D1084177

THE BEHAVIOR OF ORGANISMS

An Experimental Analysis

THE BEHAVIOR OF ORGANISMS

An Experimental Analysis

BY

(Burrhus Frederic), 1904

B. F. SKINNER

ASSISTANT PROFESSOR OF PSYCHOLOGY

UNIVERSITY OF MINNESOTA

Copley Publishing Group

Acton, Massachusetts 01720

P.O. Box 825, Cambridge, MA 02138
Internet address: www.bfskinner.org

Printed in the United States of America

ISBN 1-58390-007-1

TO YVONNE

ACKNOWLEDGMENTS

This publication of *The Behavior of Organisms* initiates a series of books sponsored by the B. F. Skinner Foundation. This first volume reproduces the seventh printing of the original 1938 work, and includes the preface written by Skinner for that 1966 edition.

The significance of this classic is only now being recognized. In it, Skinner presents a radical departure from previous explanations of behavior, including the behavioral approaches current at the time. Though Skinner did not use the word "selection" until later, this first book presents the framework and experimental evidence for selection by consequences as the basic mechanism for changes in operant behavior during the lifetime of the individual.

A project such as this involves many people. I would like to express my appreciation to W. Scott Wood who first proposed establishing a foundation and was instrumental in bringing it into existence. The other members of the Board of Directors also supported the project: Jack Michael, Ernest A. Vargas, and Margaret Vaughan. I would also like to thank the many people and organizations that contributed financial support, especially Edward L. Anderson, COBA Inc., and Aubrey Daniels & Associates.

Julie S. Vargas, President
The B. F. Skinner Foundation
Cambridge, Massachusetts

March 20, 1991

ACKNOWLEDGMENTS

The program of research responsible for most of the material in this book was proposed in a semi-historical essay on the concept of the reflex which was part of a thesis for the degree of doctor of philosophy at Harvard University in 1931. Most of the research was subsequently carried out as a Fellow of the National Research Council in the Biological Sciences and, later, as a Junior Fellow of the Society of Fellows in Harvard University. The Society of Fellows has made a generous grant toward defraying the special expenses of publishing a book of this sort. To both of these foundations I express my gratitude. It is more difficult to say how much I owe to the various men with whom I have been at one time or another in contact, especially to Professor-Emeritus Albro D. Morrill of Hamilton College, Professor Walter S. Hunter, now of Brown University, Professors W. J. Crozier, Hallowell Davis, and L. J. Henderson of Harvard University, and Professor W. T. Heron of the University of Minnesota. The book is as much theirs as mine. I have also profited greatly from a frequent interchange of opinions and ideas with Professors Leonard Carmichael, Clark L. Hull, F. S. Keller, E. C. Tolman and many others. My debt is especially great to the editor of the series, Professor R. M. Elliott, who has greatly improved the manuscript and in many ways enabled me to bring it to completion.

Part of the material has previously appeared in the papers listed at the back of the book. For permission to use text and figures I have to thank the editors and publishers of the Journal of General Psychology, the Proceedings of the National Academy of Sciences, and the Psychological Record.

B. F. S.

Minneapolis, Minnesota.
May, 1938.

PREFACE TO THE SEVENTH PRINTING

Two other theoretico-experimental analyses of behavior compose the historical setting in which this book should be evaluated. Tolman's *Purposive Behavior in Animals and Men* preceded *The Behavior of Organisms* by six years; Hull's *Principles of Behavior* followed it by five. The three books differed in many ways: they undertook to solve different problems, and they sought solutions in different places.

My debt to Sherrington, Magnus, and Pavlov is obvious in my continuing use of the word "reflex." I held to the term even after I had begun to distinguish between emitted and elicited behavior—between "respondents" and "operants"—for I wanted to preserve the notion of reflex strength. In my thesis (published as References 2, 3, and 4 on page 445, below) I had argued that " 'conditioning,' 'emotion,' 'drive,' so far as they concern behavior, are essentially to be regarded as changes in reflex strength." But in my first experiment I followed changes in the rate of responding rather than in latency, magnitude of response, or after-discharge, as in traditional practice. Probability of response thus emerged as my basic dependent variable. Its importance was clarified by the use of a cumulative record, in which changes in rate over relatively long periods of time can be seen at a glance. A simple mechanical device which generates a continuous plot of this sort while the organism is responding "makes behavioral processes visible."

With the exception of a running wheel my only apparatus was the now familiar "box" containing a food- or water-dispenser operated by a lever which could be depressed by a rat. A few relays and timers turned buzzers and lights on and off and connected or disconnected the lever and the dispenser. In the chronological order in which they were studied experimentally, my independent variables consisted of deprivation and satiation, reinforcement and nonreinforcement, schedules of reinforcement, differential reinforcement with respect to properties of stimuli and responses, aversive consequences, and a few behavioral drugs.

As I have recently pointed out (in *Operant Behavior: Areas of Research and Application,* edited by Werner Honig) operant behavior is essentially the field of purpose. Tolman made a more explicit acknowledgment of the traditional significance of his subject. His experiments were designed to make purpose visible in spatial terms, in the movement of an organism toward or away from a goal object. Almost all the figures in his book which describe apparatus are maps. (It is not surprising that his experiments led to the concept of cognitive mapping and emphasized distinctions between place- and response-learning.) Tolman's docile behavior is no doubt close to operant behavior, but docility, like purpose, was a property or characteristic of behavior rather than a relation to an independent variable.

Tolman classified his independent variables as "stimuli, initiating physiological states, and the general heredity and past training of the organism." But this phrase appears near the end of his book in a chapter entitled "Summary and Conclusions for Psychologists and Philosophers" and is less a summary than a fore-shadowing of a different formulation which he was to publish three years later and which contains the expression:

$$B = f(SPHTA).$$

This is close to the

$$R = f(S,A)$$

of my 1931 paper. Discussions which I had with Tolman during the summer of 1931, when he taught summer school at Harvard University, are, I believe, relevant. I had finished my experiments on "drive and reflex strength," and I was arguing that rate of eating, or of pressing a lever when pressing was reinforced with food, was to be described as a function of a "third variable"—that is, a variable in addition to stimulus and response, in this case a history of deprivation and satiation. Although I continued to use the concept of drive for many years (J. R. Kantor eventually convinced me of its dangers), I regarded it simply as a convenient way of referring to *environmental* variables. Tolman, however, made it an *intervening* variable, apparently in his avowed concern with replacing or reinterpreting mental processes. The

"B" in his equation represents behavior, but it was to be quantified, if at all, as a "behavior ratio." The maze had been simplified to the form of a "T," but no further, and a remnant of spatial purpose thus survived. Behavior at a choice point was as close as Tolman came to the concept of probability of response.

Hull's dependent variable was Habit Strength. "Habit" came from 19th-century studies of animal behavior, but "strength" was in the reflex tradition, and Hull accordingly used such measures as latency and magnitude of response. When he and his students turned to the use of a modified "Skinner Box" (Hull seems to have been responsible for the expression, which I have never used), the notion of probability of response began to emerge, but it appears relatively late in his book and only in passing, and has not been emphasized by subsequent workers in the Hullian tradition. Hull's independent variables are exemplified by stimuli, number of reinforcements, and deprivation.

Although all three books are ostensibly concerned with explaining observed behavior in terms of observable conditions and events, both Tolman and Hull quickly became preoccupied with internal states and processes. Some such move is inevitable so long as an effort is made to characterize the interchange between organism and environment as input and output. Output can seldom if ever be related to input in any simple way, and internal activities are therefore invented to make adjustments. It is not surprising that modern cognitive psychologists should have been strongly influenced by information theory, where a system is said to convert input into output by acquiring, processing, storing, and retrieving information. Activities of this sort are modern versions of Tolman's substitutes for mental processes. Although Tolman insisted that his behavior-determinants were "to be discovered, in the last analysis, by . . . experiments" and that "they have to be inferred 'back' from behavior," he nevertheless made them his primary object of inquiry, and this has set the pattern for cognitive psychology.

For S-R psychologists, response stands for output and stimulus for input, and when the one cannot be accounted for in terms of the other, mediating activities are again invented. (How dangerous this can be is seen in the fact that two of Hull's principles, afferent neural interaction and behavioral oscillation, serve no other function than to account for failure to relate the objective terminal events in a meaningful way.) The properties of Hull's mediating system were also to be inferred from environmental measures, but Hull himself began to insert references to the nervous system.

The Behavior of Organisms is often placed, quite erroneously, in the S-R tradition. The book remains committed to the program stated in my 1931 paper in which the stimulus occupied no special place among the independent variables. The simplest contingencies involve at least three terms—stimulus, response, and reinforcer—and at least one other variable (the deprivation associated with the reinforcer) is implied. This is very much more than input and output, and when all relevant variables are thus taken into account, there is no need to appeal to an inner apparatus, whether mental, physiological, or conceptual. The contingencies are quite enough to account for attending, remembering, learning, forgetting, generalizing, abstracting, and many other so-called cognitive processes. In the same way histories of satiation and deprivation take the place of internalized drives, schedules of reinforcement account for sustained probabilities of responding otherwise attributed to dispositions or traits, and so on.

These characteristics of the present formulation become more significant as we move from contemporary setting to subsequent history. It is instructive to examine a recent issue of the *Journal of the Experimental Analysis of Behavior* or the book *Operant Behavior: Areas of Research and Application*. The cumulative records in *The Behavior of Organisms,* purporting to show orderly changes in the behavior of individual organisms, occasioned some surprise and possibly, in some quarters, suspicion. Today, most of them seem quite crude. Improved experimental control has yielded much smoother curves. Cumulative records are also now often supplemented by distributions of inter-response times and on-line computer processing of changes in rate. The "organism" of my title, the laboratory rat, has been joined by scores of other species, including man. The lever has made room for operanda appropriate to many other topographies of behavior. Independent variables are much more carefully manipulated: buzzers and lights are often replaced by stimuli controlled with the precision characteristic of human psychophysics. To food and water have been added other positive and negative reinforcers. The negative reinforcers have opened up new territory in the study of avoidance, escape, and punishment. Caffeine and benzedrine have proved to be only the beginning of a long line of "psychotropic" drugs.

As the power of the analysis has grown, more and more complex behavior has been studied, under contingencies which approach the subtlety and complexity of the contingencies to be found in the environment at large. More of what the organism is doing at any given time is analyzed. Multiple stimuli and multiple responses compose complex systems of concurrent and chained operants. Experiments may last for weeks rather than for the standard hour of *The Behavior of Organisms*. Special environments may be maintained from birth. The apparatus required for all this is necessarily much more elaborate: simple relays, timers, and counters have given way to solid-state circuitry and computers. All these advances were facilitated by a formulation which emphasized behavior rather than supposed precursors of behavior and observable variables rather than inferred causal states or processes.

Much current basic research is essentially a technological application. The study of stimulus control, for example, has yielded a kind of nonverbal psychophysics. An extensive application to physiology seems to confirm the argument of Chapter 12 that the behavior "mediated" by the nervous system needs to be rigorously described before a neurological account can be seen to be adequate. Operant research plays an important role in psychopharmacology. The techniques are also applied with increasing frequency to what might be called the biology of behavior, sometimes to the surprise of ethologists. Many of these applications of the techniques of an experimental analysis have meant generous return support for basic research.

Other applications are more in the spirit of a behavioral technology. Toward the end of the book I pointed out that extrapolations to human affairs had been avoided ("Let him extrapolate to will"), but I was already at work on an extension to verbal behavior. My book on that subject, published nearly twenty years later, has not always been understood by linguists and psycholinguists, but it seems to me an essential step in extending the analysis to human behavior. One important application—both verbal and nonverbal—has been made in education. If teaching may be defined as the arrangement of contingencies of reinforcement under which students learn, then the various pieces of apparatus used in the experimental analysis of behavior are teaching machines. Many processes analyzed in *Verbal Behavior,* particularly in the chapter on "Supplemental Stimulation," are basic to programmed verbal instruction. An early example of "shaping" behavior by arranging a program of increasingly complex contingencies appears on pages 339–340 below.

The application to psychotherapy is not so far advanced, though O. R. Lindsley's pioneer work has been followed by many other studies of psychotic and retarded subjects. Lindsley has developed the notion of a "prosthetic environment," an environment so designed that a person may live in it effectively in spite of behavioral deficiencies. But all designed environments are prosthetic, in the sense that they facilitate or otherwise encourage specific kinds of behavior. My interest in the design of a culture —in government in the broadest sense—led to a fictional treatment in *Walden Two,* in which many issues raised by the possibility of a successful science of behavior are discussed.

So far as the facts are concerned, *The Behavior of Organisms* is out of date. It still seems to me a viable book, however, for it presents a useful formulation of behavior supported by a selection of illustrative experiments. It may also serve as a reminder that a promising conception of human behavior has been derived from an analysis which began with simple organisms in simple situations and moved on, but only as its growing power permitted, to the complexities of the world at large.

B. F. Skinner

Cambridge, Massachusetts
June, 1966

CONTENTS

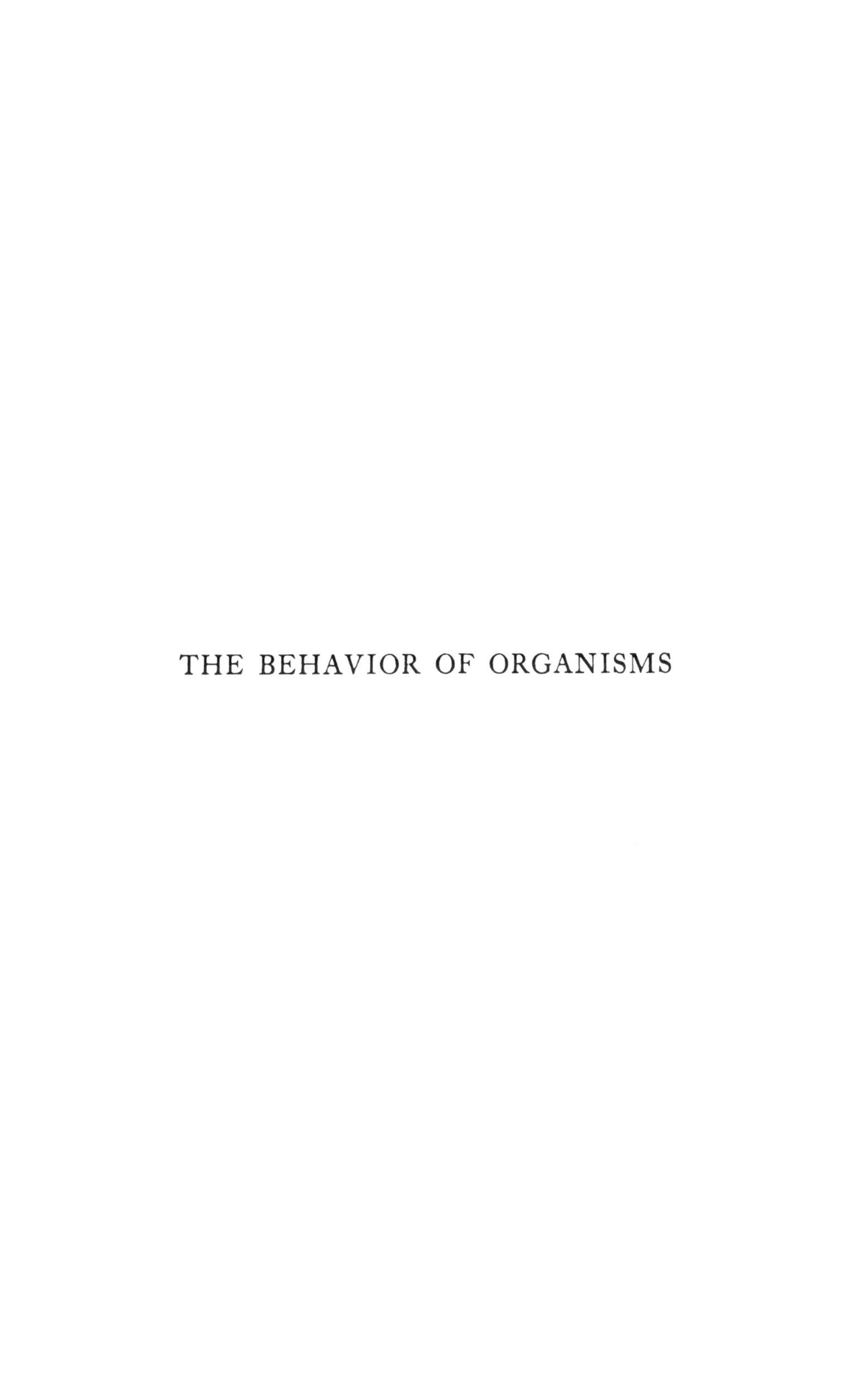

THE BEHAVIOR OF ORGANISMS

Chapter One

A SYSTEM OF BEHAVIOR

Behavior as a Scientific Datum

Although the kind of datum to which a science of behavior addresses itself is one of the commonest in human experience, it has only recently come to be regarded without reservation as a valid scientific subject matter. It is not that man has never talked about behavior nor tried to systematize and describe it, but that he has constantly done so by indirection. Behavior has that kind of complexity or intricacy which discourages simple description and in which magical explanatory concepts flourish abundantly. Primitive systems of behavior first set the pattern by placing the behavior of man under the direction of entities beyond man himself. The determination of behavior as a subject of scientific inquiry was thus efficiently disposed of, since the directing forces to which appeal was made were by hypothesis inscrutable or at least called only faintly for explanation. In more advanced systems of behavior, the ultimate direction and control have been assigned to entities placed within the organism and called psychic or mental. Nothing is gained by this stratagem because most, if not all, of the determinative properties of the original behavior must be assigned to the inner entity, which becomes, as it were, an organism in its own right. However, from this starting point three courses are possible. The inner organism may in resignation be called free, as in the case of 'free will,' when no further investigation is held to be possible. Or, it may be so vaguely defined as to disturb the curiosity of no one, as when the man in the street readily explains his behavior by appeal to a directing 'self' but does not ask nor feel it necessary to explain why the self behaves as it does. Or, it may become in turn the subject matter of a science. Some conceptions of the 'mind' and its faculties, and more recently the 'ego,' 'super-ego,' and 'id,' are examples of inner agents or organisms, designed to account for behavior, which have remained the subject of scientific investigation.

The important advance from this level of explanation that is made by turning to the nervous system as a controlling entity has unfortunately had a similar effect in discouraging a direct descriptive attack upon behavior. The change is an advance because the new entity beyond behavior to which appeal is made has a definite physical status of its own and is susceptible to scientific investigation. Its chief function with regard to a science of behavior, however, is again to divert attention away from behavior as a subject matter. The use of the nervous system as a fictional explanation of behavior was a common practice even before Descartes, and it is now much more widely current than is generally realized. At a popular level a man is said to be capable (a fact about his behavior) because he has brains (a fact about his nervous system). Whether or not such a statement has any meaning for the person who makes it is scarcely important; in either case it exemplifies the practice of explaining an obvious (if unorganized) fact by appeal to something about which little is known. The more sophisticated neurological views generally agree with the popular view in contending that behavior is in itself incomprehensible but may be reduced to law if it can be shown to be controlled by an internal system susceptible to scientific treatment. Facts about behavior are not treated in their own right, but are regarded as something to be explained or even explained away by the prior facts of the nervous system. (I am not attempting to discount the importance of a science of neurology but am referring simply to the primitive use of the nervous system as an explanatory principle in avoiding a direct description of behavior.)

The investigation of behavior as a scientific datum in its own right came about through a reformation of psychic rather than neurological fictions. Historically, it required three interesting steps, which have often been described and may be briefly summed up in the following way. Darwin, insisting upon the continuity of mind, attributed mental faculties to some subhuman species. Lloyd Morgan, with his law of parsimony, dispensed with them there in a reasonably successful attempt to account for characteristic animal behavior without them. Watson used the same technique to account for human behavior and to reestablish Darwin's desired continuity without hypothesizing mind anywhere. Thus was a science of behavior born, but under circumstances which can scarcely be said to have been auspicious. The science appeared in the form of a remodeled psychology with ill-concealed evidences of its earlier frame. It accepted an organization of data based upon ancient concepts which were not an essential part of its own structure. It inherited a language so infused with metaphor and implication that it was frequently impossible merely to talk about behavior without raising the ghosts of dead systems. Worst of all, it carried on the practice of seeking a solution for the problems of behavior elsewhere than in behavior itself. When a science of behavior had once rid itself of psychic fictions, it faced these alternatives: either it might leave their places empty and proceed to deal with its data directly, or it might make replacements. The whole weight of habit and tradition lay on the side of replacement. The altogether too obvious alternative to a mental science was a neural science, and that was the choice made by a non-mentalistic psychology. The possibility of a directly descriptive science of behavior and its peculiar advantages have received little attention.

The need for a science of behavior should be clear to anyone who looks about him at the rôle of behavior in human affairs. Indeed, the need is so obvious and so great that it has acted to discourage rather than to stimulate the establishment of such a science. It is largely because of its tremendous consequences that a rigorous treatment of behavior is still regarded in many quarters as impossible. The goal has seemed wholly inaccessible. What the eventual success of such a science might be, probably no one is now prepared to say; but the preliminary problems at least are not beyond the reach of existing scientific methods and practices, and they open up one of the most interesting prospects in modern science.

The two questions which immediately present themselves are: What will be the structure of a science of behavior? and How valid can its laws be made? These questions represent sufficiently well the double field of the present book. I am interested, first, in setting up a system of behavior in terms of which the facts of a science may be stated and, second, in testing the system experimentally at some of its more important points. In the present chapter I shall sketch what seems to me the most convenient formulation of the data at the present time, and in later chapters I shall consider some factual material fitting into this scheme. If the reader is primarily interested in facts and experimental methods, he may go directly to Chapter Two, using the index here and there to clarify the terms defined in what follows. If he is interested in the structure of a science of behavior and wishes to understand why the experiments to be reported were performed, the theoretical treatment in the rest of this chapter is indispensable.

A Definition of Behavior

It is necessary to begin with a definition. Behavior is only part of the total activity of an organism, and some formal delimitation is called for. The field might be defined historically by appeal to an established interest. As distinct from the other activities of the organism, the phenomena of behavior are held together by a common conspicuousness. Behavior is what an organism is *doing*—or more accurately what it is observed by another organism to be doing. But to say that a given sample of activity falls within the field of behavior simply because it normally comes under observation would misrepresent the significance of this property. It is more to the point to say that behavior is that part of the functioning of an organism which is engaged in acting upon or having commerce with the outside world. The peculiar properties which make behavior a unitary and unique subject matter follow from this definition. It is only because the receptors of other organisms are the most sensitive parts of the outside world that the appeal to an established interest in what an organism is doing is successful.

By behavior, then, I mean simply the movement of an organism or of its parts in a frame of reference provided by the organism itself or by various external objects or fields of force. It is convenient to speak of this as the action of the organism upon the outside world, and it is often desirable to deal with an effect rather than with the movement itself, as in the case of the production of sounds.

A Set of Terms

In approaching a field thus defined for purposes of scientific description we meet at the start the need for a set of terms. Most languages are well equipped in this respect but not to our advantage. In English, for example, we say that an organism *sees* or *feels* objects, *hears* sounds, *tastes* substances, *smells* odors, and *likes* or *dislikes* them; it *wants, seeks,* and *finds;* it *has a purpose, tries* and *succeeds* or *fails;* it *learns* and *remembers* or *forgets;* it is *frightened, angry, happy,* or *depressed; asleep* or *awake;* and so on. Most of these terms must be avoided in a scientific description of behavior, but not for the reasons usually given. It is not true that they cannot be defined. Granted that in their generally accepted usages they may not stand analysis, it is nevertheless possible to agree on what is to be meant by 'seeing an object' or 'wanting a drink' and to honor the agreement from that point forward. A set of conventional definitions could be established without going outside behavior, and was in fact so established by the early behaviorists, who spent a great deal of time (unwisely, I believe) in translating into behavioristic terms the concepts of traditional psychology, most of which had been taken from the vernacular. Vigorous attempts to redefine some terms of the popular vocabulary with reference to behavior have been made, for example, by Tolman (71).

It is likewise not true that the behavior referred to by a vocabulary of this sort cannot be dealt with quantitatively. The terms usually refer to continua (an organism sees clearly or vaguely or not at all, it likes or dislikes more or less intensely, it tries hard or feebly, it learns quickly or slowly, and so on) and once defined with reference to behavior they may be expressed in units no more arbitrary than, and perhaps even reducible to, centimeters, grams, and seconds.

The important objection to the vernacular in the description of behavior is that many of its terms imply conceptual schemes. I do not mean that a science of behavior is to dispense with a conceptual scheme but that it must not take over without careful consideration the schemes which underlie popular speech. The vernacular is clumsy and obese; its terms overlap each other, draw unnecessary or unreal distinctions, and are far from being the most convenient in dealing with the data. They have the disadvantage of being historical products, introduced because of everyday convenience rather than that special kind of convenience characteristic of a simple scientific system. It would be a miracle if such a set of terms were available for a science of behavior, and no miracle has in this case taken place. There is only one way to obtain a convenient and useful system and that is to go directly to the data.

This does not mean that we must entirely abandon ordinary speech in a science of behavior. The sole criterion for the rejection of a popular term is the implication of a system or of a formulation extending beyond immediate observations. We may freely retain all terms which are descriptive of behavior without systematic implications. Thus, the term 'try' must be rejected because it implies the relation of a given sample of behavior to past or future events; but the term 'walk' may be retained because it does not. The term 'see' must be rejected but 'look toward' may be retained, because 'see' implies more than turning the eyes toward a source of stimulation or more than the simple reception of stimuli. It is possible that some popular systematic terms will apply to the scientific system finally established. We might want to establish a relation similar to that referred to by 'try' and in that case we might reintroduce the term. But the points of contact between a popular and a scientific system will presumably not be many, and in any event the popular term must be omitted until its systematic justification has been established.

With this criterion it is possible to save a considerable part of the vernacular for use in describing the movements of organisms. Where the vernacular is vague, it may be supplemented with terms from anatomy and superficial physiology, and additional terms may be invented if necessary.

Narration and the Reflex

Once in possession of a set of terms we may proceed to a kind of description of behavior by giving a running account of a sample of behavior as it unfolds itself in some frame of reference. This is a typical method in natural history and is employed extensively in current work—for example, in child and infant behavior. It may be described as narration. It presents no special problem. If there is an objection to the use of a verbal description, the investigator may resort to sound-films and multiply them at will; and the completeness of the transcription will be limited only by an eventual unwillingness to increase the number of recording devices any further. From data obtained in this way it is possible to classify different kinds of behavior and to determine relative frequencies of occurrence. But although this is, properly speaking, a description of behavior, it is not a science in the accepted sense. We need to go beyond mere observation to a study of functional relationships. We need to establish laws by virtue of which we may predict behavior, and we may do this only by finding variables of which behavior is a function.

One kind of variable entering into the description of behavior is to be found among the external forces acting upon the organism. It is presumably not possible to show that behavior as a whole is a function of the stimulating environment as a whole. A relation between terms as complex as these does not easily submit to analysis and may perhaps never be demonstrated. The environment enters into a description of behavior when it can be shown that a given *part* of behavior may be induced at will (or according to certain laws) by a modification in part of the forces affecting the organism. Such a part, or modification of a part, of the environment is traditionally called a *stimulus* and the correlated part of the behavior a *response*. Neither term may be defined as to its essential properties without the other. For the observed relation between them I shall use the term *reflex,* for reasons which, I hope, will become clear as we proceed. Only one property of the relation is usually invoked in the use of the term—the close coincidence of occurrence of stimulus and response—but there are other important properties to be noted shortly.

The difference between the demonstration of a reflex and mere narration is, not that part of the environment may not be mentioned in the narration, but that no lawful relation between it and the behavior is asserted. In the narrative form, for example, it may be said that 'at such and such a moment the ape picked up a stick.' Here there is no reference to other instances of the same behavior either past or future. It is not asserted that all apes pick up sticks. The story is told simply of something that has once happened. The isolation of a reflex, on the other hand, is the demonstration of a predictable uniformity in behavior. In some form or other it is an inevitable part of any science of behavior. Another name may be used, and the degree of rigor in the demonstration of lawfulness may fall short of that required in the case of the reflex, but the same fundamental activity must go on whenever anything of a scientific nature is to be said about behavior that is not mere narration. Current objections to the reflex on the ground that in the analysis of behavior we destroy the very thing we are trying to understand, scarcely call for an answer. We always analyze. It is only good sense to make the act explicit—to analyze as overtly and as rigorously as possible.

So defined a reflex is not, of course, a theory. It is a fact. It is an analytical unit, which makes an investigation of behavior possible. It is by no means so simple a device as this brief account would suggest, and I shall return later to certain questions concerning its proper use. Many traditional difficulties are avoided by holding the definition at an operational level. I do not go beyond the observation of a correlation of stimulus and response. The omission of any reference to neural events may confuse the reader who is accustomed to the traditional use of the term in neurology. The issue will, I think, be clarified in Chapter Twelve, but it may be well to anticipate that discussion by noting that the concept is not used here as a 'neurological explanation' of behavior. It is a purely descriptive term.

That the reflex as a correlation of stimulus and response is not the only unit to be dealt with in a description of behavior will appear later when another kind of response that is 'emitted' rather than 'elicited' will be defined. The following argument is confined largely to elicited behavior.

The Collection of Reflexes

One step in the description of behavior is the demonstration of the relationships that are called reflexes. It leads to considerable power of prediction and control, but this is not, as has often been claimed, the aim and end of the study of behavior. Watson, for example, defined the goal of psychological study as "the ascertaining of such data and laws that, given the stimulus, psychology can predict what the response will be; or, given the response, it can specify the nature of the effective stimulus [(75), p. 10.]." But a little reflection will show that this is an impracticable program. In the field of behavior a science must contend with an extraordinary richness of experimental material. The number of stimuli to which a typical organism may respond originally is very great. The number of stimuli to which it may come to respond through a process to be described below is indefinitely large, and to each of them it may be made to respond in many ways. It follows that the number of possible reflexes is for all practical purposes infinite and that what one might call the botanizing of reflexes will be a thankless task.

Nevertheless, there is no way of reaching the goal set by this quotation, taken literally, except to botanize. The sort of prediction that it proposes would require the compilation of an exhaustive catalogue of reflexes, by reference to which predictions could be made. The catalogue would be peculiar to a single organism and would require continual revision as long as the organism lived. It is obviously unpractical. Quite aside from any question of completeness, it could not reach any degree of usefulness before becoming unmanageable from its sheer bulk. No one has seriously attempted to construct a catalogue for this purpose, and I have probably misconstrued the quotation. Those who regard wholesale prediction as the essence of the description of behavior have usually supposed it to be possible to reduce the size of the field and to reach detailed predictions by a shorter route. But no method has ever been demonstrated that would make this possible. Generally the attempt is made to reduce the total number of required terms by making each term more comprehensive (as by resorting to classes of reflexes). But the more comprehensive the term the less complete and less accurate the descriptive reference upon any given occasion when it is used. I shall show later that the level of analysis of the reflex is uniquely determined with respect to its usefulness as an analytical instrument and that it cannot be altered for the sake of reducing the number of terms to be taken into account.

We have no reason to expect, either from theoretical considerations or from a survey of what has already been done experimentally, that any wholesale prediction of response or identification of stimulus will become possible through the discovery of principles that circumvent the routine of listing reflexes. Confronted with the sheer expansiveness of the topography of behavior, we must concede the impossibility of any wholesale prediction of stimulus or response that could be called exact. The number of items to be dealt with is very great and does not seem likely to be reduced, and there is at present no reason to believe that a new order may some day be discovered to resolve the difficulty. This view may appear somber to those who believe that the study of behavior is concerned primarily with the topographical prediction of stimulus and response. But this is a mistaken, and fatal, characterization of its aim. Actually there may be little interest in the continued demonstration of reflex relationships. The discovery of a reflex was historically an important event at a time when the field of behavior was encroached upon by many other (usually metaphysical) descriptive concepts. It may still conceivably be of importance whenever there seems to be a special reason for questioning the 'reflex nature' of a given bit of behavior. But when a large number of reflexes have once been identified and especially when it has been postulated that all behavior is reflex, the mere listing of reflexes has no further theoretical interest and remains important only for special investigations (as, for example, the analysis of posture). No interest in the description of behavior itself will prompt us to press the botanizing of reflexes any further.

The Static Laws of the Reflex

I have restricted the preceding paragraphs to the *topographical* prediction of behavior in order to allow for another kind of prediction to which a science of behavior must devote itself—the prediction of the quantitative properties of representative reflexes. In limiting the concept of the reflex to the coincidence of occurrence of a stimulus and a response a considerable simplification is introduced; and the supposition that the relation so described is invariable involves another. It is only by virtue of these simplifications that the mere collection of reflexes may be shown to possess any predictive value whatsoever, and the argument against collection as the aim of a science of behavior might have been greatly strengthened by considering how reduced in value a catalogue would become after these simplifying assumptions had been lifted.

The quantitative properties arise because both stimulus and response have intensive and temporal dimensions in addition to their topography and because there is a correlation between the values assumed in the two cases. Given a stimulus over which we have quantitative control and a measure of the magnitude of the response, we are in a position to demonstrate the following laws.

THE LAW OF THRESHOLD. *The intensity of the stimulus must reach or exceed a certain critical value (called the threshold) in order to elicit a response.* A threshold follows from the necessarily limited capacity of the organism to be affected by slight external forces. The values obtained in typical reflexes of this sort are usually considerably above values for the basic receptive capacity of the organism determined in other ways (*e.g.*, in the 'discrimination' of Chapter Five).

THE LAW OF LATENCY. *An interval of time (called the latency) elapses between the beginning of the stimulus and the beginning of the response.* A latency is to be expected from the usual spatial separation of receptor and effector and from the difference in form of energy of stimulus and response. The values obtained vary greatly between reflexes, following to some slight extent a classification of receptors (*cf.* visual and thermal reflexes) and of effectors (*cf.* the responses of skeletal muscle and smooth muscle or gland).

No measure of the magnitude of stimulus or response is needed to determine the latency, provided these magnitudes may be held constant, and it is therefore a useful measure when the dimensions of either term are in doubt. One important property of latency is that it is usually a function of the intensity of the stimulus, as Sherrington originally showed [(68), pp. 18 ff.]. The stronger the stimulus the shorter the latency.

THE LAW OF THE MAGNITUDE OF THE RESPONSE. *The magnitude of the response is a function of the intensity of the stimulus.* Although there are exceptional cases which show an apparently all-or-none character, the magnitude of the response is in general graded, and there is a corresponding gradation in the intensity of the stimulus. The two magnitudes are measured on separate scales appropriate to the form of each term, but this does not interfere with the demonstration of a relation. The ratio of the magnitudes will be referred to hereafter as the *R/S* ratio.[1]

THE LAW OF AFTER-DISCHARGE. *The response may persist for some time after the cessation of the stimulus.* The term after-discharge is usually not applied to the time alone but to the total amount of activity taking place during it. In general the after-discharge increases with the intensity of the stimulus. In measuring the difference between the times of cessation the latency may be subtracted as a minor refinement.

The preceding statements regard the intensity of the stimulus as the only property of which the response is a function, but the duration must not be ignored. The laws are subject to the following elaboration:

THE LAW OF TEMPORAL SUMMATION. *Prolongation of a stimulus or repetitive presentation within certain limiting rates has the same effect as increasing the intensity.* Summation is often restricted to near-threshold values of the stimulus, when the effect is to obtain a response that is not elicited without summation, but the law applies to the magnitude of response, its latency, and so on, as well as to its mere occurrence. Thus, a sub-threshold value of a stimulus

[1] An increase in the intensity of the stimulus may result not only in an increase in the magnitude of the response but in an apparent change of topography. Thus, a mild shock to the foot may bring about simple flexion, while a stronger shock will lead to vigorous postural and progressive responses. These effects may be treated by dealing with each response separately and noting that the thresholds differ. A single stimulus is correlated with all of them but elicits any given one only when it is above its particular threshold value.

may elicit a response if it is prolonged or repeated within a certain time or at a certain rate, and the magnitude of the response and the after-discharge are functions of the duration of the stimulus as well as the intensity. The latency is frequently too short to be affected by prolongation of the stimulus, but at near-threshold values of the stimulus an effect may be felt. With repetitive presentation of a weak stimulus, the latency is a function of the frequency of presentation.[2]

The properties of latency, threshold, after-discharge and the R/S ratio are detected by presenting a stimulus at various intensities and durations and observing the time of occurrence, duration, and magnitude of the response. They may be called the *static* properties of a reflex. They supplement a topographical description in an important way and cannot be omitted from any adequate account. They are to be distinguished from a more extensive group of laws which concern changes in the state of the static properties. Changes begin to be observed when we repeat the elicitation of a reflex, as we cannot help doing if we are to check our measurements or if we are to give a description of behavior over any considerable period of time. The values of the static properties of a reflex are seldom, if ever, exactly confirmed upon successive elicitations. Important changes take place either in time or as a function of certain operations performed upon the organism. They are described by invoking another kind of law which I shall distinguish from the preceding by calling it *dynamic.*

The Dynamic Laws of Reflex Strength

In an example of a dynamic law given below (the Law of Reflex Fatigue) it is stated that, if a reflex is repeatedly elicited at a certain rate, its threshold is raised, its latency is increased, and the R/S ratio and the after-discharge are decreased [Sherrington (68)]. The operation performed upon the organism is in this case merely the

[2] The summated stimulus may appear to bring about topographical changes similar to those following an increase in the intensity of the stimulus. If the pinna of a sleeping dog is touched, the ear may be flicked; if it is repeatedly touched the dog may stir and change its position. What has happened may be treated as in the case of increased intensity. The reflex of changing position has a higher threshold than the pinna reflex and may be exhibited only with a stronger stimulus or through the summation of stimuli.

repeated elicitation of the reflex. Its effect is a simultaneous change in the values of all the static properties. The law should describe the relation between each property and the operation, but it is convenient to have a single term to describe the state of the reflex with respect to all its static properties at once. Various terms are currently in use for this purpose, such as 'intensity,' 'force,' and 'strength.' I shall use 'strength.' The value of the strength of a reflex is arbitrarily assigned to it from the values of the static properties and is never measured directly.

The strength of a reflex is not to be confused with the magnitude of the response. The latter is a function of the intensity of the stimulus, to which the strength of a reflex has no relation. A strong reflex may exhibit a response of small magnitude if the stimulus is of low intensity; conversely, a weak reflex may exhibit a fairly intense response to a very intense stimulus. I shall reserve the term strength exclusively for the meaning here assigned to it and use 'intensity' or 'magnitude' in referring to the values of stimuli and responses.

In two laws of reflex strength the state of the reflex is a function of the operation of *elicitation*. The changes are in the same direction but differ in their temporal properties.

The Law of the Refractory Phase. *Immediately after elicitation the strength of some reflexes exists at a low, perhaps zero, value. It returns to its former state during subsequent inactivity.* The time during which the value is zero is called the 'absolute refractory phase,' that during which it is below normal, the 'relative refractory phase.' The durations vary greatly between reflexes but may be of the order of a fraction of a second for the absolute phase and of a few seconds for the relative.

The refractory phase applies only to a kind of reflex in which the response utilizes an effector in opposing ways at different times. Such a response is either rhythmic or phasic. The classical examples are the scratch reflex in the dog and the lid reflex in man, both of which involve alternate flexion and extension. Mere flexion of the leg or closure of the eye involve only the first of these opposed movements and have no refractory phase; the second presentation of the stimulus only reinstates or prolongs or intensifies the response. The refractory phase may be regarded as a special mechanism for the production and support of rhythms and of responses which must cease and begin again in order to perform their functions. For data on the refractory phase see (35).

THE LAW OF REFLEX FATIGUE. *The strength of a reflex declines during repeated elicitation and returns to its former value during subsequent inactivity.* The rate of decline is a function of the rate of elicitation and of the intensity of the stimulus (and hence of the response) and varies greatly between reflexes. Because of the conflicting processes of decline and recovery the strength may stabilize itself at a constant value as a function of the rate of elicitation. At high rates the strength may reach zero. Some reflexes are practically indefatigable, as, for example, the postural reflexes evoked from the head, as has been shown by Magnus (61).

The law of reflex fatigue is directly opposed to the notion of 'canalization' in which elicitation is said to *increase* the strength of the reflex. The concept of canalization has been associated with various theories of learning and the so-called Law of Exercise. There will be no occasion to introduce it here, and the law of fatigue may therefore stand without exception. There are operations through which the strength of a reflex is increased, but simple elicitation is not one of them.

When fatigue takes place quickly and recovery is slow, the process is often referred to as 'adaptation.' Thus, certain reflexes to loud sounds may fatigue ('adapt out') fairly quickly and remain at zero or some very low strength for considerable periods. The time required to build up the original strength through inactivity may be of the order of months or years. Adaptation and fatigue are distinguished only by their temporal properties and will be considered here as instances of the same phenomenon.

The next dynamic laws to be considered involve the operation of presenting a second stimulus. The extraneous stimulus itself has no control over the response but it affects the strength of the reflex of which the response is a part.

THE LAW OF FACILITATION. *The strength of a reflex may be increased through presentation of a second stimulus which does not itself elicit the response.*

In the original experiment of Exner (41) the strength of a flexion reflex in a rabbit (observed as the magnitude of the response to a stimulus of constant intensity) was increased by a loud sound and by other intense stimuli. A loud sound is a common facilitating stimulus, but its facilitative action is confined to certain kinds of reflexes, especially skeletal. Upon such a reflex as salivation it may have an opposite effect (see next paragraph). The qualifying clause in the law, that the stimulus does not itself elicit the response, is needed to distinguish between facilitation and a process called spatial summation, to be described later. Facilitation is sometimes defined (27) so that it applies only to raising the strength from zero—*i.e.*, to producing a response where none was previously obtainable. But this is only a special case of what we have defined more comprehensively.

THE LAW OF INHIBITION. *The strength of a reflex may be decreased through presentation of a second stimulus which has no other relation to the effector involved.*

The term inhibition has been loosely used to designate any decline in reflex strength or the resulting diminished state. Two of the laws already listed (the refractory phase and reflex fatigue) have been cited as instances of inhibition by Sherrington, and examples from other authors will be given later. In the present system, where many of the more important phenomena of behavior are defined in terms of changes in strength, so broad a term is useless. All changes in strength have negative, but also positive, phases; the property of the mere direction of the change does not establish a useful class of data. The various changes included in the class may easily be distinguished on the basis of the operations that produce them, and there is little to be gained by giving them a common name.

One kind of negative change in strength has some historical right to the term inhibition. The operation is the presentation of a stimulus which does not itself affect the response in question, and the law is therefore identical with that of facilitation except for sign. Grouping the two laws together we may say that one kind of change in strength is due to the simple presentation of extraneous stimuli and that it may be either positive (facilitation) or negative (inhibition). This is a narrow sense of the term inhibition, but it is the only one in which it will be used here. The difference between it and the traditional usage may perhaps be made clear by contrasting the two pairs of terms 'inhibition-facilitation' and 'inhibition-excitation.' It is in the second pair that inhibition refers to any low state of strength or the process of reaching it. We do not need the term because we do not need its opposite. Excitation and inhibition refer to what is here seen to be a continuum of degrees of reflex strength, and we have no need to designate its two extremes. In the first pair, on the other hand, inhibition refers to a negative change in strength produced by a kind of operation that would yield a positive change under other circumstances.

There is an obvious danger in paired concepts of this sort, for they place a system under the suspicion of being designed to catch a datum no matter which way it falls. But it is not as if we were saying that extraneous stimuli affect reflex strength either one way or the other. The laws of facilitation and inhibition refer to specific stimuli and specific reflexes, and it is implied that the direction of change is capable of specification in each case.

The four dynamic laws that I have just given are classical examples. They could be supplemented with laws expressing the effects of the administration of drugs, of changes in oxygen pressure, and so on, which have also been dealt with in the classical treatises (39). They will suffice to exemplify the structure of the system at this point. I begin with the reflex as an empirical description of the topographical relation between a stimulus and a response. The static laws make the description quantitative. The statement of a reflex and its static laws predicts a certain part of the behavior of the organism by appeal to the stimulating forces that produce it. The dynamic laws enter to express the importance of other kinds of operations in affecting the same behavior, to the end that the description shall be valid at all times.

In the course of this book I shall attempt to show that a large body of material not usually considered in this light may be expressed with dynamic laws which differ from the classical examples only in the nature of the operations. The most important instances are conditioning and extinction (with their subsidiary processes of discrimination), drive, and emotion, which I propose to formulate in terms of changes in reflex strength. One type of conditioning and its corresponding extinction may be described here.

THE LAW OF CONDITIONING OF TYPE S. *The approximately simultaneous presentation of two stimuli, one of which (the 'reinforcing' stimulus) belongs to a reflex existing at the moment at some strength, may produce an increase in the strength of a third reflex composed of the response of the reinforcing reflex and the other stimulus.*

THE LAW OF EXTINCTION OF TYPE S. *If the reflex strengthened through conditioning of Type S is elicited without presentation of the reinforcing stimulus, its strength decreases.*

These laws refer to the Pavlovian type of conditioned reflex, which will be discussed in detail in Chapter Three. I wish to point out here simply that the observed data are merely changes in the strength of a reflex. As such they have no dimensions which distinguish them from changes in strength taking place during fatigue, facilitation, inhibition, or, as I shall show later, changes in drive, emotion, and so on. The process of conditioning is distinguished by what is done to the organism to induce the change; in other words, it is defined by the operation of the simultaneous presentation of the reinforcing stimulus and another stimulus. The type is called Type S to distinguish it from conditioning of Type R (see below) in which the reinforcing stimulus is contingent upon a response.

Before indicating how other divisions of the field of behavior may be formulated in terms of reflex strength, it will be necessary to consider another kind of behavior, which I have not yet mentioned. The remaining dynamic laws will then be taken up in connection with both kinds at once.

Operant Behavior

With the discovery of the stimulus and the collection of a large number of specific relationships of stimulus and response, it came to be assumed by many writers that all behavior would be accounted for in this way as soon as the appropriate stimuli could be identified. Many elaborate attempts have been made to establish the plausibility of this assumption, but they have not, I believe, proved convincing. There is a large body of behavior that does not seem to be *elicited,* in the sense in which a cinder in the eye elicits closure of the lid, although it may eventually stand in a different kind of relation to external stimuli. The original 'spontaneous' activity of the organism is chiefly of this sort, as is the greater part of the conditioned behavior of the adult organism, as I hope to show later. Merely to assert that there *must* be eliciting stimuli is an unsatisfactory appeal to ignorance. The brightest hope of establishing the generality of the eliciting stimulus was provided by Pavlov's demonstration that part of the behavior of the adult organism could be shown to be under the control of stimuli which had *acquired* their power to elicit. But a formulation of this process will show that in every case the response to the conditioned stimulus must first be elicited by an unconditioned stimulus. I do not believe that the 'stimulus' leading to the elaborate responses of singing a song or of painting a picture can be regarded as the mere substitute for a stimulus or a group of stimuli which originally elicited these responses or their component parts.

Most of the pressure behind the search for eliciting stimuli has been derived from a fear of 'spontaneity' and its implication of freedom. When spontaneity cannot be avoided, the attempt is made to define it in terms of unknown stimuli. Thus, Bethe (28) says that the term 'has long been used to describe behavior for which the stimuli are not known and I see no reason why the word should be stricken from a scientific vocabulary.' But an event may occur without any observed antecedent event and still be dealt with adequately in a descriptive science. I do not mean that there are no originating forces in spontaneous behavior but simply that they are not located in the environment. We are not in a position to see them, and we have no need to. This kind of behavior might be said to be *emitted* by the organism, and there are appropriate techniques for dealing with it in that form. One important independent variable is time. In making use of it I am simply recognizing that the observed datum is the appearance of a given identifiable sample of behavior at some more or less orderly rate. The use of a rate is perhaps the outstanding characteristic of the general method to be outlined in the following pages, where we shall be concerned very largely with behavior of this sort.

The attempt to force behavior into the simple stimulus-response formula has delayed the adequate treatment of that large part of behavior which cannot be shown to be under the control of eliciting stimuli. It will be highly important to recognize the existence of this separate field in the present work. Differences between the two kinds of behavior will accumulate throughout the book, and I shall not argue the distinction here at any length. The kind of behavior that is correlated with specific eliciting stimuli may be called *respondent* behavior and a given correlation *a respondent*. The term is intended to carry the sense of a relation to a prior event. Such behavior as is not under this kind of control I shall call *operant* and any specific example *an operant*. The term refers to a posterior event, to be noted shortly. The term reflex will be used to include both respondent and operant even though in its original meaning it applied to respondents only. A single term for both is convenient because both are topographical units of behavior and because an operant may and usually does acquire a relation to prior stimulation. In general, the notion of a reflex is to be emptied of any connotation of the active 'push' of the stimulus. The terms refer here to correlated entities, and to nothing more. All implications of dynamism and all metaphorical and figurative definitions should be avoided as far as possible.

An operant is an identifiable part of behavior of which it may be said, not that no stimulus can be found that will elicit it (there may be a respondent the response of which has the same topography), but that no correlated stimulus can be detected upon occasions when it is observed to occur. It is studied as an event appearing spontaneously with a given frequency. It has no static laws comparable with those of a respondent since in the absence of a stimulus the concepts of threshold, latency, after-discharge, and the R/S ratio are meaningless. Instead, appeal must be made to frequency of occurrence in order to establish the notion of strength. The strength of an operant is proportional to its frequency of occurrence, and the dynamic laws describe the changes in the rate of occurrence that are brought about by various operations performed upon the organism.

Other Dynamic Laws

Three of the operations already described in relation to respondent behavior involve the elicitation of the reflex and hence are inapplicable to operants. They are the refractory phase, fatigue, and conditioning of Type S. The refractory phase has a curious parallel in the rate itself, as I shall note later, and a phenomenon comparable with fatigue may also appear in an operant. The conditioning of an operant differs from that of a respondent by involving the correlation of a reinforcing stimulus with a *response*. For this reason the process may be referred to as of Type R. Its two laws are as follows.

THE LAW OF CONDITIONING OF TYPE R. *If the occurrence of an operant is followed by presentation of a reinforcing stimulus, the strength is increased.*

THE LAW OF EXTINCTION OF TYPE R. *If the occurrence of an operant already strengthened through conditioning is not followed by the reinforcing stimulus, the strength is decreased.*

The conditioning is here again a matter of a change in strength.

The strength cannot begin at zero since at least one unconditioned response must occur to permit establishment of the relation with a reinforcing stimulus. Unlike conditioning of Type S the process has the effect of determining the form of the response, which is provided for in advance by the conditions of the correlation with a reinforcing stimulus or by the way in which the response must operate upon the environment to produce a reinforcement (see Chapter Three).

It is only rarely possible to define an operant topographically (so that successive instances may be counted) without the sharper delineation of properties that is given by the act of conditioning. This dependence upon the posterior reinforcing stimulus gives the term operant its significance. In a respondent the response is the result of something previously done to the organism. This is true even for conditioned respondents because the operation of the simultaneous presentation of two stimuli precedes, or at least is independent of, the occurrence of the response. The operant, on the other hand, becomes significant for behavior and takes on an identifiable form when it acts upon the environment in such a way that a reinforcing stimulus is produced. The operant-respondent distinction goes beyond that between Types S and R because it applies to unconditioned behavior as well; but where both apply, they coincide exactly. Conditioning of Type R is impossible in a respondent because the correlation of the reinforcing stimulus with a response implies a correlation with its eliciting stimulus. It has already been noted that conditioning of Type S is impossible in operant behavior because of the absence of an eliciting stimulus.

An operant may come to have a relation to a stimulus which seems to resemble the relation between the stimulus and response in a respondent. The case arises when prior stimulation is correlated with the reinforcement of the operant. The stimulus may be said to set the occasion upon which a response will be reinforced, and therefore (through establishment of a discrimination) upon which it will occur; but it does not elicit the response. The distinction will be emphasized later.

One kind of operation that affects the strength of reflexes (both operant and respondent) falls within the traditional field of drive or motivation. It would be pointless to review here the various ways in which the field has been formulated (81). In a description of behavior in terms of the present system the subject presents itself simply as a class of dynamic changes in strength. For example, suppose that we are observing an organism in the presence of a bit of food. A certain sequence of progressive, manipulative, and ingestive reflexes will be evoked. The early stages of this sequence are operants, the later stages are respondents. At any given time the strengths may be measured either by observing the rate of occurrence in the case of the former or by exploring the static properties in the case of the latter. The problem of drive arises because the values so obtained vary between wide extremes. At one time the chain may be repeatedly evoked at a high rate, while at another no response may be forthcoming during a considerable period of time. In the vernacular we should say that the organism eats only when it is hungry. What we observe is that the strengths of these reflexes vary, and we must set about finding the operations of which they are a function. This is not difficult. Most important of all are the operations of feeding and fasting. By allowing a hungry organism, such as a rat, to eat bits of food placed before it, it is possible to show an orderly decline in the strength of this group of reflexes. Eventually a very low strength is reached and eating ceases. By allowing a certain time to elapse before food is again available it may be shown that the strength has risen to a value at which responses will occur. The same may be said of the later members of the chain, the strengths of which (as respondents) must be measured in terms of the static properties. Thus, the amount of saliva secreted in response to a gustatory stimulus may be a similar function of feeding and fasting. A complete account of the strengths of this particular group of reflexes may be given in terms of this operation, other factors being held constant. There are other operations to be taken into account, however, which affect the same group, such as deprivation of water, illness, and so on.

In another important group of changes in reflex strength the chief operation with which the changes are correlated is the presentation of what may be called 'emotional' stimuli—stimuli which typically elicit changes of this sort. They may be either unconditioned (for example, an electric shock) or conditioned according to Type S where the reinforcing stimulus has been emotional (for example, a tone which has preceded a shock). Other operations which induce an emotional change in strength are the restraint of a response, the interruption of a chain of reflexes through the removal of a reinforcing stimulus (see later), the administration of certain drugs, and so on. The resulting change in behavior is again in the strength of reflexes, as I shall show in detail in Chapter Eleven.

The operations characterizing drive and emotion differ from the others listed in that they effect concurrent changes in *groups* of reflexes. The operation of feeding, for example, brings about changes in all the operants that have been reinforced with food and in all the conditioned and unconditioned respondents concerned with ingestion. Moreover, a single operation is not unique in its effect. There is more than one way of changing the strength of the group of reflexes varying with ingestion or with an emotional stimulus. In addition to the formulation of the effect upon a single reflex, we must deal also with *the* drive or *the* emotion as the 'state' of a group of reflexes. This is done by introducing a hypothetical middle term between the operation and the resulting observed change. 'Hunger,' 'fear,' and so on, are terms of this sort. The operation of feeding is said to affect the hunger and the hunger in turn the strength of the reflex. The notion of an intermediate state is valuable when (a) more than one reflex is affected by the operation, and (b) when several operations have the same effect. Its utility may perhaps be made clear with the following schemes. When an operation is unique in its effect and applies to a single reflex, it may be represented as follows:

Operation I---- ( ) ----Strength of Reflex I,

where no middle term is needed. When there are several operations having the same effect and affecting several reflexes, the relation may be represented as follows:

Operation I ——— 'State' ——— Strength of Reflex I
Operation II ——— 'State' ——— Strength of Reflex II
Operation III ——— 'State' ——— Strength of Reflex III

In the present system hypothetical middle terms ('states') will be used in the cases of drive and emotion, but no other properties will be assigned to them. A dynamic law always refers to the change in strength of a single reflex as a function of a single operation, and the intermediate term is actually unnecessary in its expression.

An observation of the state of a reflex at any given time is limited to its strength. Since the data are changes in strength and therefore the same in all dynamic laws, the system emphasizes the great importance of defining and classifying operations. The mere strength of the reflex itself is an ambiguous fact. It is impossible to tell from a momentary observation of strength whether its value is due especially to an operation of drive, conditioning, or emotion. Suppose, for example, that we have been working with an operant that has been reinforced with food and that at a given time we observe that the organism does not respond (*i.e.*, that the strength is low). *From the state of the reflex itself,* it is impossible to distinguish between the following cases. (1) The organism is hungry and unafraid, but the response has been extinguished. (2) The response is conditioned and the organism is hungry but afraid. (3) The response is conditioned, and the organism is unafraid but not hungry. (4) The response is conditioned, but the organism is both not hungry and afraid. (5) The organism is hungry, but it is afraid, and the response has been extinguished. (6) The organism is not afraid, but it is not hungry and the response has been extinguished. (7) The response has been extinguished, and the organism is afraid but not hungry. We can decide among these possibilities by referring to other behavior. If we present the stimulus of an *unconditioned* reflex varying with hunger and fear (say, if we present food), the question of conditioning is eliminated. If the organism eats, the first case listed above is proved. If it does not eat, the possibilities are then as follows. (1) The organism is hungry but afraid. (2) It is unafraid but not hungry. (3) It is both not hungry and afraid. If we then test another reflex, the strength of which decreases in a state of fear but which does not vary with hunger, and find it strong, the organism is not afraid and must therefore not be hungry.

The strength of a reflex at any given time is a function of all the operations that affect it. The principal task of a science of behavior is to isolate their separate effects and to establish their functional relationships with the strength.

The development of dynamic laws enables us to consider behavior which does not invariably occur under a given set of circumstances as, nevertheless, reflex (*i.e.*, as lawful). The early classical examples of the reflex were those of which the lawfulness was obvious. It was obvious because the number of variables involved was limited. A flexion reflex could be described very early because it was controlled by a stimulus and was not to any considerable extent a function of the operations of drive, emotion, or conditioning, which cause the greatest variability in strength. The discovery of conditioning of Type S brought under the principle of the reflex a number of activities the lawfulness of which was not evident until the conditioning operation was controlled. Operants, as predictable entities, are naturally isolated last of all because they are not controlled through stimuli and are subject to many operations. They are not *obviously* lawful. But with a rigorous control of all relevant operations the kind of necessity that naturally characterizes simple reflexes is seen to apply to behavior generally. I offer the experimental material described later in this book in support of this statement.

The Reflex Reserve

It has already been noted that one kind of operation (for example, that involved in fatigue and conditioning) is unique in its effect and changes the strength of a single reflex, while another kind (for example, that of drive or emotion) has an effect that is common to other operations and is felt by a group of reflexes. In the latter case the notion of a middle term (such as a 'state' of drive or emotion) is convenient, but in the former a different conception is suggested. An operation affecting the strength of a single reflex always involves elicitation. In reflex fatigue, for example, the strength is a function of repeated elicitation. And this relation between strength and previous elicitation is such that we may speak of a certain amount of *available activity,* which is exhausted during the process of repeated elicitation and of which the strength of the reflex is at any moment a function.

I shall speak of the total available activity as the *reflex reserve,* a concept that will take an important place in the following chapters. In one sense the reserve is a hypothetical entity. It is a convenient way of representing the particular relation that obtains between the activity of a reflex and its subsequent strength. But I shall later show in detail that a reserve is clearly exhibited in all its relevant properties during the process that exhausts it and that the momentary strength is proportional to the reserve and therefore an available direct measure. The reserve is consequently very near to being directly treated experimentally, although no local or physiological properties are assigned to it. The notion applies to all operations that involve the elicitation of the reflex and to both operant and respondent behavior, whether conditioned or unconditioned.

One distinction between an unconditioned and a conditioned reflex is that the reserve of the former is constantly being restored spontaneously, when it is not already at a maximum. In the particular case of reflex fatigue, a spontaneous flow into the reserve is evident in the complete recovery from fatigue that takes place during rest and in the possibility of reaching a stable intermediate state when the rate of restoration just equals the rate of exhaustion at a given frequency of elicitation. In many unconditioned reflexes the reserve is very great and is replaced at so high a rate that exhaustion is difficult. On the other hand the reserve may be very low and slowly restored, as in the response of starting at a sound, which may occur once or twice at the first presentation but require a very long period of inactivity to recover its strength (the phenomenon of 'adaptation'). In conditioned reflexes the reserve is built up by the act of reinforcement, and extinction is essentially a process of exhaustion comparable with fatigue. The conception applies to both types of conditioning and leads to a much more comprehensive formulation of the process than is available in terms of mere change in strength. The relation of the reserve to the operations of reinforcement and extinction will be dealt with in detail in Chapters Three and Four.

Since the strength of a reflex is proportional to its reserve, it may be altered in two ways. Either the size of the reserve or the proportionality between it and the strength may be changed. All operations that involve elicitation affect the reserve directly, either to increase or to decrease it. Conditioning increases it; extinction and fatigue decrease it. The other operations (which are not unique in their action and affect groups of reflexes) change the proportionality between the reserve and the strength. Facilitation and certain kinds of emotion increase the strength, while inhibition and certain other kinds of emotion decrease it without modifying the reserve. The operations that control the drive also affect the proportionality factor. Without altering the total number of available responses, a change in drive may alter the rate of elicitation of an operant from a minimal to a maximal value. Several demonstrations of the distinction between altering the reserve and altering the proportionality will appear later.

In a phasic respondent the refractory phase suggests a smaller subsidiary reserve which is either completely or nearly completely exhausted with each elicitation. This subsidiary reserve is restored from the whole reserve, but the rate of restoration depends upon the size of the latter. Thus, during the fatigue of such a respondent, the refractory phase is progressively prolonged. The rate of elicitation of an operant exhibits a similar effect, as I have already noted. The total reserve of an operant does not pour out at once as soon as opportunity arises; the rate of elicitation is relatively slow and presumably depends upon a similar subsidiary reserve exhausted at each single occurrence. We may regard the emission of an operant response as occurring when the subsidiary reserve reaches a critical value. A second response cannot occur until the subsidiary reserve has been restored to the same value. The rate of restoration is again a function of the total reserve. I shall not need to refer again to the subsidiary reserve. It is relatively unimportant in the case of respondent behavior, because only a few respondents are phasic. In operant behavior the notion is carried adequately by that of a rate.

The notion of a reserve and of the varying proportionality between it and the strength is something more than a mere definition of dynamic properties in terms of reflex strength. It is a convenient way of bringing together such facts as the following (examples of which will appear later): there is a relation between the number of responses appearing during the extinction of an operant and the number of preceding reinforcements (that is to say, the number of responses that can be obtained from the organism is strictly limited by the number that has been put into it); changes in drive do not change the total number of available responses, although the rate of responding may vary greatly; emotional, facilitative, and inhibitory changes are compensated for by later changes in strength; and so on.

The Interaction of Reflexes

An actual reduction in the number of variables affecting an organic system (such as is achieved by turning off the lights in a room) is to be distinguished from a hypothetical reduction where the constancy or irrelevance of a variable is merely assumed. In the latter case the resulting analytical unit cannot always be demonstrated *in fact,* and the analysis can be regarded as complete only when a successful return has been made through synthesis to the original unanalyzed system. Now, the kind of variable represented by the stimulus may be controlled and even in many cases eliminated. Many of our techniques of analysis are devoted to this end. The reflex as an analytical unit is actually obtained *in practice*. The unit is a fact, and its validity and the validity of the laws describing its changes do not depend upon the correctness of analytical assumptions or the possibility of a later synthesis of more complex behavior.

The preceding laws have applied to a single unit isolated in this practical way. They are valid so far as they go regardless of the fate of the unit when other stimuli are allowed to enter. We are under no obligation to validate the unit through some kind of synthesis. But a description of behavior would be inadequate if it failed to give an account of how separate units exist and function together in the ordinary behavior of the organism. In addition to processes involving reflex strength, a description of behavior must deal with the interaction of its separate functional parts. Interaction may be studied in a practical way by deliberately combining previously isolated units and observing their effect upon one another. In this way we obtain a number of laws which enable us to deal with those larger samples of behavior sometimes dubiously if not erroneously designated as 'wholes.' That great pseudo-problem—Is the whole greater than the sum of its parts?—takes in the present case this intelligible form: What happens when reflexes interact? The effects of interaction are in part topographical and in part intensive.

The Law of Compatibility. *Two or more responses which do not overlap topographically may occur simultaneously without interference.* The responses may be under the control of separate stimuli (as when the patellar tendon is tapped and a light is flashed into the eyes at the same time, so that both the knee-jerk and contraction of the pupil occur simultaneously) or of a single stimulus (as when a shock to the hand elicits flexion of the arm, vasoconstriction, respiratory changes, and so on). The law seems to hold without exception for respondents but requires some qualification elsewhere. We cannot combine operants quite so deliberately because we cannot *elicit* them. We can only present discriminative stimuli simultaneously and build up appropriate drives. When the required discriminative stimuli are few, so that a topographical overlap of stimuli is avoided, and when the drive is strong, there may be no interference up to the limit set by the topography of the responses. Thus, a machinist may stop his lathe by pressing a clutch with his foot, slow the chuck with one hand, loosen the tailstock with the other hand, turn his head in order to catch the light on a certain part of his work, and call to his helper, all at the same time. But this will occur only if the drive is strong—if the machinist is working rapidly. Usually responses make way for each other and occur in some sort of serial order, even when there is little or no topographical overlap.

THE LAW OF PREPOTENCY. *When two reflexes overlap topographically and the responses are incompatible, one response may occur to the exclusion of the other.* The notion of prepotency has been extensively investigated by Sherrington (68). The effect is not inhibition as defined above because both stimuli control the same effector. Various theories of inhibition have appealed to prepotency —to the competitive activity of incompatible responses—as an explanatory principle, but with the limited definition here adopted the two processes are clearly distinct.[3] Prepotency applies to both operant and respondent behavior, but as in the case of the Law of Compatibility it is not easily demonstrated with operants because the precise moment of elicitation is not controlled.

THE LAW OF ALGEBRAIC SUMMATION. *The simultaneous elicitation of two responses utilizing the same effectors but in opposite directions produces a response the extent of which is an algebraic resultant.* If one reflex is much stronger, little or no trace of the weaker may be observed, and the case resembles that of prepotency. When the two exactly balance, either no response is observed, or, as is often the case in systems of this sort, a rapid oscillation of greater or less amplitude appears. When the strengths differ slightly a partial or slower response occurs in the direction of the stronger. A familiar example is the operant behavior of a squirrel in approaching a novel object. There are two responses—one toward the object, the other away from it. If either is relatively strong, the resulting behavior is simply approach or withdrawal. If approach is strong, but withdrawal not negligible, a slow approach takes place. When the two balance (as they often do at some point in the approach), a rapid oscillation may be observed. A respondent example, which has been demonstrated by Magnus (61), is the position of the eye of the rabbit during changes of posture, which is due to the algebraic summation of reflexes from the labyrinths and from the receptors in the

[3] For a recent exposition of inhibition as prepotency see (77).

muscles of the neck. When one response is wholly obscured, the effect must be distinguished from inhibition, as in the case of prepotency, on the grounds that both stimuli here control the effector.

THE LAW OF BLENDING. *Two responses showing some topographical overlap may be elicited together but in necessarily modified forms.* In playing the piano while balancing a wine glass on the back of the hand, the usual movements of the fingers are modified by the balancing behavior. The result is a mechanical interaction of the musculature and resembles the external modification of a response as when one plays with a weight attached to the hand. In the former case the usual form of the behavior is modified by other reflexes, in the latter by an external force. Most of the normal behavior of an organism shows blending of this sort. Many respondent examples are given by Magnus (61).

THE LAW OF SPATIAL SUMMATION. *When two reflexes have the same form of response, the response to both stimuli in combination has a greater magnitude and a shorter latency.* Spatial summation differs from temporal summation in raising a topographical problem. A reflex is defined in terms of both stimulus and response. Two stimuli define separate reflexes even though the response is the same. In spatial summation we are dealing with the interaction of reflexes rather than with the intensification of the stimulus of a single reflex as in temporal summation. This is obvious when the stimuli are distantly located or in separate sensory fields. A familiar example may easily be demonstrated on infants. A movement of the hand before the eyes, and a slight sound, neither of which will evoke winking if presented alone, may be effective in combination. As is most frequently the case in dealing with summation, the example applies to threshold values; but where the stimuli are strong enough to evoke the response separately, the effect of their combination may presumably also be felt on the magnitude of the response. When the stimuli are in the same sensory field, and especially when they are closely adjacent, this formulation in terms of interaction may seem awkward, but it is, I believe, in harmony with the actual data and is demanded by the present system. The fundamental observation is that the response to the combined stimulus is stronger than that to either stimulus separately. The result should be included under algebraic summation, taken literally, but a separate class is usually set up for the case in which the responses not only involve the same effectors but utilize them in the same direction. Much of the behavior of an organism is under the control of more than one stimulus operating synergically. Excellent examples may be found in the work of Magnus on the multiple control of posture (61, 62).

A distinction between spatial summation and facilitation may be pointed out. In facilitation the strength of a reflex changes according to a dynamic law, where the defining operation is the presentation of a stimulus. The relation between the stimulus and the change in strength is like that of inhibition except for the direction of the change. The single condition which distinguishes facilitation from summation is that the facilitating stimulus must not of itself be capable of eliciting the response. If this distinction were not maintained, the two phenomena would be indistinguishable. In summation we cannot regard one stimulus as raising the strength of the relation between the second and the response because of the observed direct effect upon the response of the first. In facilitation, with this effect lacking, we must formulate the change as an example of a dynamic law. Facilitation involves one reflex and a stimulus, but summation involves two reflexes.

Two other laws which come under the broad heading of interaction have a slightly different status, but may be listed here.

THE LAW OF CHAINING. *The response of one reflex may constitute or produce the eliciting or discriminative stimulus of another.* The stimuli may be proprioceptive (as in the serial reaction of throwing a ball) or produced externally by a change in the position of receptors (as when the organism looks to the right and then responds to a resulting visual stimulus or reaches out and then seizes the object which touches its hand). The Law of Chaining is considered again in the following chapter.

THE LAW OF INDUCTION. *A dynamic change in the strength of a reflex may be accompanied by a similar but not so extensive change in a related reflex, where the relation is due to the possession of common properties of stimulus or response.* The dynamic changes are limited to those which affect the reserve. An example of induction is the fatigue of a flexion reflex from one locus of stimulation through repeated elicitation of a reflex from another locus. This is not the meaning of induction given by Sherrington or Pavlov. In Sherrington's usage the term refers both to summation from adjacent stimuli (immediate induction) and to the 'post-inhibitory' strengthening of a related reflex (successive induction). The latter is, as Sherrington points out, in several ways the reverse of the former, and the use of a single term is misleading. Pavlov adopts the term from Sherrington but uses only the second meaning. Neither case matches the present definition, for a fuller explanation of which Chapter Five should be consulted.

The Generic Nature of the Concepts of Stimulus and Response

The preceding system is based upon the assumption that both behavior and environment may be broken into parts which retain their identity throughout an experiment and undergo orderly changes. If this assumption were not in some sense justified, a science of behavior would be impossible. But the analysis of behavior is not an act of arbitrary subdividing. We cannot define the concepts of stimulus and response quite as simply as 'parts of behavior and environment' without taking account of the natural lines of fracture along which behavior and environment actually break.

If we could confine ourselves to the elicitation of a reflex upon a single occasion, the problem would not arise. The complete description of such an event would present technical difficulties only; and if no limit were placed upon apparatus, an adequate account of what might be termed the stimulus and the response could in most cases be given. We should be free of the question of *what* we were describing. But a reproducible unit is required in order to predict behavior, and an account of a single elicitation, no matter how perfect, is inadequate for this purpose. It is very difficult to find a stimulus and a response which maintain precisely the same topographical properties upon two successive occasions. The identifiable unit is something more or something less than such a completely described entity.

In the traditional field of reflex physiology this problem is dealt with by main force. An investigation is confined to a reflex in which the response is originally of a very simple sort or may be easily simplified (flexion, for example, or salivation) and in which the stimulus is of a convenient form and may be localized sharply. It is easier to restrict the stimulus than the response, since the stimulus presents itself as the independent variable, but it is possible by surgical or other technical means to control some of the properties of the response also. In this way a sort of reproducibility is devised, and a restricted preparation is frequently obtained in which a stimulus is correlated with a response and all properties of both terms are capable of specification within a very narrow range. For many purposes a preparation of this kind may be an adequate solution of the problem of reproducibility. Some degree of restriction is probably always required before successful experimentation can be carried on. But severe restriction must be rejected as a general solution, since it implies an arbitrary unit, the exact character of which depends upon the selection of properties and does not fully correspond to the material originally under investigation. The very act of restriction suppresses an important characteristic of the typical reflex, and it is, moreover, not a practical solution that can be extended to behavior as a whole.

An example of the problem is as follows. In the relatively simple flexion reflex, the exact location of the stimulus is unimportant; a correlated response may be demonstrated even though the stimulus is applied elsewhere within a rather wide range. The form of energy also need not be specific. Similarly, on the side of the response we cannot specify the exact direction of the flexion if we have not simplified, or, having simplified, we cannot justify the selection of one direction as against the other. So far as the mere elicitation of the reflex is concerned, most of the properties of the two events in the correlation are, therefore, irrelevant. The only relevant properties are flexion (the reduction of the angle made by adjacent segments of a limb at a given joint) and a given ('noxious') kind of stimulation applied within a rather large area. It will be seen, then, that in stating the flexion reflex as a unit the term 'stimulus' must refer to a *class* of events, the members of which possess some property in common, but otherwise differ rather freely, and the term 'response' to a similar class showing a greater freedom of variation but also defined rigorously with respect to one or more properties. The correlation that is called the reflex is a correlation of classes, and the problem of analysis is the problem of finding the right defining properties.

The level of analysis at which significant classes emerge is not determinable from the mere demonstration of a correlation of stimulus and response but must be arrived at through a study of the dynamic properties of the resulting unit. This is obvious in the case of the operant since the correlation with a stimulus does not exist, but it holds as well for respondent behavior. In order to show this it will be necessary to review the procedure of setting up a reflex.

The first step toward what is called the discovery of a reflex is the observation of some aspect of behavior that occurs repeatedly under general stimulation. The control over the response is almost exclusively that of specification. We have the refusal of all responses not answering to the criteria that we have selected. When the defining property of a class has been decided upon, the stimuli that elicit responses possessing it are discovered by exploration. One stimulus may be enough to demonstrate the sort of correlation sought for, but either deliberately or through lack of control the properties are usually varied in later elicitations and other members of the stimulus class thus added. Subsequently the defining property of the stimulus is identified as the part common to the different stimuli found to be effective.

There must be defining properties on the sides of both stimulus and response or the classes will have no necessary reference to real aspects of behavior. If the flexion reflex is allowed to be defined simply as a reflex having for its response a class defined by flexion, there is nothing to prevent the definition of an infinite number of reflexes upon similar bases. For example, we could say that there is a reflex or class of reflexes defined by this property: that in elicitation the center of gravity of the organism moves to the north. Such a class is experimentally useless, since it brings together quite unrelated activities, but we must be ready to show that all flexions are related in a way in which all geographical movements of the center of gravity are not, and to do this we must appeal to the observed fact that all flexions are elicitable by stimuli of a few classes. As soon as this relation is apparent our tentative response-class begins to take on experimental reality as a characteristic of the behavior of the organism.[4]

Since we are completely free in this first choice, it is easy to select a wrong property, but this is soon detected in our inability to show a correlation with a single stimulus-class. However, a certain freedom in specifying the response remains. By including other properties in our specification we may set up less comprehensive classes, for which correspondingly less comprehensive stimulus-classes may be found. For example, if we begin with 'flexion in a specific direction only,' we obtain a stimulus-class embracing a smaller stimulating area. There is nothing to prevent taking such a restricted unit at

[4] The impossibility of defining a functional stimulus without reference to a functional response, and *vice versa*, has been especially emphasized by Kantor (53).

the start, so long as for any such class a stimulus-class may be found, and if a restricted unit is taken first the very broadest term can be arrived at only by removing restrictions.

Within the class given by a first defining property, then, we may set up subclasses through the arbitrary restriction of other properties. This procedure yields a series of responses, generated by progressive restriction, each member of which possesses a corresponding stimulus in a more or less parallel series. We approach as a limit the correlation of a completely specified response and a stimulus which is not necessarily strictly constant but may be held so experimentally. At this stage the unit is unpractical and never fully representative.

Usually the first restrictions are designed to protect the defining property by excluding extreme cases. They clarify the definition and add weight to the expressed correlation with a stimulus-class. In general, as we progressively restrict, the descriptive term assigned to the reflex comes to include more and more of the two events and is consequently so much the more useful. At the same time a greater and greater restriction of the stimulus-class is demanded, so that the increase in the validity and completeness of the correlation is paid for with added experimental effort.

If we now examine the dynamic properties of this series of correlations, we find that with progressive restriction the dynamic changes in strength become more and more regular. The changes in question are those affecting the reserve, not the proportionality of reserve and strength. If we are measuring fatigue, for example, we shall not obtain too smooth a curve if our stimulus varies in such a way as to produce at one time one direction of flexion and at another time another; but as we restrict the stimulus to obtain a less variable response, the smoothness of the curve increases. This is essentially a consequence of the Law of Induction. In such a process as fatigue or extinction we are examining the effect of one elicitation upon another following it. We look for this effect to follow the main rule of induction: it will be a function of the degree of community of properties. In a completely restricted preparation we should, therefore, have a sort of complete induction, since two successive elicitations would be identical. Each elicitation would have its full effect, and the curve for the dynamic change would be smooth. But if we are using only a partially restricted entity, successive elicitations need not have identically the same properties, and dynamic processes may or may not be advanced full steps. An improvement in data follows from any change that makes successive elicitations more likely to resemble each other.

The generic nature of the concepts of stimulus and response is demonstrated by the fact that complete induction obtains (and the dynamic changes therefore reach an optimal uniformity) *before* all the properties of stimulus or response have been fully specified in the description and respected in each elicitation. I am not prepared to demonstrate this fact in the case of respondent behavior because the present degree of orderliness of the dynamic changes is inadequate, but in operant behavior the same argument holds. The appeal to dynamic laws is especially significant in the case of operant behavior because the other kind of operation that establishes the properties of a class (the presentation of a stimulus) is lacking. At a qualitative level the definition of an operant depends upon the repetition of a sample of behavior with greater or less uniformity. Before we can see precisely what a given act consists of, we must examine the changes it undergoes in strength. Here again we merely specify what is to be counted as a response and refuse to accept instances not coming up to the specification. A specification is successful if the entity which it describes gives smooth curves for the dynamic laws. Since the laws that apply here are those that affect the reserve, the proof that the response is a class of events and that any given instance possesses irrelevant properties reduces essentially to a proof that responses which possess an irrelevant property contribute or subtract from the reserve with the same effectiveness as responses that do not. Many examples will be described later which concern the behavior of a rat in pressing a lever. The number of distinguishable acts on the part of the rat that will give the required movement of the lever is indefinite and very large. They constitute a class, which is sufficiently well-defined by the phrase 'pressing the lever.' It will be shown later that under various circumstances the rate of responding is significant. It maintains itself or changes in lawful ways. But the responses which contribute to this total number-per-unit-time are not identical. They are selected at random from the whole class—that is, by circumstances which are independent of the conditions determining the rate. The members of the class are quantitatively mutually replaceable in spite of their differences. If only such responses as had been made in a very special way were counted (that is, if the response had been restricted through further specification), the smoothness of the resulting curves would have been *decreased*. The curves would have been destroyed through the elimination of many responses that contributed to them. The set of properties that define 'pressing a lever' is thus uniquely determined; specifying either fewer or more would destroy the consistency of the experimental result. It may be added that in the case of conditioned operant behavior the defining property of a class is exactly that given by the conditions of the reinforcement. If the reinforcement depended, for example, upon making the response with a certain group of muscles, the class would change to one defined by that property. Such a class might vary in other ways, but by restricting the reinforcements further and further we could approach a response answering to very rigorous specifications (see Chapter Eight). The present point is that when the reinforcement depends upon such a property as 'pressing a lever,' *other properties of the behavior may vary widely, although smooth curves are still obtained.*

It is true that the non-defining properties are often not wholly negligible and that the members of classes are consequently not exactly mutually replaceable. On the side of the response, the data will not show this in most cases because of the present lack of precision. But it is certain that there are outlying members of a class which have not a full substitutive power; that is to say, there are flexions and pressings that are so unusual because of other properties that they do not fully *count as such*. It ought to be supposed that lesser differences would be significant in a more sensitive test. If we should examine a large number of responses leading to the movement of the lever, most of them would be relatively quite similar, but there would be smaller groups set off by distinguishing properties and a few quite anomalous responses. It is because of the high frequency of occurrence of the similar ones that they are typical of the response 'pressing the lever,' but it is also because of this frequency that any lack of effectiveness of atypical responses is not at present sufficiently strongly felt to be noted.

On the side of the stimulus, small differences may be demonstrated, since we here control the values of the non-defining properties and may mass the effect of a given property. Thus, it can be shown that in the flexion reflex fatigue from one locus of stimulation does not result in complete fatigue of the reflex from another locus. Here particular stimuli have been segregated into two groups on the basis of the property of location, and the relevance of the property to the course of a secondary change appears. In this case we are justified in speaking of different reflexes from the two loci. Similarly, in the example of pressing a lever the reinforcement is made in the presence of an indefinitely large class of stimuli arising from the lever and the surrounding parts of the apparatus. It is possible to control some of the properties of these members. For example, the lever may be made to stimulate either in the light or in the dark, so that all properties which arise as visible radiation can be introduced or removed at will. It will later be shown that these are not wholly irrelevant in the subsequent extinction of the operant. In either of these cases if we had allowed the stimulus to vary at random with respect to the non-defining property, we should have obtained reasonably smooth curves for the dynamic processes, according to present standards of smoothness. It is only by separating the stimuli into groups that we can show their lack of complete equivalence. Once having shown this, we can no longer disregard the importance of the property, even in the absence of grouping.

This demonstration of the generic nature of the stimulus and response does not pretend to go beyond the limits set by the present degree of experimental precision, but its main features are too well marked to be seriously disturbed by limiting conditions. A practical consistency in the dynamic laws may appear at such a relatively unrestricted level—and, as one might say, so suddenly—in the series of progressively restricted entities, that extrapolation to complete consistency appears to fall far short of complete restriction. It would be idle to consider the possibility of details that have at present no experimental reality or importance. It may be that the location of the stimulus to flexion or the forces affecting the organism as the lever is pressed are somehow significant up to the point of complete specification; but we are here interested only in the degree of consistency that can be obtained while they are still by no means completely determined. This consistency is so remarkable that it promises very little improvement from further restriction.

The preceding argument may be summarized as follows. A preparation may be restricted for two quite different reasons—either to obtain a greater precision of reference (so that the description of a response, for example, will be more nearly complete and accurate) or to obtain consistent curves for dynamic processes. The increase in precision gives a greater authority to the statement of a correlation, which is desirable; but it will not help in deciding upon a unit. It leads ultimately to a completely restricted entity, which is usually unreproducible and otherwise unpractical, so that it is necessary to stop at some arbitrary level—for example, at a compromise between precision of reference and the experimental effort of restriction. The second criterion yields, on the other hand, a unit which is by no means arbitrary. The appearance of smooth curves in dynamic processes marks a unique point in the progressive restriction of a preparation, and it is to this uniquely determined entity that the term reflex may be assigned. *A respondent,* then, *regarded as a correlation of a stimulus and a response and an operant regarded as a functional part of behavior are defined at levels of specification marked by the orderliness of dynamic changes.*

In deciding upon this definition we choose simplicity and consistency of data as against exact reproducibility as our ultimate criterion, or perhaps we temper the extent to which exact reproducibility is to be demanded and use the consistency of the data in our defense. This would be good scientific method if we were not forced to it for other reasons. To insist upon the constancy of properties that can be shown not to affect the measurements in hand is to make a fetish of exactitude. It is obvious why this has so often been done. In the case of a respondent what is wanted is the 'necessary and sufficient' correlation of a stimulus and a response. The procedure recommended by the present analysis is to discover the defining properties of a stimulus and a response and to express the correlation in terms of classes. The usual expedient has been to hold all properties of a given term constant so far as this is possible. In a successful case all properties seem to be relevant because they invariably occur upon all occasions. It is almost as if, faced with the evident irrelevance of many properties, the reflex physiologist had invented the highly restricted preparation to *make* them relevant. In giving a complete account of an arbitrarily restricted preparation, we describe at the same time too little and too much. We include material that is irrelevant to the principal datum, so that part of the description is superfluous, and we deliberately ignore the broader character of the stimulus and the response. The same argument applies to operant behavior. The complete description of one act of pressing the lever would have very little usefulness, since most of the information would be irrelevant to the fact of emission, with which we are chiefly concerned, and would tell us nothing about the set of properties that yield a consistent result.

Some amount of restriction is practically indispensable. It has the merit of holding a defining property constant even though the property has not been identified. Until we have discovered a defining property, it is necessary to resort to restriction to guarantee ultimate validity. And since it is often difficult to designate defining properties clearly, especially where extreme values of other properties interfere, some measure of precautionary restriction is usually necessary. It is often not obvious that it is being used. We should find it very difficult to define the class 'pressing a lever' without considerable precautionary restriction of essentially non-defining properties—concerning the size of the lever and so on. The use of a uniform lever from experiment to experiment is in itself a considerable act of restriction and is apparently necessary to assure a consistent result.

Freedom from the requirement of complete reproducibility broadens our field of operation immeasurably. We are no longer limited to the very few preparations in which some semblance of complete reproducibility is to be found, for we are able to define 'parts of behavior and environment' having experimental reality and reproducible in their own fashion. In particular the behavior of the intact organism is made available for study with an expectation of precision comparable with that of the classical spinal preparation. Indeed, if smoothness of the dynamic changes is to be taken as an ultimate criterion, the intact organism often shows much greater consistency than the spinal preparation used in reflex physiology, even though the number of uncontrolled non-defining properties is much smaller in the latter case. Evidence of this will be forthcoming later.

The generic nature of stimuli and responses is in no sense a justification for the broader terms of the popular vocabulary. No property is a valid defining property of a class until its experimental reality has been demonstrated, and this rule excludes a great many terms commonly brought into the description of behavior. For example, when it is casually observed that a child hides when confronted with a dog, it may be said in an uncritical extension of the terminology of the reflex that the dog is a stimulus and the hiding a response. It is obvious at once that the word 'hiding' does not refer to a unique set of movements nor 'dog' to a unique set of stimulating forces. In order to make these terms validly descriptive it is necessary to define the classes to which they refer. It must be shown what properties of a stimulus give it a place in the class 'dog' and what properties of a response make it an instance of 'hiding.' (It will not be enough to dignify the popular vocabulary by appealing to essential properties of 'dogness' or 'hidingness' and to suppose them intuitively known.) The resulting classes must be shown to be correlated experimentally and it ought also to be shown that dynamic changes in the correlation are lawful.[5] It is not at all certain that the properties thus found to be significant are those now supposedly referred to by the words 'dog' and 'hiding' even after allowing for the inevitable vagueness of the popular term.

The existence of a popular term does create some presumption in favor of the existence of a corresponding experimentally real concept, but this does not free us from the necessity of defining the class and of demonstrating the reality if the term is to be used for scientific purposes. It has still to be shown that most of the terms borrowed from the popular vocabulary are validly descriptive—that they lead to consistent and reproducible experimentation. We cannot, with Watson (76), define a response as 'anything the animal does, such as turning toward or away from a light, jumping at a sound, and more highly organized activities such as building a skyscraper, drawing plans, having babies, writing books, and the like.' There is no reason to expect that responses of the latter sort will obey simple dynamic laws. The analysis has not been pressed to the point at which orderly changes emerge.

This restriction upon the use of the popular vocabulary is often not felt because the partial legitimacy of the popular term frequently results in some experimental consistency. The experimenter is more likely than not to hit upon experimentally real terms, and he may have some private set of properties resulting from his own training that will serve. The word 'hiding' may always be used *by him* in connection with events having certain definite prop-

[5] It will appear later that this example is actually a discriminated operant rather than a respondent.

erties, and his own results will be consistent by virtue of this definition *per accidens.* But it is a mistake for him to suppose that these properties are communicated in his use of the popular term. If no more accurate supplementary specification is given, the difficulty will become apparent whenever his experiments are repeated by someone with another set of private defining properties and will be the greater the wider the difference in background of the two experimenters.

This raises a problem in epistemology, which is inevitable in a field of this sort. The relation of organism to environment must be supposed to include the special case of the relation of scientist to subject matter. If we contemplate an eventual successful extension of our methods, we must suppose ourselves to be describing an activity of which describing is itself one manifestation. It is necessary to raise this epistemological point in order to explain why it is that popular terms so often refer to what are later found to be experimentally real entities. The reason is that such terms are in themselves responses of a generic sort: they are the responses of the population of which the experimenter is a member. Consequently, when the organism under investigation fairly closely resembles man (for example, when it is a dog); the popular term may be very close to the experimentally real entity. The experimenter may hit immediately upon the right property of the stimulus, not because he has manipulated it experimentally, but because he himself reacts in a measure similarly to the dog. On the other hand if the organism is, let us say, an ant or an amoeba, it is much more difficult to detect the real stimulus-class without experimentation. If it were not for this explanation, the partial legitimacy of the popular term would be a striking coincidence, which might be used (and indeed has been used) as an argument for the admission of a special method, such as 'empathy' or 'identification' or 'anthropomorphizing' (71), into the study of behavior. In insisting that no amount of reality in the popular terms already examined will excuse us from defining a new term experimentally if it is to be used at all, I am rejecting any such process. The rule that the generic term may be used only when its experimental reality has been verified will not admit the possibility of an ancillary principle, available in and peculiar to the study of behavior, leading to the definition of concepts through some other means than the sort of experimental procedure here outlined.

Chapter Two

SCOPE AND METHOD

The Direction of Inquiry

So far as scientific method is concerned, the system set up in the preceding chapter may be characterized as follows. It is positivistic. It confines itself to description rather than explanation. Its concepts are defined in terms of immediate observations and are not given local or physiological properties. A reflex is not an arc, a drive is not the state of a center, extinction is not the exhaustion of a physiological substance or state. Terms of this sort are used merely to bring together groups of observations, to state uniformities, and to express properties of behavior which transcend single instances. They are not hypotheses, in the sense of things to be proved or disproved, but convenient representations of things already known. As to hypotheses, the system does not require them —at least in the usual sense.

It is often objected that a positivistic system offers no incentive to experimentation. The hypothesis, even the bad hypothesis, is said to be justified by its effect in producing research (presumably even bad research), and it is held or implied that some such device is usually needed. This is an historical question about the motivation of human behavior. There are doubtless many men whose curiosity about nature is less than their curiosity about the accuracy of their guesses, but it may be noted that science does in fact progress without the aid of this kind of explanatory prophecy. Much can be claimed for the greater efficiency of the descriptive system, when it is once motivated.

Granted, however, that such a system does possess the requisite moving force, it may still be insisted that a merely descriptive science must be lacking in direction. A fact is a fact; and the positivistic system does not seem to prefer one to another. Hypotheses are declared to solve this problem by directing the choice of facts (what directs the choice of hypotheses is not often discussed), and without them a distinction between the useful and the useless fact is said to be impossible. This is a narrow view of a descriptive science. The mere accumulation of uniformities is not a science at all. It is necessary to organize facts in such a way that a simple and convenient description can be given, and for this purpose a structure or system is required. The exigencies of a satisfactory system provide all the direction in the acquisition of facts that can be desired. Although natural history has set the pattern for the collection of isolated bits of curious behavior, there is no danger that a science of behavior will reach that level.

The research to be described in this volume has been dictated by the formulation of the system described in the preceding chapter, and the general direction of inquiry may be justified by appeal to the system in the following way. There is a lack of balance, at the present time, in favor of respondent as against operant behavior. This is explicable on historical grounds. The discovery of the stimulus as a controlling variable was the first great advance in reducing behavior to some kind of order—a discovery that naturally encouraged research in bringing to light stimulus-response relationships. The investigation of the lower reflexes which began with Marshall Hall and reached its height with Sherrington established the reflex as a valid concept and set the pattern for analogous research on higher behavior. Pavlov's discovery of the conditioned reflex of Type S emerged from the study of unconditioned alimentary respondents, and the greater part of the work on conditioning has kept to this type. The early contention that the concepts applicable to spinal respondents and to conditioned reflexes of Type S could be extended to behavior in general has delayed the investigation of operant behavior. There is, therefore, good reason to direct research toward obtaining a better balance between the two fields, especially since the greater part of the behavior of the intact organism is operant.

An important historical phase of the investigation of respondents was topographical. New reflexes were discovered and named. But with the extension of the field through the discovery of conditioning and the realization that reflexes could be endlessly multiplied, the isolation and naming of reflexes lost much of its importance. No comparable phase has tended to arise in the operant field. The general topography of operant behavior is not important, because most if not all specific operants are conditioned. I suggest that the dynamic properties of operant behavior may be studied with a single reflex (or at least with only as many as are needed to assure the general applicability of the results). If this is true, there should be no incentive to 'botanize.' The present work is accordingly confined to a single reflex—the behavior of pressing downward a horizontal bar or lever. The only topographical problem to be considered is the legitimacy of defining a response in terms of certain properties. This is the problem of a unit of behavior, discussed at the end of the preceding chapter, which extends in the present work principally to the field of discrimination and to the necessity of formulating a discrimination in terms of related reflexes. The problem may be treated conveniently with a single operant.

The principal problems in the field of behavior lie in the direction of the laws of reflex strength, and this is the chief burden of the following work. The remaining part of the field—the interaction of separate reflexes—I have felt could not (except for induction and chaining) be successfully investigated until the laws of strength applying to the single reflex were better known. The stage of combining two reflexes in order to observe the resultant behavior has not been reached. The formulation and classification of the kinds of operations inducing changes in strength are chiefly an observational rather than an experimental problem, but the quantitative study of the laws governing the relation between the operations and the changes is clearly experimental. The central problem of the book is the formulation of behavior, the treatment of which is intended to be reasonably exhaustive. The experimental work fills in the formulative framework so far as I have been able to perform the experiments up to the present time.

The order in which the dynamic laws are considered is determined by experimental convenience. The study of variables which are most easily held constant may conveniently be postponed in favor of an experimental attack upon the less tractable. Age, sexual cycles, and health affect reflex strength, but they may be made relatively unimportant by using healthy male organisms in the least rapidly changing part of their life span. Drugs and surgical techniques that affect strength may simply be avoided. Emotion can for the most part be eliminated by careful handling and adaptation to the apparatus and procedure. Drive and conditioning, then, remain as the two principal factors to be investigated.

They are perhaps of the same order of tractableness. Either may be held fairly constant when the other is being studied. Conditioning is taken up first in the following pages because there is more to say about it now. Such a plan of procedure does not imply that the untreated factors are not important in the description of behavior but simply that being assumed to be susceptible of control they may be left until a later stage of the investigation.

It should be clearly understood that this book is not a survey or summary of the literature on behavior. It avoids topographical exploration. In the chapter on conditioning I shall not review all the kinds of reflexes that have been conditioned; in the chapter on drive I shall not try to cite the available material on all drives; and so on. My concern is with the formulation of typical material only. If I fail to discuss any *kind* of process about which there is at the present time any reliable information, it will be a serious omission, but I shall deliberately pass over many important investigations which are topographically parallel to those mentioned. Perhaps I should say something about the almost exclusive use of my own experimental material in connection with operant behavior. There is no implication whatever that this is the only important work that has been done in the field, but simply that I have had little luck in finding relevant material elsewhere because of differences in basic formulations and their effect upon the choice of variables to be studied.

The Organism

In the broadest sense a science of behavior should be concerned with all kinds of organisms, but it is reasonable to limit oneself, at least in the beginning, to a single representative example. Through a certain anthropocentricity of interests we are likely to choose an organism as similar to man as is consistent with experimental convenience and control. The organism used here is the white rat. It differs from man in its sensory equipment (especially in its poorer vision), in its reactive capacities (as of hands, larynx, and so on), and in limitations in certain other capacities such as that for forming discriminations. It has the advantage over man of submitting to the experimental control of its drives and routine of living. There are other organisms differing less widely in capacity that would serve as well in this respect, such as the ape, dog, or cat, but the rat has the following added advantages. It is cheap and cheaply kept; it occupies very little laboratory space; and it is amazingly stable in the face of long and difficult treatment. Some of the procedures to be described later could not have been used with dogs or apes because of the tendency of such organisms to develop 'neuroses' (64, 59). The rat is also easily adapted to confinement and has an advantage in this respect, especially over the cat. This does not mean that a result obtained with a rat is not applicable to a cat, but that the cat possesses strong reflexes conflicting with a particular experimental procedure which are weak or lacking in the case of the rat. Confinement is used as a means of excluding extraneous stimuli, and it is well to choose an organism for which this exclusion is as simply arranged as possible.

The rats used in the following experiments were in part members of a long inbred strain of albinos and an inbred hooded strain,[1] and in part commercial albinos of unidentified stock. With a very few exceptions all were males. Experimentation was usually begun at about 100 days of age. Experimental groups were practically always made up of litter mates. The rats were healthy at the beginning of the experiments and were discarded if any illness developed.

The Operant

The operant that I have used is the behavior of pressing downward a small lever. A typical lever is made of ⅛-inch brass rod and is shown in its place in the apparatus in Figure 1. The part available to the rat is a horizontal section 8 cm. long parallel to and approximately 1 cm. from the wall of the experimental box and 8 to 10 cm. above the floor. In order to press the lever down the rat must lift its forelegs from the floor, put one or both of them on the bar, and press downward with about 10 grams of pressure. The vertical movement of the bar is through a distance of about 1.5 cm.

The selection of this sample of operant behavior is based upon the following considerations.

(a) Either it is a practically universal unconditioned response (if it is to be regarded as unconditioned investigatory behavior) or it does not presuppose conditioned manipulatory behavior that is at all extraordinary for the species. Less than one per cent of the

[1] Kindly supplied by Dr. Gregory Pincus of the Harvard Biological Laboratories.

rats I have used have failed to make the response at some time or other.

(b) It has a convenient frequency of occurrence before conditioning takes place. An untrained rat placed in a small box with the lever will press it from one to ten or more times per hour, depending upon hunger, the presence of other stimuli, and so on. This is

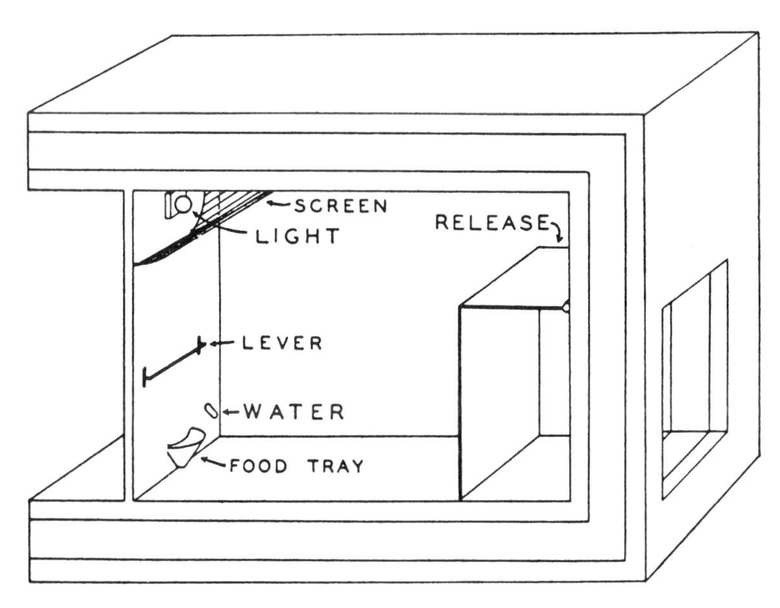

FIGURE 1

A TYPICAL EXPERIMENTAL BOX

One side has been cut away to show the part occupied by the animal. The space behind the panel at the left contains the rest of the lever, the food magazine, and other pieces of apparatus.

an adequate amount of 'spontaneous' activity for the conditioning of an operant.

(c) At the same time it does not occur so frequently without conditioning that the effect of reinforcement is obscured. In this respect it may be contrasted with running, lifting the fore part of the body into the air, and so on.

(d) It is not included in any other significant behavior. The response of flexing a foreleg, for example, might be a component part in the responses of scratching, eating, cleaning the face, running, climbing, and so on. A description of its changes in strength would need to take all these various behaviors into account. Two difficulties of this sort arise in the case of pressing a lever, but they may be eliminated in the following way. (1) The lever may be pressed when the rat is exploring the wall space above the lever, but this may be corrected by projecting the wall or a screen forward for a short distance above the lever (see the figure). (2) In certain types of experiments (involving an emotional reaction) the rat may gnaw the lever and incidentally move it up and down. When necessary this may be avoided by using a lever of larger diameter (say, 1.5 cm.) so that gnawing is either impossible or quickly discouraged.

(e) The response is relatively unambiguous. There is no difficulty in deciding whether or not a given movement is to be counted as an elicitation, as would be the case if the response were defined as a given movement of a leg, for example.

(f) It is made in approximately the same way upon each occasion. The differences actually observed will be discussed later.

(g) Lastly, the response requires external discriminative stimulation (provided by the lever). The nature of this will be considered later. Its presence is necessary for two reasons. If the response did not need what Tolman (71) calls external support, it might be made by the organism outside the experimental situation, but no record would be taken and no reinforcement provided. Experiments which extended over a number of experimental periods would be seriously disturbed. Secondly, the sample must be discriminative in order to be typical. It would be quite possible experimentally to use such an 'unsupported' response as flexing a leg or flicking the tail, but it is only in verbal behavior that such non-mechanically effective responses are reinforced (*i.e.*, when they become gestures). In general, a response must act upon the environment to produce its own reinforcement. Although the connection between the movement of the lever and reinforcement is in one sense artificial, it closely parallels the typical discriminated operant in the normal behavior of the rat.

The response of pressing a lever meets these several requirements with reasonable success and is perhaps nearly optimal in this respect. It does not follow that the laws arrived at in this case cannot be demonstrated with other kinds of responses. The analysis necessary for the demonstration would simply be more difficult.

For example, if the response were part of many different kinds of conditioned and unconditioned behavior [see (d), page 49], the curves obtained during various changes in strength would be composite and highly complex but not for that reason less lawful.

The reinforcement of the response is accomplished automatically with a food magazine which discharges pellets of food of uniform size into a tray immediately beneath the lever. The pellets are made with a device similar to a druggist's 'pill-machine' and are composed of the standard food with which the rats are normally fed. When the lever has been pressed and a pellet eaten, the rat is left in approximately the same position relative to the lever as at the beginning.

The entire behavior of lifting up the fore part of the body, pressing and releasing the lever, reaching into the tray, seizing the pellet of food, withdrawing from the tray, and eating the pellet is, of course, an extremely complex act. It is a chain of reflexes, which for experimental purposes must be analyzed into its component parts. The analysis is most easily carried out by observing the way in which the behavior is acquired. In describing material of this sort an abbreviated system of notation is practically indispensable, and I shall insert here a table of the symbols to be used, not only in this discussion, but during the course of the book.

A System of Notation

The symbols to be used and their definitions are as follows:

$S =$ a stimulus

$R =$ a response

$S . R =$ a respondent

$s . R =$ an operant (The originating force in an operant is specified in the formulation as s, although it is not under experimental control.)

The properties of a term are indicated with lower case letters. Thus, $Sabc \ldots =$ a stimulus with the properties $a, b, c \ldots$ (wave-length, pattern, intensity . . .). The absence of a property is indicated with the corresponding Greek lower case letter. Thus $R\,\alpha =$ a response lacking the property a.

Superscripts comment upon the term, refer to its place in a formula, and so on, without specifying properties. Specific examples (to be defined later) are: S^1 (a reinforcing stimulus), S^D (a discriminative stimulus correlated with reinforcement), and S^Δ (a discriminative stimulus negatively correlated with reinforcement).

A composite stimulus is indicated by juxtaposition of its parts without punctuation, as in $sS^{D'}S^{D''}$.

When examples are given, they are inserted after the symbols with a separating colon, as in the expression *S: shock . R: flexion.*

[ ] = 'the strength of' the enclosed reflex. Thus, $[sS^D\lambda . R]$ will later be seen to represent 'the strength of an operant in the presence of a stimulus which is correlated with a reinforcement and is characterized as such by the absence of the property *l*.'

$\rightarrow$ = 'is followed by.' For example, we may later use the expression $s . R\, a \rightarrow S^1$ for 'a response made without respect to discriminative stimulation which is reinforced provided it does not possess the property *a*.'

Analysis of the Chain

The Law of Chaining was stated in Chapter One in this way: the response of one reflex may produce the eliciting or discriminative stimulus of another. This is a principle of extraordinarily wide application in the integration of behavior. Most of the reflexes of the intact organism are parts of chains. The sample of behavior used here is typical in this respect, and its 'molecular' nature (71) must be carefully stated.

The later stages of the behavior of pressing the lever and eating the pellet are unconditioned respondents. A hungry rat will respond to the touch of food upon the lips by seizing the food with its teeth (and hands), chewing, moistening with saliva, and swallowing. The response of seizing the food in the mouth has been studied by Magnus (61). The behavior of chewing and swallowing may be analyzed into a rather intricate series of reflexes, which are of only slight interest here.[2] The chain is important in the

[2] Wright (79) lists the following steps in the case of human deglutition:

'1. After mastication, the food is rolled into a bolus and is moved to the back of the mouth by elevation of the front of the tongue. The mylohyoid contracts, and the bolus is thrown back between the pillars of the fauces.

'2. The sensory nerves at the entrance to the pharynx are stimulated, and reflexly through the medullary centres the following complex co-ordinated movement results:

study of behavior only up to the point of the seizing of food, which may be written *S: food . R: seizing.*

At the beginning of an experiment a hungry rat is placed in the experimental box containing the tray for several periods of, say, one hour. Investigatory responses to the walls of the box, the tray, and so on, are elicited but soon adapt out to a fairly low strength. The tray contains food, however, and certain movements made by the rat in the presence of stimulation from the tray and adjacent parts of the box are reinforced by the action of *S: food*. Such movements become fully conditioned and are made with considerable frequency by a hungry rat. The response is a discriminated operant, the nature of which will be more or less thoroughly investigated later. It may be written sS^D*: tray . R: approach to tray*. The chain stands at this point as follows:

sS^D*: tray . R: approach* → *S: food . R: seizing.*

The next step in building up the total sample is the establishment of a 'remote' discrimination (see Chapter Five), in which sS^D*: tray . R: approach* is reinforced only when a discriminative stimulus supplied by the sound of the food magazine is presented. The tray is empty except after the magazine has dropped a pellet of food into it. The rat comes to respond to the tray when the magazine sounds but not frequently otherwise. The reinforced reflex may be written sS^D*: tray* S^D*: sound . R: approach.* As will be shown in Chapter Six the sound of the magazine now acquires reinforcing power and a further member of the chain may be added, namely, pressing the lever, which produces the sound of the magazine because of the arrangement of the apparatus. The response to the lever depends upon discriminative stimulation supplied tactually by the lever itself and the reflex may be written

sS^D*: tactual lever . R: pressing.*

---

i. The soft palate is elevated and the post-pharyngeal wall bulges forward to shut off the posterior nares.

ii. The posterior pillars of the fauces approximate to shut off the mouth cavity.

iii. The larynx is pulled up under cover of the root of the tongue, and the vocal cords are approximated.

iv. The epiglottis serves as a sloping ledge to guide the bolus past the laryngeal opening.

v. Respiration is inhibited.

vi. The superior constrictor muscles are relaxed to receive the bolus.

'3. A peristaltic wave propels the food along the oesophagus.'

A further link is added because the tactual stimulation from the lever reinforces all responses toward the lever which produce it. Such a response is also a discriminative operant, which may be written sS^D*: visual lever (or stimulation from adjacent parts of the apparatus) . R: lifting hands and fore part of body*. The completed chain may be written:

sS^D*: visual lever . R: lifting* → sS^D*: tactual lever . R: pressing* →.
sS^D*: tray* S^D*: sound of magazine . R: approach to tray* →
S: food . R: seizing,

where the second arrow is understood to connect the response with S^D*: sound* only. Dropping out the names of the terms, numbering the parts, and omitting the discriminative stimulation supplied by the tray in the absence of the sound of the magazine, we have:

$$sS^{D\,IV} . R^{IV} \rightarrow sS^{D\,III} . R^{III} \rightarrow sS^{D\,II} . R^{II} \rightarrow S^{I} . R^{I}$$

which represents the final structure of the behavior.

Of these four reflexes only Reflex III will be recorded and studied in what follows, but it is possible to give a fair account of the whole chain through this one member. Any occurrence of Reflex IV is almost invariably followed by Reflex III (if the lever is touched, it is pressed), and any occurrence of Reflex III implies the occurrence of Reflex IV (if the lever is pressed, it must have been touched). When the chaining is intact, Reflexes II and I practically always follow. The chaining is under experimental control, however, and the effect of breaking the chain will be studied in certain cases. The use of a chain cannot be avoided in dealing with operant behavior because the very act of reinforcement implies it. A simpler example, as I have said, could have been used by making Reflex III independent of external discriminative stimuli, for example, by using mere flexion of a limb, but it would have been less typical of the normal behavior of the organism.

Chaining is not peculiar to operant behavior. The example of swallowing described above (footnote, p. 52) is almost entirely respondent. Excellent examples of the analysis of chains of respondents are given by Magnus (61) in his work on posture and progression. The principle is the same, with the slight exception that in the case of operants the responses produce discriminative rather than eliciting stimuli.

In every case what we have is a chain of reflexes, not a 'chained reflex.' The connections between parts are purely mechanical and may be broken at will. Any section of a chain may be elicited in isolation with the same properties which characterize it as part of the total chain. There is no reason to appeal to any unique property of the whole sample as an 'act.' I make these statements as explicitly as possible in view of prevailing opinions to the contrary. Experimental justification for the present 'molecular' view will accumulate during the rest of this work.

Control of Extraneous Factors

Whatever success the experiments to be described later may have in revealing uniformities in reflex strength is due to the procedures through which the reflex is isolated and through which extraneous factors affecting the strength are controlled. A first precaution is the removal of stimuli which elicit other reflexes, the necessity of which follows obviously enough from the formulation given in Chapter One. Not all such stimuli can be removed, but a nearly maximal isolation can be achieved by conducting the experiments in a sound-proof, dark, smooth-walled, and well-ventilated box, such as is shown in Figure 1. The stimuli that remain are chiefly the sounds produced by the rat's own movements and the tactual stimulation from the box. By placing the ceiling beyond reach of the rat one wall is effectually removed. The size of the base should be a compromise between a minimal size representing the smallest possible stimulating surface and a maximal size eliciting no responses to restraint. The boxes that I have used vary from 10 cm. x 20 cm. to 30 cm. x 35 cm. at the base. Reflexes in response to the walls and the incidental stimuli produced by the rat itself adapt out quickly, and the rat remains in a relatively inactive state until specific reflexes in response to the tray and lever are established. The ventilation of the box is achieved by drawing air out through a small tube not shown in the figure.

The control of the drive (in this case hunger) is not so easily arranged. For most purposes the same degree of drive must be reproduced upon successive days, and in many cases the degree must be varied in a known way. The use of different periods of fast is open to the objection that the organism does not eat continuously. If it were true that food is ingested at a stable (necessarily low) rate, it would be possible by cutting off the supply of food at different times before the experiment to obtain an array of degrees of hunger having some relation (not necessarily linear) to the lengths of fast. But the rat ordinarily confines its eating activity to a few periods during the day, and much depends upon the state of the organism just before the fast is begun. The method is valid only where the irregularity due to this factor does not matter, which means during fasts of the order of several days. These induce extreme degrees of hunger, which are complicated by other factors, and the method is inadequate for most purposes.

The feeding of limited amounts of food daily will also produce progressive changes in hunger unless the amounts are happily chosen. Modifying the amounts as the weight of the rat changes will avoid prolonged progressive changes but is probably too slow an adjustment to give a uniform hunger from day to day.

It has been possible to reproduce a given degree of drive upon successive days through the following procedure. The rat is placed on a schedule of daily feeding, according to which it is allowed to eat freely once a day for a definite length of time. With dry food a feeding period of one or two hours may be advisable; with a mash as little as ten minutes will suffice. The food I have used is a standard dog biscuit (Purina Dog Chow), which is capable of maintaining rats in good health for several months without a supplementary diet. After about a week of this procedure a high and essentially constant degree of hunger is reached each day just before the time of feeding. Proof of the constancy will be given in Chapters Four and Ten. From this essentially maximal value various lower states may be reached by feeding uniform amounts of food, as will also be shown later. Since thirst affects hunger, a supply of water is made available at all times. The 'hunger cycle' obtained in this way is a function of external stimuli which act as a sort of clock. It can be successfully maintained only if the living conditions of the rat are held constant. The cycle is also a function of the temperature and perhaps also of the humidity. In most of the present cases the rats remained between experiments in a dark sound-proof room at a temperature of from 75° to 80°, varying for any one experiment less than one degree. The humidity was not controlled.

The experiments are conducted at the usual feeding time in order to take advantage of the constancy of the peak of the cycle. When small amounts of food are used in the experiment, however, the major part of the daily ration must be given later, and this introduces a difficulty. If an experiment is repeated for several successive days, the peak of the cycle may shift from the beginning to the end of the experimental hour. Evidence for such a shift may be obtained independently from a study of the spontaneous activity of the rat (see Chapter Nine). In order to eliminate the shift, the experiments may be conducted on alternate days only, the animals being fed on the intervening days in their living-cages at the beginning of the hour of the experiment. In many of the experiments here described in which a process extended over a period of several days this procedure was used. The intervening days are neglected in the description, where such a phrase as 'the preceding day' is to be understood to mean the preceding experimental day, with an intervening non-experimental day omitted.

The elimination of extraneous stimuli has the added effect of avoiding most sources of emotional change. The principal precaution that must be taken is in the handling of the animals at the beginning of an experiment. The effects of handling may be minimized by confining the rat behind a release door when it is put into the experimental box and allowing it to remain there for a minute or two after the box is closed and before the experiment proper begins. The release door should be reasonably silent in operation and out of reach of the rat when open. The drawing in Figure 1 shows such a door in place. It is operated by a projection of the shaft upon which it is mounted and is held against the ceiling when open.

While I have never repeated experiments without precautions of this sort, I believe that the regularity of the data obtained testifies to their advisability.

The Measurement of the Behavior

The problem of recording behavior is in general easily solved. Compared with the data of many other sciences behavior is macroscopic and slow. A single moving picture camera will provide most of the required information in a convenient form, although a battery of cameras and a sound-recorder will, of course, do even better. But a mere record, in the sense of a portrait or representation, is not to be confused with measurement. No matter how complete, a representation is only the beginning of a science. What it does—

whether obtained with a system of notation or with photography and phonography—is to permit leisurely inspection and measurement. The most complete and accurate record of behavior possible would differ from the behavior itself only in that it could be run slowly, or held still, or repeated at will.

So far as measurement is concerned, much of the detail of a complete representation is unnecessary and even inconvenient. I am not speaking here of the wholesale measurement of behavior which yields a sort of quantified narrative but of measurement which presupposes an analytic and selective system. The need for quantification in the study of behavior is fairly widely understood, but it has frequently led to a sort of opportunism. The experimenter takes his measures where he can find them and is satisfied if they are quantitative even if they are trivial or irrelevant. Within a system exhibiting reasonable rigor the relative importance of data may be estimated and much useless measurement avoided. With a systematic formulation of behavior it is usually possible to know in advance what aspect of behavior is going to vary during a given process and what must, therefore, be measured. In the present case the following aspects of the system bear upon the problem of the measure to be taken: (1) the definition of behavior as that part of the activity of the organism which affects the external world; (2) the practical isolation of a unit of behavior; (3) the definition of a response as a class of events; and (4) the demonstration that the rate of responding is the principal measure of the strength of an operant. It follows that the main datum to be measured in the study of the dynamic laws of an operant is the length of time elapsing between a response and the response immediately preceding it or, in other words, the rate of responding.

It may be objected that information other than the rate could surely do no harm and might be interesting or even valuable. It is true that simultaneous photographic records would have been useful in perhaps one per cent of the following cases; beyond that, opposing practical considerations must be taken into account. By recording the rate of responding only, it has been possible for one person to study approximately two million responses within six years. With a few exceptions only one thing is known of each response—how long a period of time elapsed between it and the preceding response. This single datum is enough for the purpose of the present formulation, and the result is, I believe, more valuable than a more complete description of a small part of the same material would have been.

The movement of the lever is recorded electrically as a graph of the total number of responses plotted against time. The required apparatus consists of a slow kymograph and a vertically moving writing point. At each response the point is moved a uniform distance by an electrically operated ratchet. A step-like line is

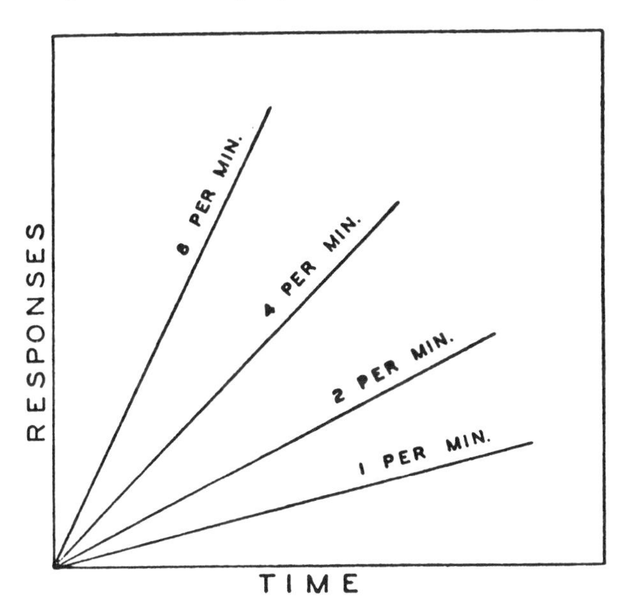

FIGURE 2

SAMPLE SLOPES OBTAINED WITH THE COORDINATES MOST FREQUENTLY USED IN THE FOLLOWING CHAPTERS

The number of responses per minute represented by each slope is indicated. The actual records are step-like.

obtained, the slope of which is proportional to the rate of responding. The speed of the kymograph and the height of the step are chosen to give a convenient slope at the more frequent rates of responding. In Figure 2 some representative slopes are given for the coordinate values used in the greater part of the following account. The step-like character is not shown in the figure. [The movement of the lever operates the recorder by closing a mercury switch on the other side of the panel bearing the lever. In the first experiments with this method a needle attached to the lever-arm dipped into a small cup of mercury. When the lever was moved slowly there was a tendency for the contact to chatter, and this was corrected by inserting into the circuit to the recorder a device which made it impossible for a second contact to be recorded within, say, one second. It has been found that a commercial mercury tube switch does not require this precaution.]

With a record of this sort it is possible to survey at a glance the state of a reflex and its various changes in strength during an experimental period. The form of the record is especially adapted to the study of dynamic laws. We are often interested, it is true, in the course of the change in *rate* rather than in the total number of responses, but it is much easier to record the responses of the rat in a cumulative or integral curve than in a differential. When we are interested in the rate, the curves must be read with respect to their slopes. It is often convenient to have a plot showing rate against time, and examples are given in many cases below. I have not converted all records to this form, however, partly because I wish to remain as close to the experimental data as possible but also because the cumulative curve has a special advantage in dealing with the notion of a reserve and with its subsidiary effects (such as compensation for temporary deviations).

Records of this sort are easily classified and filed, and they supply a permanent first-hand account of the behavior. It may be noted that at no point does the experimenter intervene for purposes of interpretation. All the curves given in this book (except those obtained by averaging or those extending over a number of days) are photographic reproductions of records made directly by the rats themselves. The presence of the experimenter is not required after the experiment has begun. Many of the figures reproduced later were taken *in absentia*. Because of the automatic character of the apparatus it is possible to conduct several experiments simultaneously. I have usually worked with sets of four, although in certain cases as many as twelve animals have been studied at the same time.

Chapter Three

CONDITIONING AND EXTINCTION

The Process of Conditioning

'The term "conditioned" is becoming more and more generally employed, and I think its use is fully justified in that, compared with the inborn reflexes, these new reflexes actually do depend on very many conditions, both in their formation and in the maintenance of their physiological activity. Of course the terms "conditioned" and "unconditioned" could be replaced by others of arguably equal merit. Thus, for example, we might retain the term "inborn reflexes," and call the new type "acquired reflexes"; or call the former "species reflexes" since they are characteristic of the species, and the latter "individual reflexes" since they vary from animal to animal in a species, and even in the same animal at different times and under different conditions.'

This quotation from Pavlov [(64), p. 5] will serve to explain the use of the term 'conditioned.' It denotes a class of reflexes which are *conditional* upon a certain operation performed upon the organism (called reinforcement). The French term (*les réflexes conditionnels*) is in better accord with this meaning. A conditioned reflex may be identified as such by showing, not that it does not exist at birth (an unconditioned reflex may 'mature' later and a conditioned reflex develop earlier), but that it did not exist until the operation of reinforcement had been performed. It may also be distinguished by showing that through elicitation without reinforcement it is removed from the repertory of the organism.

The emphasis in the quotation upon a *kind* of reflex is unfortunate. Except for its dependence upon reinforcement the conditioned reflex behaves with respect to other operations just as any other reflex. The important thing is the process of conditioning and its reciprocal process of extinction. The changes in strength effected by reinforcement continue after the reflex has been acquired and after the distinction between innate and acquired has become trivial. It can even be shown that conditioning may take place when the creation of a topographically new reflex is not involved, as when the effect of reinforcement is to increase the strength of a reflex which parallels one already existing. (For example, let the introduction of a mild solution of acid into the mouth precede the administration of food. Eventually the secretion following the acid is predominantly conditioned, but the reflex *S: acid . R: salivation* is not topographically new.) All reflexes are subject to experimental modification in strength. The change in strength called conditioning is distinguished merely by the specific operation that brings it about. The study of conditioning is not the study of a kind of reflex but of the operation of reinforcement and its effect upon reflex strength.

The operation of reinforcement is defined as the presentation of a certain kind of stimulus in a temporal relation with either a stimulus or a response. A reinforcing stimulus is defined as such by its power to produce the resulting change. There is no circularity about this; some stimuli are found to produce the change, others not, and they are classified as reinforcing and non-reinforcing accordingly. A stimulus may possess the power to reinforce when it is first presented (when it is usually the stimulus of an unconditioned respondent) or it may acquire the power through conditioning (see Chapter Six).

Conditioning of Type S

In Chapter One it was pointed out that there are two types of conditioned reflex, defined according to whether the reinforcing stimulus is correlated with a stimulus or with a response. The case involving a correlation with a stimulus (Type S) may be represented as follows:

where S^0 is a stimulus which elicits the irrelevant response R^0 but does not (in the typical case) elicit R^1 prior to conditioning, and where S^1 is a reinforcing stimulus eliciting R^1. In a typical example S^0 is a tone, S^1 the introduction of food into the mouth, and R^1 salivation. The requirements for conditioning are some strength of $S^1 . R^1$ at the moment and the approximately simultaneous presentation of the two stimuli. The result is an increase in $[S^0 . R^1]$. The reflex may exist at zero strength prior to the conditioning. The change is in one direction only—an increase in strength—and differs in this respect from that in Type R which could involve a decrease. The present type is confined to respondents and is the type originally studied by Pavlov.

Since conditioning is a change in the strength of $S^0 . R^1$, which is a respondent, it is measured in terms of static properties. Pavlov uses the R/S ratio (including after-discharge) and the latency—that is, the effect of the reinforcement is observed as an increase in the magnitude of R^1 in response to an S^0 of constant intensity and duration or as a reduction in the time elapsing between presentation of S^0 and the beginning of R^1. The strength can be measured only when S^1 is not presented.

Some of the factors affecting the rate of conditioning may be listed. Quantitative determinations of the rate as a function of these variables are in general lacking.

Properties of S^0. In the usual experiment S^0 is a mild stimulus eliciting no unconditioned response of any importance. According to Pavlov [(64), p. 29], 'conditioned reflexes are quite readily formed to stimuli to which the animal is more or less indifferent at the outset, though strictly speaking no stimulus within the animal's range of perception exists to which it would be absolutely indifferent.' Stimuli may be arranged in order according to their corresponding rates of conditioning. For example, with a common reinforcement (food) a conditioned salivary reflex was established to visual stimulation from a rotating object in five combined presentations but to a buzzer in only one. With a different reinforcement (acid) the odor of amyl acetate required 20 presentations. Stimuli which belong to strong reflexes may interfere with the process of conditioning, probably because they produce emotional changes, but conditioning may nevertheless be effected in many cases. Pavlov reports the conditioning of an alimentary response to a strong electric current which originally elicited a violent 'defensive' response. The defensive response adapted out completely during the conditioning, but, as Hull's summary of the evidence shows, (48) this is not a universal result.

Properties of $S^1 . R^1$. The provision that $S^1 . R^1$ should have some considerable strength should be particularly noted. The mere simultaneous presentation of two stimuli, when neither of them evokes a response, is not asserted by this law to have any effect. The provision accounts for the failure to apply the formula successfully to many examples of learning through temporal contiguity. It also means that the organism must be awake and the basic drive underlying $S^1 . R^1$ strong. A conditioned alimentary reflex is easily established in a hungry dog but slowly or not at all in one recently fed.

Temporal relation of S^0 and S^1. The required relation of the stimuli is expressed by Pavlov as follows: 'The fundamental requisite is that any external stimulus which is to become the signal in a conditioned reflex must overlap in point of time with the action of an unconditioned stimulus. . . . It is equally necessary that the conditioned stimulus should begin to operate before the unconditioned comes into action [(64), pp. 26, 27]. Pavlov's own work has shown that overlap in time is not a necessary condition, although it may greatly aid the development of conditioning. The temporal order has been disputed. Hull (48) has summarized the evidence on this point and has concluded that 'backward' conditioning (conditioning in which S^0 follows S^1) is possible. However, the two examples he cites (experiments by Switzer and Wolfle) are not unequivocally of this type, as will be shown later (Chapter Six).

The course of the change. Surprisingly little work has been done on the curve for the 'acquisition' of a conditioned reflex of Type S. Hull (48) has summarized evidence to show that the increase in strength is positively accelerated at the beginning of the process. Since the strength eventually reaches a maximum, an S-shaped curve probably characterizes the process. It is difficult to obtain smooth curves because of the complicating condition that S^1 must be withheld in measuring the strength, an act which itself affects the strength. To delay the presentation of S^1 until the response to S^0 has taken place is not a solution of this difficulty because it introduces a temporal element of some importance for conditioning (see Chapter Seven).

EXTINCTION OF TYPE S

Extinction of a conditioned reflex of Type S occurs when S^0 is presented without S^1. The resulting change is in the strength of $S^0 . R^1$ and is identical with that of conditioning except for sign. It is measured in the same way—usually in terms of the R/S ratio or of latency. Pavlov's statement that 'the rate of experimental extinction is measured by the period of time during which a given stimulus must be applied at definite regular intervals without reinforcement before the reflex response becomes zero (64),' is unsatisfactory. The momentary rate of extinction is the rate at which the strength is falling. It is probably a function of the strength at the moment but not necessarily of the strength at which the process began, although the latter is included in Pavlov's 'rate.'

The rate of extinction is rather easily disturbed by extraneous factors. The very act of withholding reinforcement may be an example; not only may it cause the reduction in strength called extinction, but it may also set up an emotional state causing fluctuation in the state of the reflex. The total time required for extinction at a given rate of elicitation is a function of the amount of conditioning that has previously taken place and of the momentary state of the drive. It is also obviously a function of the rate of elicitation. In an experiment by Pavlov extinction was obtained in 15 minutes when the interval between presentations of S^0 was 2 minutes, in 20 minutes when it was 4 minutes, in 54 minutes when it was 8 minutes, and in more than two hours when it was 16 minutes. But this is a misleading way of presenting the data. When the strength of the reflex (in terms of the amount of secretion during one minute) is plotted against presentations of the stimulus, the four curves are of the same order.

An extinguished reflex may be reconditioned by restoring the correlation of S^0 and S^1. Successive extinction curves taken after comparable amounts of reconditioning show a progressively more rapid decline in rate. If an extinguished reflex is left inactive for, say, a few hours, its strength increases slightly through what is called spontaneous recovery. These various properties will be taken up later in comparing the extinction of an operant.

Pavlov has interpreted extinction as a case of inhibition and has offered cases of what he calls 'disinhibition' which seem to affirm the inhibitory nature of extinction. According to the present system this is a gratuitous interpretation, which will be considered later.

Conditioning of Type R

The second type of conditioning, in which the reinforcing stimulus is correlated with a response, may be represented as follows:

$$s \,.\, R^0 \rightarrow S^1 \,.\, R^1$$

where $s \, . \, R^0$ is some part of the unconditioned operant behavior of the organism and S^1 is again a reinforcing stimulus. The requirements for conditioning are some considerable strength of $S^1 \, . \, R^1$ and the connection indicated by $\rightarrow$. The effect is a change in $[s \, . \, R^0]$, which may be either an increase or, possibly, a decrease. In the present example of pressing a lever the strength may increase if S^1 is, for example, food, and it may decrease if it is, for example, a shock.[1] There are thus two kinds of reinforcing stimuli—positive and negative. The cessation of a positive reinforcement acts as a negative, the cessation of a negative as a positive. Differences between the two types of conditioning will be summarized later.

EXPERIMENTAL PROCEDURE

The chain of reflexes involved in pressing the lever and obtaining food is set up through conditioning of Type R. The order in which the four principal members are added was described in the preceding chapter. I shall now turn directly to the conditioning of the single member of pressing the lever. An unconditioned hungry rat placed in the experimental box will exhibit a certain low strength of this operant. That is to say, it will press the lever a few times during an hour at a rather irregular rate. If the response is now correlated with the reinforcing stimulus of the sound of the magazine, the rate should increase according to the present law. The required datum is the rate at which the change in strength takes place.

In order to obtain a maximal reinforcement of the first response to the lever, the discriminative response to the sound of the magazine must be well established. The rat is placed in the experimental box without food and pellets are discharged from the magazine periodically. The rat comes to respond to the tray only when the magazine has sounded, or in other words it discriminates between the tray alone and the tray plus the sound of the magazine. The process is analogous to a discrimination described in detail in Chapter Five. For the moment I may simply note that in the typical case the rat comes to respond to the tray immediately after the sound within ten applications. In the following experiments from 50 to 200 pellets were discharged in this way in order to give the reflex a considerable strength. I have not been able to detect any

[1] But see the sections on negative reinforcement in this and the following chapter.

difference in the rate of conditioning of the response to the lever due to the number of these reinforcements of the subsequent member of the chain, beyond, of course, the number required to establish the discrimination. A relation presumably exists, but the rate of conditioning is too rapid to permit a demonstration of a significant difference with the numbers of animals here used.

During this preliminary training the lever may be in place in the box (but disconnected from the magazine) and unconditioned responses recorded. Or it may be in place but fixed, or it may be absent. Adaptation to the lever as a source of novel stimulation as

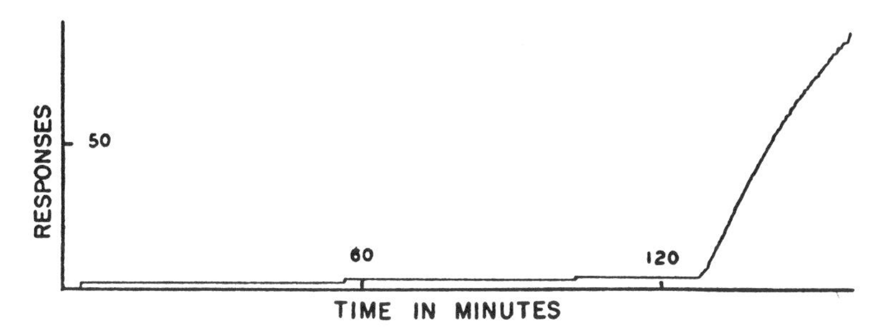

FIGURE 3

ORIGINAL CONDITIONING

All responses to the lever were reinforced. The first three reinforcements were apparently ineffective. The fourth is followed by a rapid increase in rate.

well as adaptation of the actual response of pressing will presumably influence the rate of conditioning, but I have again not been able to detect any significant difference.

On the day of conditioning the rat is placed in the box as usual. The lever is present, and for the first time in the history of the rat its movement downward will operate the magazine. Figure 3 gives a record of the resulting change in behavior. In this case the lever had been present during the preliminary training, but it was then resting at its lowest point and no movement downward was possible. On the day of conditioning a first response was made five minutes after release. The reinforcement had no observable effect upon the behavior. A second response was made 51½ minutes later, also without effect. A third was made 47½ minutes later and a fourth 25 minutes after that. The fourth response was followed by an appreciable increase in rate showing a swift acceleration to a maximum. The intervals elapsing before the fifth, sixth, and following responses were 43, 21, 28, 10, 10, and 15 seconds respectively. From that point on the rat responded at an essentially constant rate. A negative acceleration as the result of a change in hunger due to the ingestion of the pellets (see Chapter Nine) is shown later in the record.

This example is unusual in that conditioning does not take place until the fourth reinforcement. Five records showing a quicker

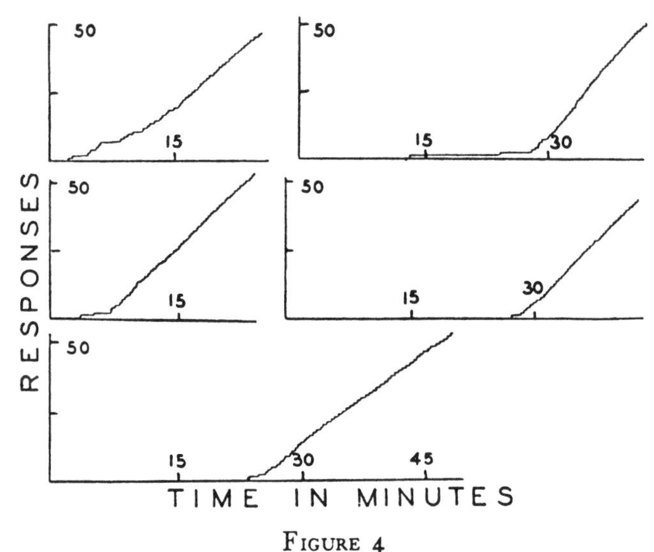

FIGURE 4

ORIGINAL CONDITIONING

All responses were reinforced. The change in rate here occurs more rapidly than in Figure 3.

effect are given in Figure 4, where conditioning occurs with the first or second reinforcement, although the rate is not immediately maximal. In the optimal case the first reinforcement produces a complete and maximal change, as shown in the four records in Figure 5.

These three figures give a very fair representation of the result for the 78 rats tested with this method. Three of these rats showed the very low unconditioned rate of about one response per hour. Conditioning did not occur within three hours, and no further experimentation was carried out. Of the remaining 75 records the slowest conditioning is shown in that in Figure 3. Twenty cases gave the instantaneous result shown in Figure 5, and the remaining cases were similar to those in Figure 4.

I think it may be concluded from the high frequency of occurrence of the instantaneous change that a single reinforcement is capable of raising the strength of the operant to essentially a maximal value. Apparently the only qualification of the experimental result is the possible effect of unconditioned strength. When every response to the lever is reinforced, as in this procedure, the appearance of a sudden change in rate is equivocal so far as the effect of

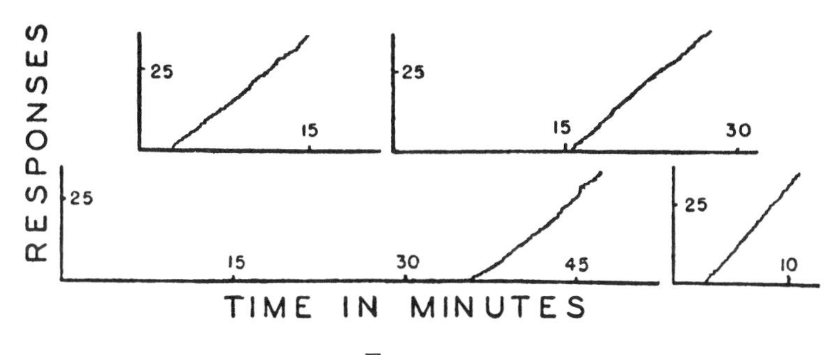

FIGURE 5

ORIGINAL CONDITIONING

The reinforcement of all responses produces an instantaneous change to a practically maximal rate of responding.

a single reinforcement is concerned. If the unconditioned rate is fairly high, the second response may be due to unconditioned strength rather than to the first reinforcement. Since the second response is also reinforced, the subsequent maximal rate may be due to both reinforcements. But the examples of an instantaneous change occur too frequently to be explained by the unconditioned rates actually observed in control records. In at least twenty-five per cent of the cases the second response follows the first as rapidly as any later response follows the preceding, but this can hardly be due to an unconditioned rate of responding of, say, ten responses per hour. A procedure in which this equivocality is lacking will be described shortly.

On the other hand, the failure to obtain an instantaneous change in every case may be accounted for in various ways, such as the following:

(1) The fact that the reflex correlated with reinforcement is the second member of a chain means that the first member—the response of lifting up and touching the lever—must also be reinforced by the sound of the magazine in order to obtain a maximal rate. But this response may have been made some time before pressing, and, as I shall show in the following chapter, an interval of as little as two seconds may reduce the effect of a reinforcement by one-third. In a certain number of cases we must suppose that the pressing is reinforced but that touching the lever is not (at least not so extensively). The rat must touch the lever again before this initial member is conditioned. Since the experiments are performed in the dark the discriminative stimuli controlling the response do not act in concert with directive stimuli for exploration and hence are not optimal for quick conditioning.

(2) The response of pressing the lever is defined as any movement causing the lever to move downward. Some instances may be of unusual form. For example, the rat may stand on its hind legs, lose its balance, and fall against the lever. Cases of this sort are too rare to affect the rate of responding after conditioning has taken place but some allowance must be made for them, especially before any particularly strong behavior has been developed.

(3) If a rat is fairly active, the sound of the magazine may reinforce other behavior immediately preceding the pressing. If such behavior has a greater unconditioned strength, its total strength after conditioning may be enough to cause a significant conflict with the repetition of the precise form of response producing the reinforcement.

(4) When the lever has not been present prior to the day of conditioning, its movement may have an emotional effect, one result of which is a depression in rate.

(5) Perhaps the most important factor interfering with an instantaneous change in rate is the dropping out of intermediate members of the chain. According to the conditions of the experiment mere lifting up and touching the lever is followed by the sound of the magazine when the lever is pressed. But proprioceptive stimuli from the act of lifting up and exteroceptive stimuli from the touch of the lever thus become discriminative stimuli correlated with the presence of food in the tray. An occasional effect upon the behavior is

clearly observable. In the 78 cases here described the rats were not watched during the process of conditioning, since the apparatus does not permit visual observation without interfering with the animals or at least without the additional distracting stimulation from a light. However, in a few other cases the process was observed by leaving the box open. (None of these cases gave instantaneous conditioning.) After the first reinforcement it was occasionally observed that the rat touched the lever, perhaps moved it slightly, and then responded to the tray without pressing and without being stimulated by the sound of the magazine. The response to this partial discriminative stimulation is never reinforced and is therefore extinguished (see Chapter Five), but it has an effect upon the recorded rate in the early stages of conditioning. In some of the present cases of a non-instantaneous change we may suppose that between the recorded responses there occurred some instances of incomplete elicitation of the initial member of the chain. That incidental stimulation from the response to the lever may acquire the same discriminative function as the sound of the magazine may be shown by the behavior of the rat when the magazine is disconnected after conditioning has occurred. A response to the lever is not followed by the sound but a response to the tray almost invariably follows.

Although these five factors operate to produce less than a maximal change in rate, they are all compatible with the assertion of the actual instantaneity of the process. They affect the observed rate but not the rate of conditioning. That they are not wholly responsible for the failure to obtain instantaneous changes in rate in every case, however, is indicated by the behavior of the rat during reconditioning after extinction, which frequently shows some initial acceleration in spite of the fact that the definition of the reinforced response should in such cases be fully achieved. Records of reconditioning are given later in Figure 10. In some of these cases the change is instantaneous, however, and I believe that the conclusion is justified that a practically complete instantaneous change is extremely common if not actually typical of the process of original conditioning.

I may point out that the result depends upon the preliminary development of the later reflexes in the chain. It could not be obtained consistently when the total act of 'pressing the lever and obtaining food' is regarded as a unit and treated experimentally as such. If reinforcing power is not first given to the sound of the magazine through the establishment of a discrimination, a certain interval of time will elapse between the response and the stimulation from the food, and the effectiveness of the reinforcement will be severely reduced. In rare cases instantaneous conditioning may be observed, but the consistent result obtained in the present case will necessarily be lacking.

The instantaneity of conditioning of Type R has several interesting consequences. When the demonstration of conditioning is based upon a mere change in rate as distinct from a change in the reflex reserve (to be discussed later), it is impossible to answer questions concerning (1) the inheritance of ability to acquire a response, (2) 'transfer of training' from one response to another, (3) 'saving' in the process of reconditioning, and so on. There is no possible improvement upon an instantaneous change. It is also probably impossible to demonstrate a special kind of conditioning called 'insight learning.' Insight has been defined in various ways and related to various preliminary conditions of the 'problem,' such as the possibility of 'perceiving the relation of the response to the reinforcement.' The actual presence of 'insight' is usually claimed only when a 'successful' response is made and readily repeated. It is difficult to see how the present case fails to qualify, although the only relation to be 'perceived' is that of the temporal contiguity of the response and the reinforcement. In Köhler's criticism of the work of Thorndike this is held to be insufficient (55).

More important for our present purposes is the difficulty raised in studying the effects of various conditions of the experiment. Examples already noted are the prior adaptation of the stimulation arising from the lever, prior responses of pressing the lever, and the degree to which the discriminative response to the sound of the magazine is established. Over the ranges so far explored no effect has been noted upon the rate of conditioning, but if the process were slower, it is reasonable to suppose that an effect might have been observed. Another variable, to be described in Chapter Ten, is the strength of $S^1 . R^1$, which was seen to be important for conditioning of Type S.

One condition which was important for Type S and which seems naturally to apply here is the time elapsing between the response (the stimulus in Type S) and the reinforcement. This question has not been thoroughly investigated, but the following results may be described. An interval is introduced by allowing the contact on the lever to release a timing pendulum, which operates the magazine at the end of the desired interval. With one group of eight rats intervals of 1, 2, 3, and 4 seconds were introduced in this way. The resulting curves are given in Figure 6. The rates of acceleration are all comparable with those obtained with simultaneous reinforce-

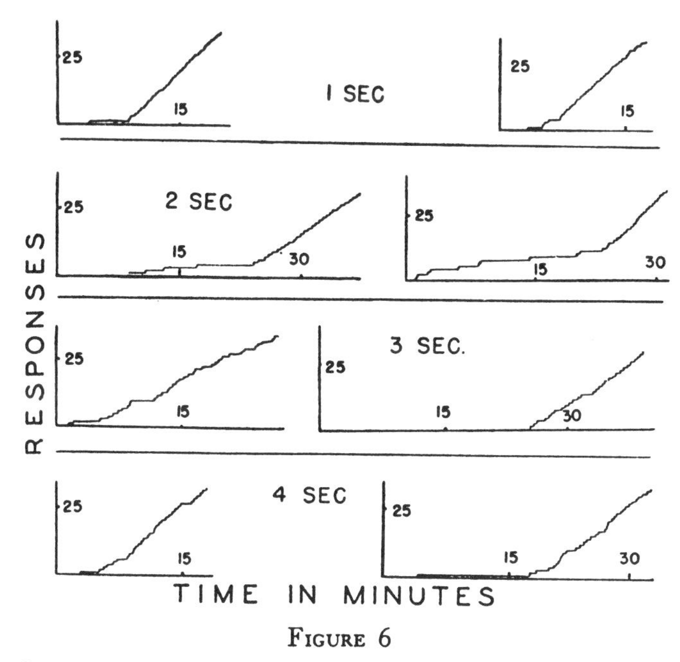

FIGURE 6

ORIGINAL CONDITIONING WITH DELAYED REINFORCEMENT

Intervals of time as marked were introduced between the response and the delivery of food.

ment with the possible exception of one case at two seconds. The slight irregularity in the rates eventually reached at three and four seconds is due to the fact that as soon as the reflex has acquired some strength a second response may occur during the interval before reinforcement. A new interval must be begun at this point or the second response will be reinforced too quickly, but this means that the first response must go unreinforced. At the longer intervals this result is seriously disturbing. I have been able to condition the response with an interval as long as eight seconds, but in view of the inclusion of unreinforced responses, it is difficult to say whether or not the acceleration is really retarded.

EXTINCTION OF TYPE R

Extinction of a conditioned reflex of Type R occurs when the response is no longer followed by the reinforcing stimulus. The change is merely a decrease in [*s*. *R*], the course of which is indicated in the following experiment. Four rats were conditioned as described above and about 100 responses were reinforced in each case. On the following day the magazines were disconnected, so that the movement of the lever was without effect, except upon the recording devices. The rats were placed in the experimental boxes and released at the usual time and in the usual way. Their responses to the lever during the following hour are recorded in Figure 7. The animals were released approximately at the beginning of the curves, except Rat D. In this experiment a release door was used which was pushed open by the rat, and Rat D showed a characteristic delay. Its record begins about five minutes after the unlatching of the door.

When the first response to the lever fails to supply the stimulus for the next member of the usual chain, the response is elicited again immediately, and a high rate of elicitation is maintained for a short time. This is soon interrupted, and the rate subsequently undergoes an extensive fluctuation. In spite of the irregularity it is possible to indicate a general course for each curve. When this is done, as in Figure 7, it is apparent that the deviations have the character of depressions in the curve, and that they are followed by compensatory increases in rate. Because of the compensation the smooth curve is eventually fairly closely approximated.

These curves are typical. Better approximations to a smooth curve could have been given, but they would have represented less accurately what an average animal does. The curves usually show marked deviations, which must be accepted as characteristic of the change in strength observed under these conditions. The deviations are not, however, haphazard. The intervals elapsing between successive responses are distributed by no means at random, and the observed fluctuations are, in that sense, real. Before giving some account of their nature, it will be well to consider certain possible sources of disturbance in the recorded curves and certain incidental changes in the behavior of the rat, which must be taken into account in determining the course of the major process. In the first place, it is possible that the *character* of the response may change during extinction, in such a way that the closing of the mercury switch at the lever will no longer serve as a reliable criterion of the behavior. But this is evidently not true. Throughout the course of the extinc-

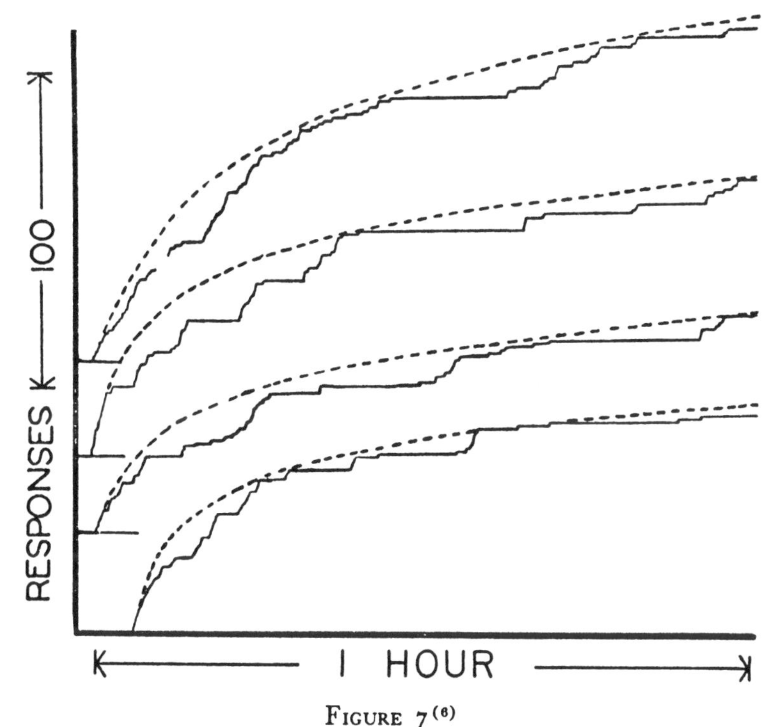

FIGURE 7[6]
FOUR TYPICAL EXTINCTION CURVES
No responses were reinforced.

tion the contacts are in general cleanly made and broken, and their duration holds rather closely to a constant average value. Another possibility is that toward the end of such a curve responses may appear which are due to unconditioned strength—the 'investigatory' responses which would be made during the hour even if no conditioning had taken place. But the number of such responses after the first day or two is negligible.

The only important kind of irregularity is characteristic of the behavior of the rat except when the reflex is at maximal strength. Under circumstances which produce a constant but sub-maximal rate of responding, the intervals occurring between successive responses vary. The constant rate is a resultant. The actual record may show short periods of no responding followed by a compensatory increase in activity, and during extinction (but very rarely otherwise) there may be an increase in rate followed by a depression. The former case is reasonably attributed to the prepotent action of incidental stimulation in the experimental box. The latter may be an emotional effect (see Chapter Eleven). Many examples of deviations with compensation will appear in later records.

It is apparent from a casual inspection of Figure 7 that the typical curve for extinction is wave-like in character. There is no very uniform wave-length or amplitude, but the rate of elicitation clearly tends to pass from a high to a low value and back again. The transitions are sometimes smoothly executed but are more often abrupt. Since an effect of this sort is presumably continuous, we may regard the smooth transition as representative of the actual change in strength and account for the abrupt transitions by appealing to the disturbing factors just discussed. The probable course of the change in strength in the four curves in Figure 7, as distinct from the disturbed course given experimentally, is represented in Figure 8. Examples which show the cyclic character more clearly will be given later.

A simple theoretical account of this cyclic fluctuation can be derived from the assumption that an interruption of the normal chain of ingestive reflexes sets up an emotional effect that lowers the rate of elicitation. That failure to reinforce is a common emotional operation has been pointed out in Chapter One. In order to account for the present observations we must assume a time-lag for the effect and must appeal also to a compensatory increase in rate. In the process of extinction the rate of elicitation begins at a maximum. Under the conditions of the experiment each response is unreinforced and an emotional effect is generated. A period of reduced rate supervenes, during which the previously generated effect disappears and (because of the lowered rate) is not replaced to any considerable extent. Through a compensatory increase the rate returns to a nearly maximal value, where it enters upon a second cycle. The properties of the resulting curve will depend upon the extent of the depressing effect of a single interrupted elicitation, upon the time-lag, and upon the nature of the process of recovery. One characteristic of such an emotional effect is that it readily adapts out, and curves for extinction may be obtained that lack this cyclic effect when time is provided for adaptation (see the following chapter).

In Figure 8 we may inspect more easily the theoretical curves that have been fitted to the data. They are logarithmic and are drawn as envelops, upon the assumption that the deviations are depressions. The above account of the fluctuation is based upon that assumption,

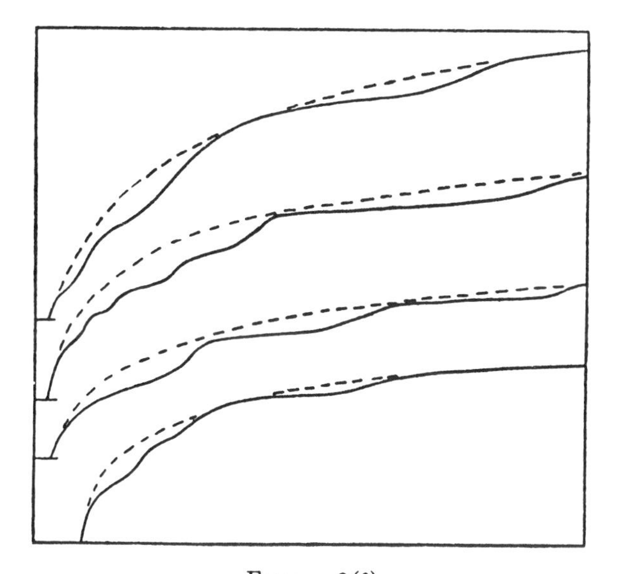

FIGURE 8[6]

INFERRED COURSE OF THE RECORDS IN FIGURE 7

Allowance has been made for expected irregularities.

since it requires the existence of a compensatory effect. Furthermore, the first short section is heavily weighted in each record, and the theoretical curves are permitted to lie considerably above the immediately following sections of the curves. This is also justified by the present interpretation of the fluctuation. Since we have assumed a time-lag, the first part of the curve must be free of the depressing effect. Moreover there is no reason to expect the first few recoveries from depression to be successful in reaching the extrapolated curve, since the generation of a depressing effect is in no way related to whether or not the rat is 'on schedule.' *Eventually* the theoretical curve is reached, because the emotional effect partially adapts out, and because the curve has meanwhile fallen off.

Other kinds of curves could doubtless be fitted as well to these data. A logarithmic curve is not necessarily required but may be used as a simple empirical description. It is perhaps at present impossible to establish the exact nature of the change more closely. Because of the cyclic fluctuation and its variable wave length it is useless to appeal to averages. The deviations are all in one direction if the present interpretation is correct. A smooth curve could doubtless be obtained by averaging a large number of cases but it would not represent the envelop approached by each curve separately.

The present interpretation of the extinction curve may be summarized as follows:

(1) The decrease in reflex strength observed during the process of extinction follows an approximately logarithmic course.

(2) The essential condition for extinction (the interruption of the chain) leads also to a temporary (emotional) change, which depresses the rate of elicitation.

(3) The depressive effect shows a time-lag, which in conjunction with a compensatory increase in rate produces a cyclic fluctuation.

(4) The depressive effect is subject to adaptation.

SPONTANEOUS RECOVERY FROM EXTINCTION

At the end of the experimental period represented in Figure 7 a low rate of elicitation has been reached. The strength of the reflex has been reduced approximately to its value prior to conditioning. But if the rat is replaced in the apparatus at a later time (no reconditioning having taken place), a small extinction curve will be obtained. A loss in degree of extinction has taken place. In Figure 9 one of the curves from Figure 7 is reproduced with a further curve obtained 48 hours later. To avoid any effect of change of daily procedure the rat was put into the experimental box as usual on the intervening day, but a supply of food was available and no responses to the lever were made. There was thus no reinforcement (or further extinction) of the reflex between the two records. Nevertheless, on the second day of extinction the rate begins at a value many times that of the final value of the first day. The record shows a somewhat greater recovery than is usually observed after twenty-four hours, but the order of magnitude is the same and the character of the curve typical.[1] Such an effect ultimately disappears if the extinction is sufficiently prolonged, and it is probable that a state of complete extinction can be reached. Since the reflex may persist

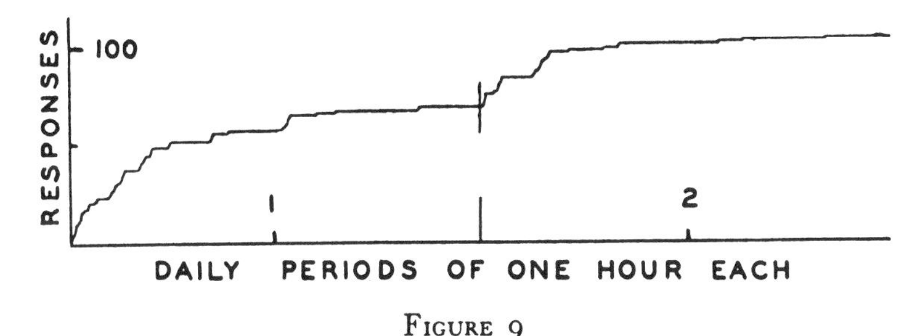

FIGURE 9

SPONTANEOUS RECOVERY FROM EXTINCTION

The first curve is from Figure 7. The second was obtained from the same rat forty-seven hours later with no intervening conditioning.

with some unconditioned strength in spite of the extinction, we are limited to the statement that the frequency of elicitation is reduced approximately to its value prior to conditioning.

RECONDITIONING

Reconditioning after extinction is effected by restoring the correlation of R^0 and S^1. Since there is no effect comparable to 'extinction below zero' in the case of an operant, reconditioning should resemble original conditioning in every respect, except that any modification due to the topographical definition of the response should be lacking. This expectation is borne out by the data. Curves for reconditioning are, in my experience, remarkably like curves for original conditioning. Figure 10 (page 80) gives two examples obtained after prolonged extinction following a procedure of periodic reconditioning to be described in the next chapter. The magazines were connected at the vertical lines, and reconditioning began at the following response. A slight positive acceleration may be observed, which resembles the acceleration often obtained in original conditioning.

A reconditioned reflex may, of course, again be extinguished. In

[1] Youtz (*J. Exper. Psychol.*, in press) has found that second extinction curves carried to an arbitrary zero show spontaneous recovery amounting to from 28% to 58% of the original extinction curve.

Figure 11 the reflex had been thoroughly extinguished on the preceding day. Some loss of extinction is shown in the small curve at A. The rate then reached a very low value. At B ten responses were reinforced. A positive acceleration in the curve for reconditioning is evident. The reinforcement was then omitted and the resulting extinction follows at C. A base-line for the curve for extinction has been added. Since this is a second curve for extinction, the emotional effect causing a cyclic fluctuation has to a considerable extent

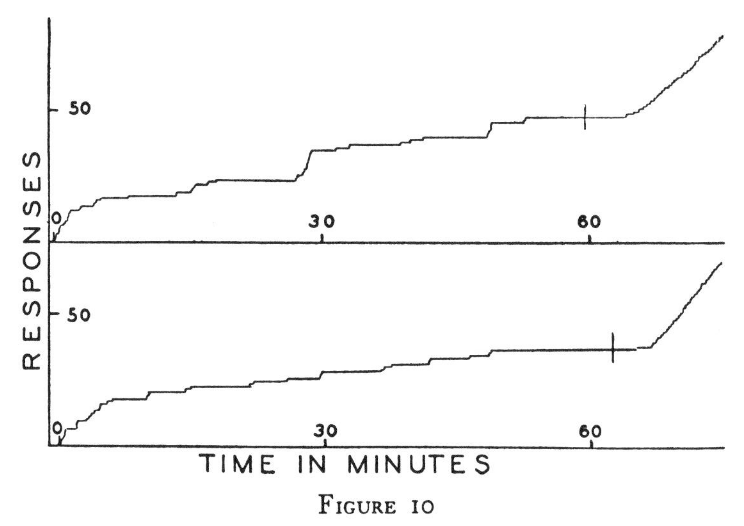

FIGURE 10

RECONDITIONING AFTER EXTINCTION

The records begin with extinction curves in a late stage. All responses were reinforced after the vertical lines.

adapted out. There is still some fluctuation, however, and the effects of minor deviations may again be observed. The area embraced by the curve in this record is exceptionally large for the reinforcement of only ten responses.

That such an increase in strength is actually reconditioning and not simply a facilitative effect due to the reception of food is shown in Figure 12. In this experiment the reflex had been very thoroughly extinguished. There is a slight loss of extinction evident at the beginning of the curve, followed by only two responses during the next 25 minutes. Three pellets of food were then discharged from the magazine by the experimenter, the lever remaining disconnected.

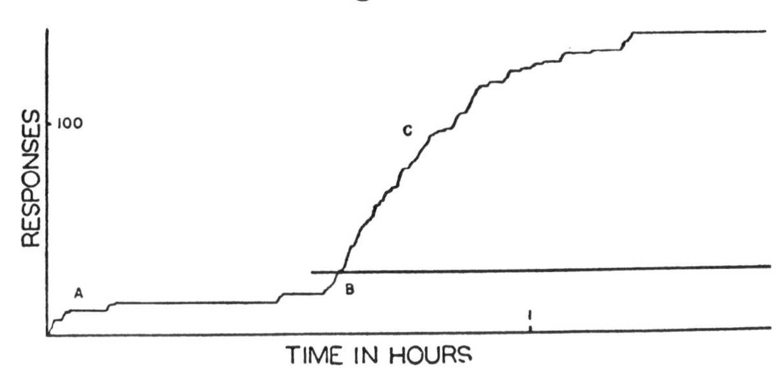

FIGURE 11 [6]

RECONDITIONING AND SUBSEQUENT EXTINCTION

At A: some recovery from extinction. At B: ten reinforced responses. At C: further extinction.

The rat pressed the lever only five times in the following 18 minutes, indicating that a facilitative effect, if present, was very slight. At the next response the magazine had been connected, and the rat obtained three pellets of food, pressing the lever once for each. The

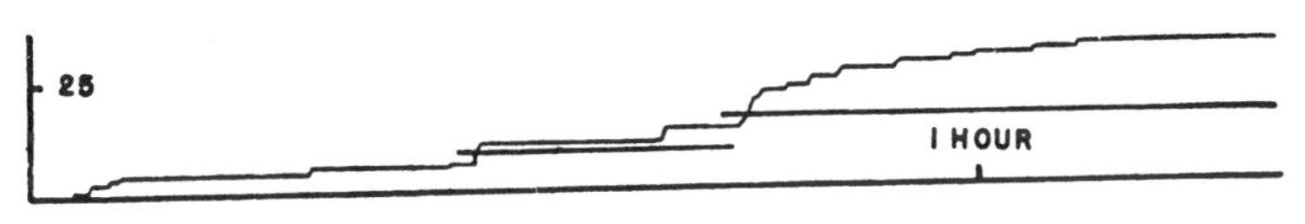

FIGURE 12 [6]

CONTROL AGAINST THE POSSIBLE FACILITATIVE ACTION OF A REINFORCING STIMULUS

The record begins with advanced extinction. Just below the first horizontal line three pellets of food were given without responses. Just below the second line three responses were reinforced.

magazine was then disconnected and a significant curve for extinction obtained.

SECONDARY CONDITIONING

Pavlov has described a process of secondary conditioning (of Type S) in which a stimulus acquires the properties of a conditioned stimulus as the result of accompanying a previously conditioned stimulus upon occasions when no ultimate unconditioned reinforcement is provided. For example, a black square presented to a dog for 10 seconds is followed 15 seconds later by the sound of a metronome, a conditioned stimulus for a salivary reflex. After ten presentations (no reinforcement being given) the square elicits a small salivary response. Such a procedure is identical with that of establishing a discrimination, in which the new stimulus should acquire 'inhibitory' rather than 'excitatory' properties (to use Pavlov's terms). The different result depends, according to Pavlov, upon the provision that in secondary conditioning the new stimulus must be withdrawn a few seconds *after* the conditioned stimulus is applied. This unusual temporal order severely limits the application of the result in normal behavior, and as I shall try to show in Chapter Six, the whole notion of secondary conditioning of Type S is probably suspect.

There is, however, a process that might be called secondary conditioning of Type R. It does not involve a conflict with the process of discrimination because it is a response rather than a stimulus that is reinforced. The process is that of adding an initial member to a chain of reflexes without ultimately reinforcing the chain. In the present example, the sound of the magazine acquires reinforcing value through its correlation with ultimate reinforcement. It can function as a reinforcing agent even when this ultimate reinforcement is lacking. Its reinforcing power will be weakened through the resulting extinction, but considerable conditioning can be effected before a state of more or less complete extinction is reached. The maximal result to be obtained from a given amount of previous conditioning of the sound of the magazine is shown in the following experiment.

The usual preliminary procedure was carried out with four rats, as the result of which they came to respond readily to the sound of the magazine by approaching the food tray. Sixty combined presentations of the sound and food were given. On the day of conditioning the magazine was connected with the lever but was empty. For the first time in the history of the rat the movement of the lever produced the (hitherto always reinforced) sound of the magazine, but no responses were ultimately reinforced with food. The four resulting curves are given in Figure 13. The sound of the magazine serves to reinforce the response to the lever so that a considerable number of responses are made. The heights of the curves are of the same order as those for extinction after a comparable amount of previous reinforcement, but the reinforcement is not effective enough to build up an initial maximal rate such as is observed during extinction. The records might be regarded as flattened extinction curves. In the third curve in the figure a period of relatively rapid responding occurs, but it is clearly a compensatory increase in rate following a period of little or no responding.

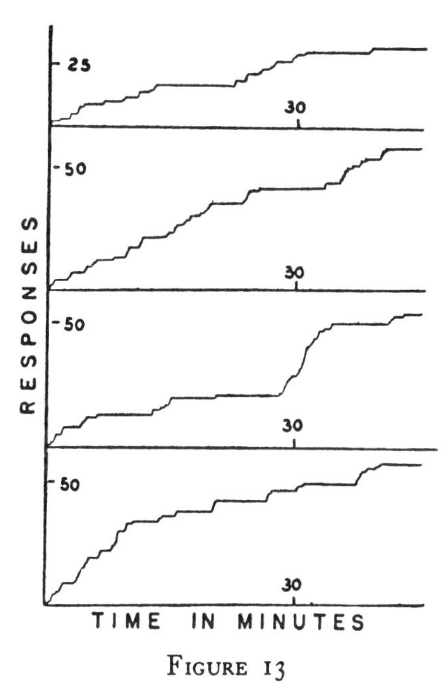

FIGURE 13

ORIGINAL CONDITIONING WITHOUT ULTIMATE REINFORCEMENT

No responses had ever been reinforced prior to this day, nor were any reinforced during the period represented. But the sound of the magazine had previously been followed by the delivery of food and here exerts a considerable reinforcing effect although no food is given.

Conditioning and the Reflex Reserve

In discussing the notion of a reflex reserve in Chapter One the extinction curve of Type R was used as an example. The process of conditioning was regarded as creating a certain number of potential responses, which could occur later without reinforcement. During extinction this reserve is exhausted. At any point the rate of responding may be assumed to be roughly proportional to the existing reserve. At the beginning of extinction the reserve and the rate are both maximal. As responses occur the reserve is drained and the rate declines. The slope of the envelop of the extinction curve gives the maximal rate of emission at any point. The significant deviations are below this envelop and suggest that incidental factors may change the proportionality in the direction of reducing the rate.

The phenomenon of spontaneous recovery does not disturb the assumption of a relation between rate and reserve. The envelop is not exceeded. If extinction had been continued during the time required for recovery, the curve would have been high enough to cover the new curve and probably much higher. Spontaneous recovery simply

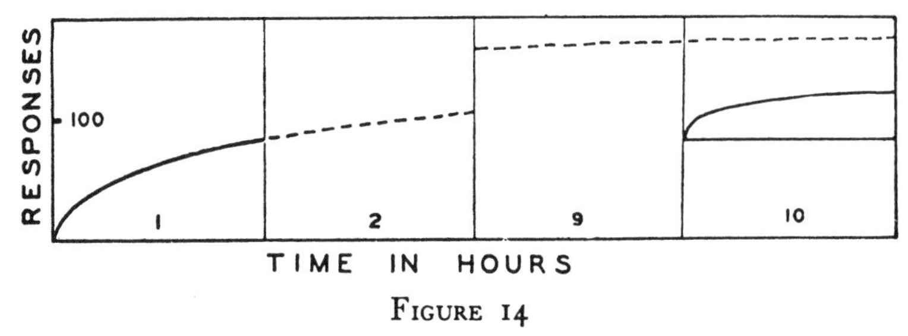

FIGURE 14

SPONTANEOUS RECOVERY AND THE EXTRAPOLATION OF AN EXTINCTION CURVE

A theoretical case to illustrate the relation of the recovery curve to the envelop of the extinction curve.

indicates that some of the responses that would have occurred during the intervening time have become immediately accessible as soon as opportunity for elicitation arises. In Figure 14 I have extrapolated a typical curve for ten hours (Hours 3–8 inclusive being omitted from the graph) and have indicated the place of a second extinction curve showing spontaneous recovery. The second curve begins at the height at which the first stopped, and it continues well below the projected envelop of the first. There is no reason to expect that the envelop will be reached, since it describes the process of unhampered emission.

The rapid compensatory increases in rate following periods of little or no activity during extinction differ from spontaneous recovery in that the factor responsible for the inactivity is not the absence of necessary external discriminative stimuli but either the prepotent activity of competing stimuli or an emotional effect. In the case of compensation apparently all the responses that would have occurred had activity been maintained are available when responding is resumed. If this is not true, the cyclic fluctuations typical of original extinction permanently lower the curve below the extrapolation of its first part. The available examples of extinction curves are too irregular to establish the fact of a full compensation but it seems probable that the envelop is very nearly, if not actually, reached in most cases. The difference between compensation and recovery may be due either to the short intervals in the case of the former or to the factors causing inactivity.

The phenomenon of compensation, like that of recovery, requires the notion of an *immediate* reserve distinct from the total reserve which determines the rate in the absence of interruption. The process is catenary. The rate is proportional to the immediate reserve, which is contributed to from the total reserve. When elicitation is continuous, the total reserve controls the process. When elicitation is interrupted, the immediate reserve is built up; and a period of increased activity is made possible when responding is resumed, until the total reserve again becomes the controlling factor. The period of time during which responding may be suspended without making the original envelop inaccessible will depend upon the size of the immediate reserve.

I do not believe that the data at present available are adequate to support a mathematical formulation of such a system, and I shall therefore leave the matter at this qualitative stage. A considerable amount of other material bearing upon the notion of a reserve will be given later.

CONDITIONING AS THE CREATION OF A RESERVE

The definition of conditioning that has been given here is in terms of a change in reflex strength, but the act of reinforcement has another distinguishable effect. It establishes the potentiality of a subsequent extinction curve, the size of which is a measure of the extent of the conditioning. There is no simple relation between these two measures. It is possible to reach a maximal rate of responding (a maximal strength) very quickly. Further reinforcement does not affect this measure, but it continues to build up the reserve described by a subsequent extinction curve. The typical effect of 'over-conditioning' is felt not upon any immediate property of the behavior but upon its subsequent changes during extinction. (The lack of a relation between the two measures is not incompatible with the assumption of a relation between rate and reserve because of the interposition of the limited 'immediate' reserve.)

So far as I know, all standard techniques for the study of conditioning deal only with the change in strength, perhaps because it is the immediately observed datum. The measure is awkward not only because it ceases to function while reinforcement continues to have an effect but because it raises technical difficulties. In order to measure the strength of a reflex we must elicit it, and if it is a conditioned reflex, the elicitation must be either reinforced or unreinforced. In either case there is an effect upon the state of the reflex, altering the condition which is being measured. This is an unavoidable dilemma, the consequences of which will be more or less serious according to the method used and the relative importance of the effect of a single elicitation. One example has already been noted in the proof of an instantaneous change in strength due to one reinforcement. The second response must be either reinforced or not, and in either case the strength is affected. The only available 'rate' is given by the interval between the first two responses, since the second elicitation contributes a further conditioning effect of its own. We need a statistical demonstration that this interval is shorter than would be observed if the first response were not reinforced. A single experiment cannot be convincing. Although a sufficient percentage of cases was observed to make an instantaneous change highly probable, the appeal to a number of cases is unfortunate, since the amount of conditioning actually observed after one elicitation varies greatly between cases, and in averaging together a group of first intervals we unavoidably obscure the exceptional examples, which are of special interest. Moreover, although a statistical demonstration in the present case does show that conditioning may take place at one elicitation, we are faced with the comparative unreliability of a single interval as a measure of the extent of the change.

The alternative method is to examine the reserve created by a single reinforcement. The procedure is simple. The first response is reinforced as usual, and the magazine is then permanently disconnected. If conditioning takes place, an extinction curve is obtained, the height of which is a measure of the extent of the conditioning.

An experiment which demonstrates a considerable amount of conditioning from the reinforcement of a single response is represented in Figure 15. In experiments of this sort it is necessary (for reasons to be noted shortly) to take control records of unconditioned responding after the preliminary training to the sound of the magazine. In Figure 15 Record A is a control taken prior to the preliminary training. The rat was hungry and no food was available.

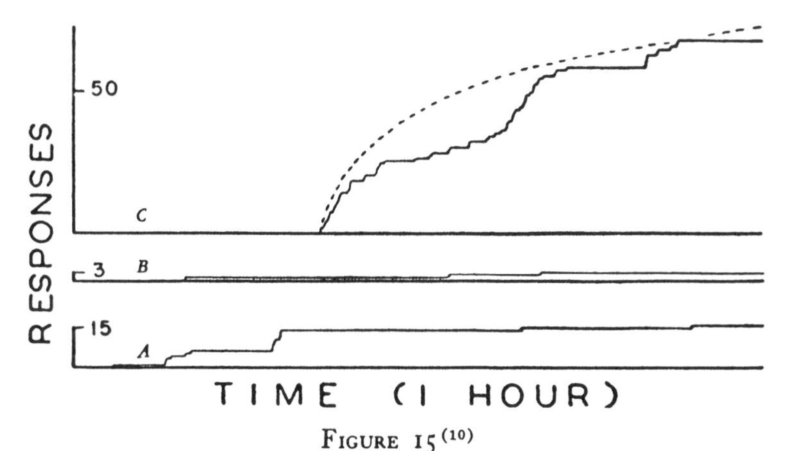

FIGURE 15 (10)

EXTINCTION CURVE FOLLOWING THE REINFORCEMENT OF ONE RESPONSE

The first response in Curve C was the only response in the history of the organism to be reinforced.

The grouping of responses in the first part of this record is probably due to the fact that when the rat responds to the lever, it is in an optimal position to respond again. The reflex adapts out quite thoroughly before the end of the hour. Record B is a control for investigatory responses in the absence of food after training to the sound of the magazine. Its significance will appear later. Record C was taken on the next day. The first response occurred in about 20 minutes. It was triply reinforced. That is to say, one pellet of food was discharged simultaneously with the response, and two more were discharged by the experimenter within three seconds. Although the rat received three pellets of food, *only one response was reinforced.* It will be seen from the record that the rat began to respond again immediately and during the next 40 minutes traced out a typical extinction curve. We are forced to conclude that this is the effect of the reinforcement of one response.

The record has the general properties of original extinction. A logarithmic envelop may be drawn above the curve (the broken line is for the equation $N = K \log t$, where N is the number of responses at time t and K is a constant), although the actual contacts with the experimental curve are no justification for such a form in this particular case. The experimental curve suffers a cyclic deviation below the envelop, which is an exceptionally good demonstration of the wave-like character first inferred from records

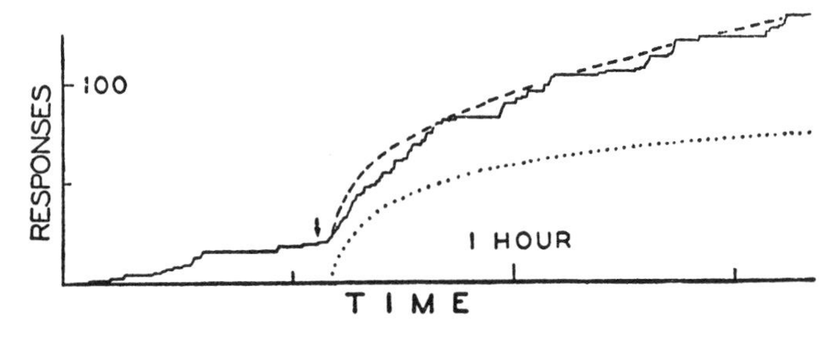

FIGURE 16 (10)

EXTINCTION CURVE FOLLOWING THE REINFORCEMENT OF ONE RESPONSE

In this case some responding occurs without specific reinforcement. At the arrow one response was reinforced.

much less regular. The curve differs in one respect from those obtained after more extensive conditioning: it lacks the initial burst of responses that lie upon the theoretical curve. To fit Figure 15 in this way it would be necessary to shift the origin of the theoretical curve about one minute to the right. This may indicate a lag in the effect of the conditioning.

The result is not always so clear-cut. In perhaps a majority of cases the control line B fails to maintain the slope to be expected from A. The training to the sound of the magazine produces an increase in the observed rate of responding that is independent of any actual reinforcement with the magazine. A new base-line appears, of which the first part of the record in Figure 16 is an example. Although no reinforcement with the magazine had yet taken place, a significant rate of elicitation is evident. A probable explanation is as follows. In the preliminary training to the magazine a discrimination is established. The rat comes to respond strongly to the tray-plus-the-sound-of-the-magazine, but not to respond to the tray alone. It can easily be shown that such a discrimination is frequently very broadly generalized (see Chapter Five). The rat will respond to the tray plus almost any sort of extra stimulation. Now, even though the magazine remains disconnected, a response to the lever will stimulate the rat in several ways, and this stimulation—auditory, tactual, and kinaesthetic—may be sufficient to elicit the discriminatory response to the tray. Although no pellet is received and the sequence of reflexes therefore not eventually reinforced, the response to the lever and the discriminatory response to the tray are present, and some conditioning of the former should take place. The base-line has the general form of the curve obtained in secondary extinction described above in Figure 13, page 83, but shows less conditioning.

When this generalization takes place, the extinction curve must be observed against the new base-line. At the arrow in Figure 16 a single response to the lever was reinforced with one pellet of food. After a short pause the curve of extinction was begun. Assuming that the base-line is approximately linear, we may subtract it from the experimental curve. The dotted curve in Figure 16 is for the above equation with very nearly the same value of K. The broken line is for the same equation plus ct on the right-hand side, where c is another constant determined from the base-line. The figure thus compares very well with Figure 15, although the wave form is much less smoothly executed. Figure 16 also lacks the initial burst. The shift in the origin required in this case would be about two minutes. (It will be understood that the extinction curve begins at the first unreinforced response, not at the arrow.)

In the case of Figure 15 an adequate control against a change in base-line is provided by Record B. Generalization of the discrimination has not occurred, and Record C cannot be regarded as including a base-line of any significant slope. The curve is due wholly to the reinforcement of the first response and is not contributed to by partial reinforcement of other responses. This is obviously the simpler result, and some way should be devised to obtain it regularly. Although there is no insuperable difficulty in accounting for the effect of generalization, it is for certain purposes objectionable.

The conditioning that occurs at the arrow in Figure 16, for example, is not original conditioning if this interpretation of the preceding base-line is correct, for the general stimulation resulting from the earlier responses has had a reinforcing effect.

It is evident that the extinction curve supplies more definite information about the change occurring at reinforcement than the mere increase in rate appealed to in the earlier part of this chapter. If some aspect of the curve (its height or area) is to be taken as a proper measure of the *amount* of conditioning, then in the present type of experiment, with a single reinforcement, the amount varies considerably among rats. The curves in Figures 15 and 16 are somewhat larger than the average, although they are by no means exceptional. In rare cases no effect may be shown. 'Good' records cannot, of course, be accidental. On the other hand, I have already noted several reasons why a first reinforcement may not be fully effective. These sources of variability seem for the moment to be beyond experimental control.

THE MAXIMAL SIZE OF THE RESERVE

There is an upper limit to the size of the reserve, and successive reinforcements are less and less effective in adding to the total as the maximal value is approached. The relatively large reserves created in the preceding experiment by single reinforcements are by no means doubled when the first two responses are reinforced. The extinction curves given in Figure 7 show the reserves created by about one hundred reinforcements. The numbers of responses are of the order of 75 to 100, and when the curves are extrapolated to a point at which responding would practically cease, the height should not exceed 200 responses. The largest extinction curve I have seen is reproduced in Figure 17. It was obtained with a similar procedure by Dr. F. S. Keller and Mr. A. Kerr of Colgate University with whose permission it is reproduced here. The rat had received 250 reinforcements prior to extinction.

In a later chapter a method of creating much larger reserves will be described, in which responses are only periodically reinforced.

LOSS OF RESERVE WITH LAPSE OF TIME

The extinction curve is sometimes unjustifiably identified with the 'forgetting curves' of Ebbinghaus and others. As Hunter (52) has noted, both processes involve a loss of strength, but, as I have already said in connection with the subject of inhibition, the mere direction of the change is not an adequate defining property. Forgetting is a loss of strength due, not to the active process of extinction (which requires elicitation without reinforcement), but to the passive lapse of time. In the usual all-or-none manner of stating the problem ('Does the organism still remember a response?') the question is whether the strength has gone below a

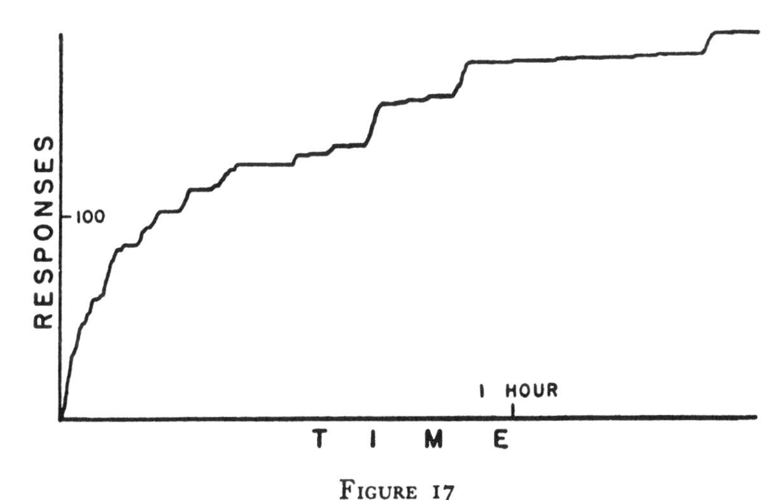

FIGURE 17

UNUSUALLY LARGE EXTINCTION CURVE FOLLOWING ORIGINAL CONDITIONING

Experiment by F. S. Keller and A. Kerr.

threshold value. This is a mere end-point and can yield only a crude account of the change. In the present case we want to know how the strength changes as a function of time when no operation is performed upon the organism to affect either the degree of conditioning or any of the other variables of which the strength is a function.

As in all cases of conditioning the question really concerns the reserve of which the strength is a function. In order to investigate 'forgetting' we need simply to introduce an interval of time between conditioning and extinction and to examine the effect upon the extinction curve. Certain precautions concerning other variables must be taken. With such an organism as the rat an important example is aging. The rat is shortlived, and changes due to age take place rapidly. In the following experiment the rats were conditioned at approximately 100 days of age, and extinction curves were taken forty-five days later. It will be shown in the following chapter that no considerable change in strength due to age occurs during a period of this length.

A second important variable that must be controlled is the drive. The records to be compared must be obtained under comparable states of drive. An internal check on the degree of drive is available in the shape of the extinction curve, as will be shown in Chapter Ten. The required constancy was obtained in the following way. The rats were put on the usual feeding routine described in Chapter Two until a daily hunger cycle had been well established. The conditioning was then carried out, and the rats were immediately taken back to the animal room, where they remained for forty days with a constant supply of food and water. At the end of that time they were brought back to the experimental room and again placed on the feeding routine for five days. As a precaution against any loss of adaptation to the experimental box they were placed in the boxes for one of the feeding periods (the third). The lever had been removed on that day. At the end of five days the hunger cycle had been adequately reestablished, and extinction curves were obtained on the sixth day.

Four rats were used which compared closely in age and preliminary training with those in Figure 7. The number of reinforced responses on the day of conditioning was the same in the two cases (*ca.* 100). Consequently, the effect of the interval of time between conditioning and extinction may be determined by a direct comparison of the two sets of results. The averaged curves for the two groups are given in Figure 18. The lower curve is for the group extinguished after a delay of forty-five days. The extinction was allowed to run for one hour and twenty-two minutes. The upper curve is the average of the curves in Figure 7. As I have already said, an averaged curve is of little use in showing the fundamental *shape* of the extinction curve, since the cyclic deviations are characteristically in one direction only and their wave-lengths vary, but it serves to summarize the general effect and is sufficient for our present purpose.

The figure shows that the numbers of responses occurring during one hour of extinction are of the same order of magnitude in the two cases. The average heights of the curves at the end of one hour are 86 and 69 responses for immediate and delayed extinction respectively, indicating only a slight loss in reserve during the forty-five days.[1] The general shape of the curve is maintained. The averaged curve indicates this only roughly because of the flattening due to deviations below the envelop, but the individual records confirm the statement. The availability of the reserve is therefore also unchanged. The mere length of time at which a reserve remains un-

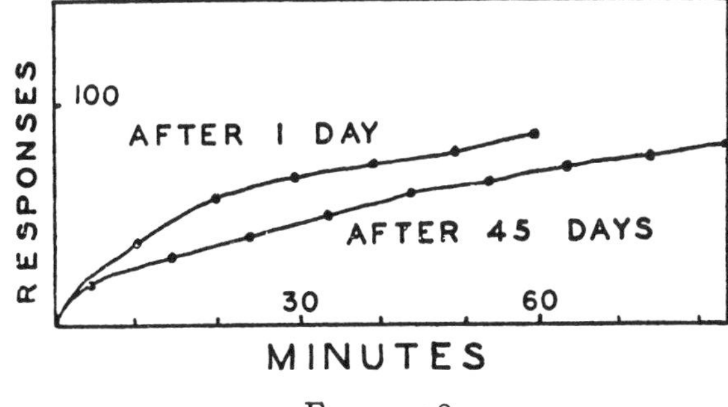

FIGURE 18

AVERAGED EXTINCTION CURVES SHOWING ONLY SLIGHT LOSS OF RESERVE DURING FORTY-FIVE DAYS

touched does not affect the relation between it and the rate of elicitation.

So far as this little experiment goes, it may be concluded that the rat 'forgets' slowly. At the end of forty-five days the strength of the operant is very far from that threshold point at which no response would be forthcoming,—or, in other words, at which the rat would have 'forgotten what to do to get food.' The mere lapse of time is an inefficient factor in reducing the reflex reserve.

An experiment by Keller (54) reveals another process involving the reserve, which also takes place during a period of inactivity. The 'spontaneous recovery' in strength that occurs after partial extinction when further extinction is later carried out was interpreted above as the effect of the intervening time in making immediately available a number of responses that would have been

[1] Youtz (*J. Exper. Psychol.*, in press) has found that with small amounts of conditioning (10 and 40 reinforcements) extinction curves obtained after fifteen days show more responses than those after one day.

emitted during that time if the extinction had been continued without interruption. Presumably the number made available is related to the length of the interruption. In the experiments described above, the period was either twenty-three or forty-seven hours. In Keller's experiment it was forty-five days.

The apparatus and procedure were similar in all essential respects to that described in Chapter Two. Two groups of four rats each were thoroughly conditioned by reinforcing from two to three hundred responses. When the drive had been suitably adjusted, extinction curves were taken for one and one-half hours. Further extinction was obtained for a similar period on the following day with Group B and after forty-five days with Group A. A third

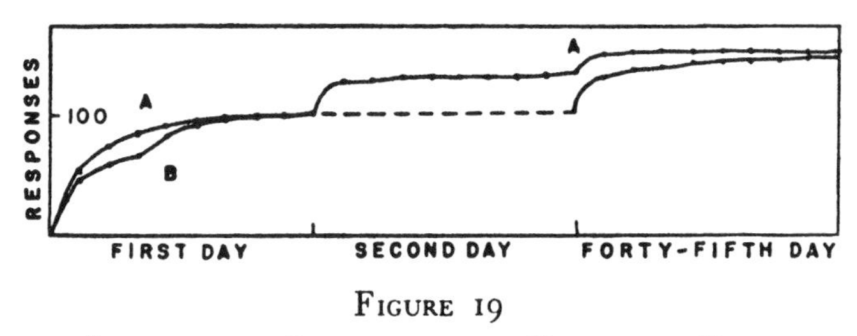

FIGURE 19

SPONTANEOUS RECOVERY AFTER FORTY-FIVE DAYS

Group A was further extinguished on the second and forty-fifth days; Group B on the forty-fifth day only. Experiment by F. S. Keller.

extinction curve was also obtained from Group B on the forty-fifth day. The result is shown in Figure 19, which gives the averaged curves for each group, where measurements were taken every ten minutes.

The curves for original extinction have the typical properties already described. The averaged curve for Group B in Figure 19 shows the effect of more extensive cyclic deviations in the first part of the curve, of such a sort as to indicate a slightly lower average degree of hunger in this group (see Chapter Ten). Both curves reach essentially zero slopes by the end of the period, and it may be concluded that practically all the available reserve has by that time been exhausted. The heights differ by only two responses (the difference being cumulative throughout the hour and a half), and the two groups are therefore well matched for the purposes of the experiment. Figure 19 also shows that the curves for further extinction have the usual properties: a concentration of responses at the beginning as evidence of 'recovery' and an essentially flat section thereafter. This is true even after forty-five days.

The crucial point concerns the numbers of responses obtained in further extinction on the second and forty-fifth days. The average for Group B on the second day was 37; that for Group A on the forty-fifth day 46. If these figures may be taken as significant, there is a twenty-five per cent increase in spontaneous recovery as the result of postponing the further extinction forty-five days. The groups are too small, of course, to give very great weight to the exact amount, but the experiment does provide a qualitative demonstration of the effect of the interval upon subsequent spontaneous recovery. The bulk of the recovery takes place during the first twenty-four hours but there is additional recovery subsequently. The third curve for Group B on the forty-fifth day shows still further recovery, although not as much as on the second day. The number of responses on this third day is more than enough to balance the totals for the two groups. This is to be expected, because Group B has been extinguished half again as long as Group A and has in the end also had the benefit of the forty-five day interval.

It might be argued that such a difference could be due to a difference in the drives of the groups on the second and forty-fifth days. During most of the intervening period both groups were fed continuously, but five days before the last test each group was again placed on a schedule of daily feedings for limited times. That this was successful in producing approximately the same degree of drive is evident from the shapes of the curves on the forty-fifth day. If there is any difference between Group A on the forty-fifth day and Group B on the second, it is in the direction of a greater degree of hunger in the latter case, which would work against the result. This is to be inferred from the experiment described in Chapter Ten and the slightly greater slope of the main part of the curve for Group A.

Although there is a definite increase in the amount of recovery, it is not true that the rats have 'forgotten the extinction while remembering the conditioning.' There is some reason to expect the contrary or at least to hold it as a possibility, but the curve after forty-five days still shows the effect of the first extinction almost as clearly as a curve taken on the second day. Here again time alone seems to have little effect upon the reserve or its state of repletion or exhaustion. The case is especially interesting because the effect of time is to produce an *increase* in the momentary strength of the reflex contrary to the usual tendency toward a decrease.

The validity of this experiment depends upon the demonstration given above that there is little loss in the reserve over a period of as much as forty-five days. If there is some slight loss, the present experiment is all the more significant in indicating further recovery during the longer period.

Extinction as 'Inhibition'

In discussing the Law of Inhibition it was noted that the term had been extended very widely to include many different kinds of negative changes in strength. The definition of a class of phenomena in terms of the mere direction of the change was held not to be useful. A stronger objection may be made to the assumption that a property observed in one case of a negative change is to be expected in another. An example of the unwarranted transfer of a property is shown in regarding extinction as a form of inhibition and of supposing that it represents the *suppression* of activity. This view was advanced by Pavlov in considerable detail but is incompatible with the present system. The metaphor of suppression is perhaps justified in the inhibition defined in Chapter One in the sense that, if it were not for the inhibiting stimulus, a certain amount of activity would be observed. But in extinction no comparable suppressing force can be pointed out. Pavlov has appealed to the conditioned stimulus itself (64). Thus he says, 'The positive conditioned stimulus itself becomes, under definite conditions, negative or inhibitory; it now evokes in the cells of the cortex [read: in the behavior of the organism] a process of inhibition instead of the usual excitation (p. 48).' And again, 'A stimulus to a positive conditioned reflex can under certain definite conditions readily be transformed into a stimulus for a negative or inhibitory conditioned reflex (p. 67).' But all that is observed is that the stimulus is now ineffective in evoking the response, as it was prior to conditioning.

Other properties of extinction attributed to inhibition by Pavlov are likewise as easily stated without the concept. To account for difficulty of extinction by saying that 'the greater the intensity of the excitatory process, the more intense must be the inhibitory process in order to overcome it' (p. 61) adds nothing to the observation that a strong reflex is extinguished slowly (as we should expect from the relation of strength to reserve). And to attribute the fluctuations observed during extinction to the 'struggle which is taking place between the nervous processes of excitation and inhibition before one or the other of them gains the mastery' (p. 60) is little more than mythology.

The principal argument against the notion of a suppressing force in extinction is logical. The number of terms needed in the system is unnecessarily increased. The interpretation of an observed state of inactivity employs the double postulate of excitatory and inhibitory forces which cancel each other. Reconditioning must be regarded as the removal of inhibition (p. 67), although it has all the external properties of original conditioning. This can hardly be

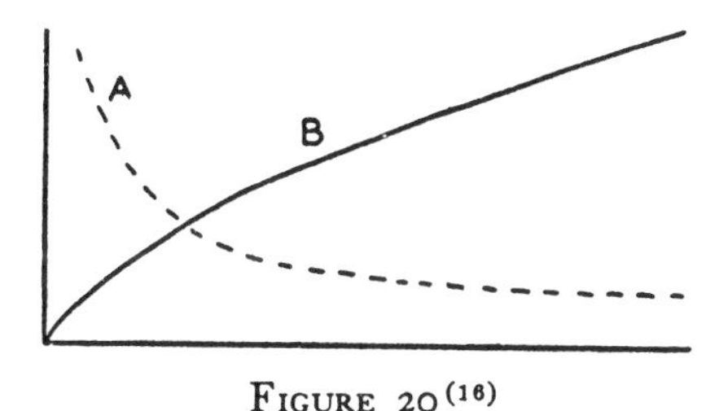

FIGURE 20[16]

EXTINCTION REPRESENTED AS (A) DECLINE IN RATE OR (B) INCREASE IN TOTAL ACTIVITY

justified by designating original conditioning also as a removal of inhibition, since we should be led to the absurd conclusion that all possible conditioned reflexes pre-exist in the organism in a state of suppressed excitability.

On the factual side some proof of the inhibitory character of extinction is claimed from the phenomenon of spontaneous recovery. But it is immaterial whether we regard the recovery as due to the spontaneous removal of inhibition or the spontaneous recovery of strength. Better support for Pavlov's view is the supposed phenomenon of 'disinhibition,' in which a reflex in the course of extinction is said to be released from inhibition by an extraneous stimulus. This effect, if true, would invalidate the formulation given above. Extinction cannot be the mere exhaustion of a reserve due to conditioning if the strength of the reflex can be restored by an event which in itself has no reinforcing value. Suppose, for example, that we are observing the extinction of a reflex along the dotted curve in Figure 20 (page 97), representing the change in strength. If we convert this into a summation curve (solid line), which represents the total amount of activity (say, the total number of elicitations under continuous stimulation) in time, we have an envelop characterizing the reserve and its gradual exhaustion. The experimental curve may fall below this envelop if the organism is disturbed in any way, but it should not go above it. The effect reported by Pavlov would be represented schematically as in Figure 21. The presentation of a stimulus at the arrow would produce an increase in the strength of the reflex, and the effect on the sum-

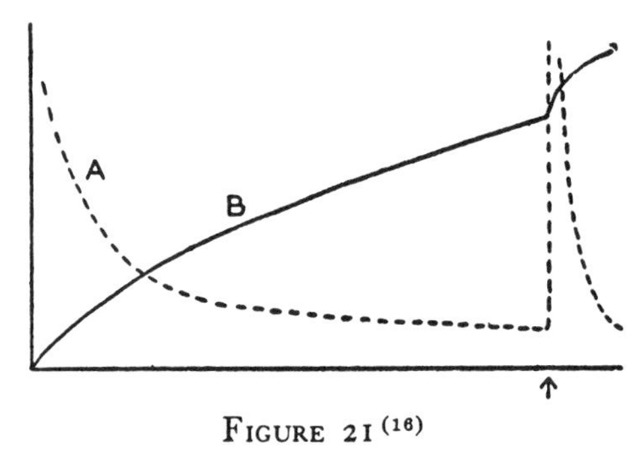

FIGURE 21 (16)

EFFECT UPON EXTINCTION TO BE EXPECTED IN 'DISINHIBITION'

mation curve would be to send it above its envelop—an embarrassing fact from the present point of view.

An experimental attempt to discover disinhibition in typical extinction curves has, however, yielded no result that would violate the present formulation. Curves were used which had been preceded by periodic reconditioning, since they are not significantly disturbed by the cyclic effect which characterizes curves of original extinction and are therefore better suited to revealing slight changes (see the following chapter). The procedure was simply to get extinction curves in progress and then to introduce the 'disinhibiting' stimulus.

Eighteen rats were used in the experiment and two records were taken for six of them, so that twenty-four cases were obtained in all. Several extraneous stimuli were used. In twelve cases the rats were quickly removed from the apparatus and tossed into the air in such a way that vigorous righting reflexes were evoked. In three cases the tails were pricked lightly with a needle. In two cases the empty food-magazine was sounded. This stimulus might have introduced some reconditioning effect; but any sound would be likely to do this through induction (see Chapter Five). In seven cases the light was turned on directly over the lever. No lesser intensities of stimulation were used deliberately, but they have occurred incidentally during many other experiments and are in fact produced inevitably by the rat itself. Since the typical curve obtained with the

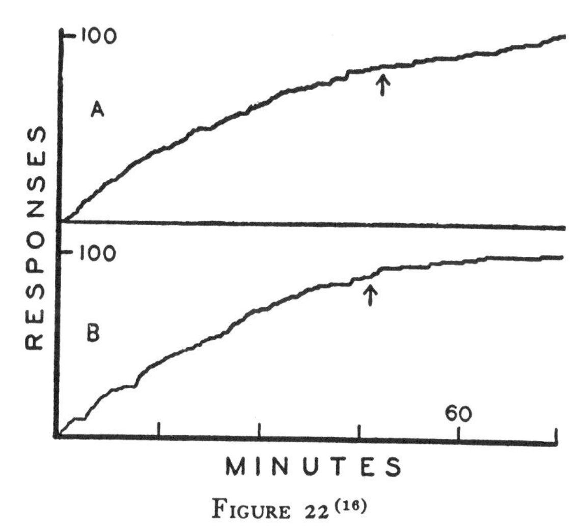

FIGURE 22 (16)

EXTINCTION CURVES SHOWING NO 'DISINHIBITORY' EFFECT

At the arrows the rats were quickly removed from the apparatus, tossed in the air, and replaced. Record A shows no effect. Record B shows one or two responses immediately after return to the apparatus, followed by a slight depression in rate.

method is either smooth or falls *below* the envelop, it may be concluded that stimuli of low orders of intensity do not produce the desired effect.

Two typical records are given in Figure 22. At the arrows the rats were removed from the apparatus and tossed into the air as described. The Record A shows no significant change, and this was true of six cases. Record B shows a very slight increase in rate (perhaps two extra responses) immediately after the return to the apparatus. This was true in one other case with a stimulus of this sort and in four other cases where the stimulus was a light. It was obvious in all these cases, however, that the stimulus was administered at a time when the record was a little below its envelop, and that these extra responses only brought it up to its proper position. In all other cases there was not only no increase in rate but an actual decrease after the 'disinhibiting' stimulus. The result may be pronounced and is precisely what we should expect when the stimulus is strong enough to have an emotional effect.

No one of the twenty-four cases gave any result which would invalidate the interpretation of the extinction curve as a description of the exhaustion of a reserve, and the experiments throw con-

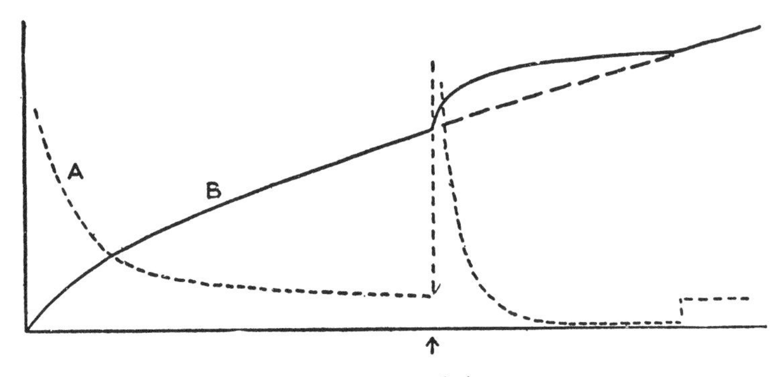

FIGURE 23 (16)

'DISINHIBITION' REPRESENTED AS A TEMPORARY CHANGE WITH SUBSEQUENT COMPENSATION

siderable doubt upon the reality of the effect reported by Pavlov. They provide several ways of disposing of Pavlov's observations without appealing to a phenomenon of disinhibition. In describing an effect of this sort it is obviously necessary to have the whole extinction curve available, or at least some considerable part before and after the disinhibition. Otherwise we may mistake certain possible increases in rate in which the stimulus is merely facilitative. An increase due to facilitation is compatible with the present interpretation and is not an evidence of disinhibition. When the curve is at the envelop any possible effect would be a momentary increase above this point, followed by full compensation during a later period of decreased rate. The existence of an increase which exceeds the envelop is doubtful, but Figure 23 shows a possible way of completing Figure 21 in agreement with this view. In the present experiment the light in particular seemed to have a facilitative effect, probably because of its position directly above the lever, by virtue of which it would bring the lever as a source of stimulation more directly before the rat, but the envelop was not exceeded in any of these cases. Whenever facilitation occurred, the curve was for some reason below its envelop, and the effect was simply to bring it to its proper position. The most striking case of this kind is reproduced in Figure 24. The justification for holding that the rat was considerably 'in arrears' at the time of the 'disinhibition' is the smoothness of the first part of the curve, which gives it enough weight to keep the extrapolation significantly above the irregular middle

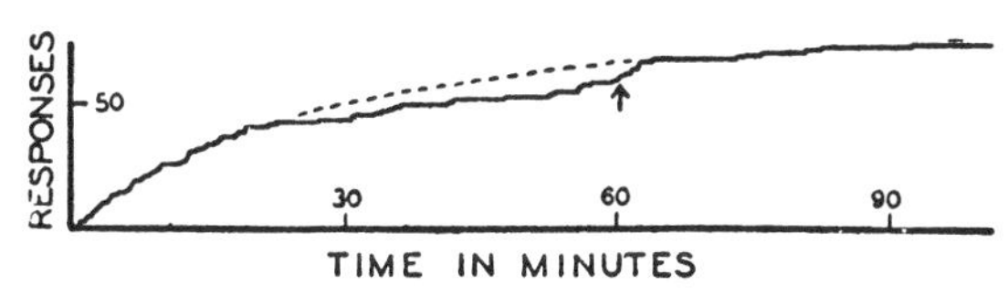

FIGURE 24[16]

A 'DISINHIBITORY' EFFECT RELATED TO THE WHOLE EXTINCTION CURVE

At the arrow a light directly above the lever was turned on. The resulting increase in rate is the greatest obtained in 24 cases. When related to the rest of the curve it is seen to be a recovery from a position considerably below the envelop of the curve.

section. This can be seen more easily by foreshortening the curve. Apparently none of the published work on disinhibition gives the curves for extinction upon which the effect was operative. Until this is done and the fact of disinhibition then confirmed, the present interpretation may be regarded as not seriously threatened.

There is one reason why we might expect an apparently positive result in the Pavlov type of experiment which would be lacking in the present case. With the salivary reflex many stimuli may elicit the response. We must deal not only with salivation as part of the ingestion of food and the expulsion of noxious substances but as part of many patterns of emotional excitement and even of investigatory responses. As Pavlov has said, 'Footfalls of a passerby, chance conversations in neighboring rooms, slamming of a door or vibration from a passing van, street-cries, even shadows cast through the windows into the room . . . set up a disturbance in the cerebral hemispheres . . . [(64), p. 20].' It would be hard to find a disinhibiting stimulus that would not produce salivation of its own accord. If the envelop is exceeded in such a case it may mean simply that other reflexes are being included—that is, that we have a superposition of envelops. The result is not disinhibition and is therefore not embarrassing for the present system.

The distinction I am here making between extinction as the suppression of activity and as the exhaustion of a reserve may be stated as follows. The notion of suppression applies to any factor altering the relation between the reserve and the rate of responding in such a way that the latter is reduced. Thus the reflex in response to the lever is suppressed (1) when the rat is not hungry (see Chapter Nine), (2) when it is frightened (see Chapter Eleven), or (3) when some other reflex takes prepotency over it. In all cases of this sort later increases in rate may be observed and the full force of the reserve ultimately exhibited. In extinction on the other hand the proportionality of rate and reserve is not changed, but the reserve itself is reduced. The metaphor of suppression is inapplicable.

The Extinction of Chained Reflexes

The extinction of the initial member of a chain of reflexes may be brought about by interrupting the chain at any point prior to the unconditioned reflex upon which the conditioning is based. In the sample of behavior considered here the initial operant of lifting the hands to the lever may be extinguished by breaking the chain at any one of the three arrows in the formula on page 54. This is done (for the three arrows respectively) by removing the lever, disconnecting the magazine, or leaving the magazine empty. In extinguishing the second member of the chain (pressing the lever) only two alternatives are available. They may be indicated as follows:

$$_sS^{D\ III} \,.\, R^{III} \overset{(A)}{\rightarrow} {_sS^{D\ II}} \,.\, R^{II} \overset{(B)}{\rightarrow} S^{I} \,.\, R^{I}$$

The following cases arise.

(1) Breaking the chain at *A* produces the simple extinction of $_sS^{D\ III} \,.\, R^{III}$, since it establishes the essential conditions by removing the reinforcing stimulus $S^{D\ II}$. The method does not supply information about $_sS^{D\ II} \,.\, R^{II}$ directly, since it measures only the rate

of responding to the lever, but it is clear from inspection that $sS^{D\ II} . R^{II}$ remains conditioned. After considerable extinction of $sS^{D\ III} . R^{III}$ the sound of the magazine will evoke R^{II} immediately.

(2) If we reestablish the connection at *A*, the first elicitation of $sS^{D\ III} . R^{III}$ reconditions the reflex, and if the connection is broken again another extinction curve is obtained (see page 81).

(3) We may also extinguish $sS^{D\ III} . R^{III}$ by breaking the chain at *B*. Examples are given below in Figures 25 B and 26 C. I have not been able to detect any difference between extinction curves obtained from *A* and *B* with the numbers of cases so far considered.

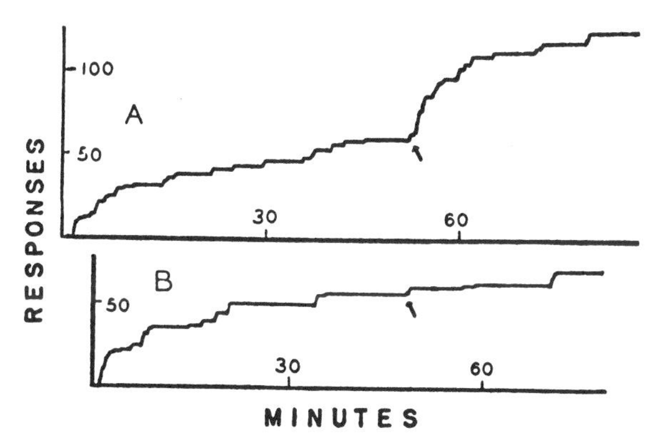

FIGURE 25 (11)

SEPARATE EXTINCTION OF THE MEMBERS OF THE CHAIN

A: the chain is broken first at *A* (see text), then reconnected at *A* and broken at *B*. B: the chain is broken first at *B*, then at *A*.

(4) If we reestablish the connection at *B*, $sS^{D\ III} . R^{III}$ will be reconditioned as soon as it is once elicited, and a subsequent extinction curve may be obtained.

(5) If after extinction from *A* we reconnect at *A* but break at *B*, we also get reconditioning of $sS^{D\ III} . R^{III}$, because $S^{D\ II}$ is an adequate reinforcement, whether or not the later stages of the chain are reinforced (page 83). But since $sS^{D\ II} . R^{II}$ will now be extinguished, it will not continue to reinforce $sS^{D\ III} . R^{III}$, and the latter will decline in strength. The effect of reconnection at *A* is shown typically in Figure 25 A. The first part of the curve shows the extinction of $sS^{D\ III} . R^{III}$ when the chain is broken at *A*. The characteristic cyclic deviation is evident. At the arrow the connection at *A* was restored

(by connecting the magazine) but that at *B* had been broken (by leaving the magazine empty). It will be seen that the response to the lever is quickly reconditioned and then goes through a second decline corresponding to the extinction of the reinforcing reflex $sS^{D\ II} . R^{II}$. More regular curves, of a slightly different shape, may be obtained by using extinction after periodic reconditioning (see

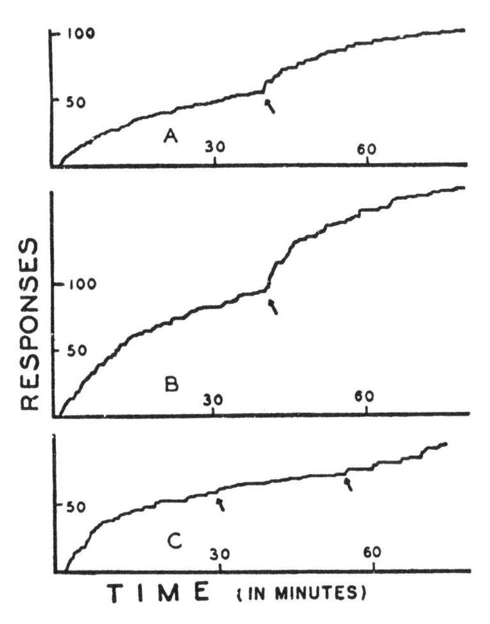

FIGURE 26 (11)

SEPARATE EXTINCTION OF THE MEMBERS OF THE CHAIN AFTER PERIODIC RECONDITIONING

A and B: The chain is broken at *A* (see text), then reconnected at *A* and broken at *B*. C: The chain is broken first at *B* then at *A*.

the following chapter), examples of which are given in Figure 26. The curves at *A* and *B* correspond to Figure 25 A. They show the effect of different initial reserves, which give different slopes to the curves.

It is apparent that these second curves are not flattened as in the case of original secondary extinction on page 83. The cases differ because the response in the present experiment has previously been conditioned. This is presumably responsible for the much more effective secondary conditioning from the sound of the magazine. The secondary conditioning in this case is really *re*conditioning. The number of responses due to the reinforcing effect of the sound of the magazine is again of the same order as in the initial extinction curve itself. It is tempting to suggest the following law: a discriminative stimulus (such as the sound of the magazine) used as a reinforcement in the absence of ultimate reinforcement creates in another reflex a reserve just equal to that of the reflex to which it belongs. The present evidence is hardly capable of establishing the law very conclusively.

(6) If we break at *A* after having extinguished from *B*, no new extinction curve for $sS^{D\ III} \, . \, R^{III}$ is obtained. This is shown typically in Figures 25 B and 26 C. The first part of each curve is for extinction at *B*. At the arrow the chain is broken at *A*. The second arrow in Figure 26 C marks the restoration of the chaining at *A* and shows no significant effect.

We may conclude from these records that the interruption of a chain extinguishes all members up to the point of interruption but not beyond. Since the interruption suppresses the elicitation of all members coming after it, a law may be stated more comprehensively as follows:

LAW OF THE EXTINCTION OF CHAINED REFLEXES. *In a chain of reflexes not ultimately reinforced only the members actually elicited undergo extinction.*

The experimental procedure does not make clear whether in Case (3) all members of a chain preceding the break decline simultaneously or whether the extinction works backward from the point of interruption in successive stages. In terms of rate of elicitation the decline is of course simultaneous, but differences in reserve might be revealed by investigating the members separately at any given point. A detailed comparison of curves for original extinction from *A* and *B* should be made. A greater reserve should appear in the case at *B* if the extinction is progressive rather than simultaneous. It seems quite clear, however, that a larger total number of responses to the lever is obtained by making successive breaks at *A* and *B* than by making an original break at *B*. Presumably, the longer the chain, the greater the difference. The greater number of responses in extinction from *A* and *B* separately is compatible with the notion of a reserve since in the case of an initial break at *A* the reinforcing effect of $S^{D\ II}$ is preserved and may be used later to create a further reserve.

This experiment shows the autonomous status of each member of the chain very clearly. It would be difficult to reconcile data of this sort with a doctrine of the 'wholeness' of the act of 'pressing the lever to get food.'

The following experiment indicates that the second curve obtained by restoring the connection at *A* while breaking at *B* is due to the reconditioning action of the sound of the magazine, although the result is ambiguous with respect to the total additional reserve that can be created in this way. The experiment involves reducing the reinforcing effect with an interval of time. The effect of an interval is not easily demonstrated in original conditioning, as I have already shown, but in the following chapter a better technique will be described with which an interval of as little as two seconds can be shown to reduce the effect by one-third. The interval is used in the present experiment in the following way. Extinction is first obtained with a break at A. The empty magazine is then connected through the time pendulum, so that every response not followed within the interval by another response is reinforced at the end of the interval. (The same difficulty raised by a second response falling within the interval is here encountered and treated as described above.) The new curve to be obtained should show the reduced efficiency of the reinforcement. When the curve has reached a low slope, the reinforcement can be made simultaneous as an additional check. It is at this latter point that an ambiguous result is obtained: in some cases a third extinction curve follows, while in others there is no effect.

The extinction curves for the eight rats tested with this procedure show the effect of considerable preceding experimentation involving prolonged periodic reconditioning (reflected chiefly in the flattening of the curves at the beginning), and the result should be accepted with caution. Of the eight cases two showed no effect upon changing to simultaneous from delayed reinforcements, four showed a slight effect (up to 50 per cent of the effect of the delayed), and two showed from 50 to 100 or more per cent of the effect of the delayed reinforcement. In Figure 27 examples of the first and last group are given. In both curves it will be seen that restoration at *A* when the sound follows after an interval is much less effective than in the simultaneous case described above. The previous prolonged periodic training may be partly responsible for this, however. In the upper record the extinction curve that is added when the reinforcement becomes simultaneous (at the second vertical line) is slightly greater than that added when the delayed reinforcement

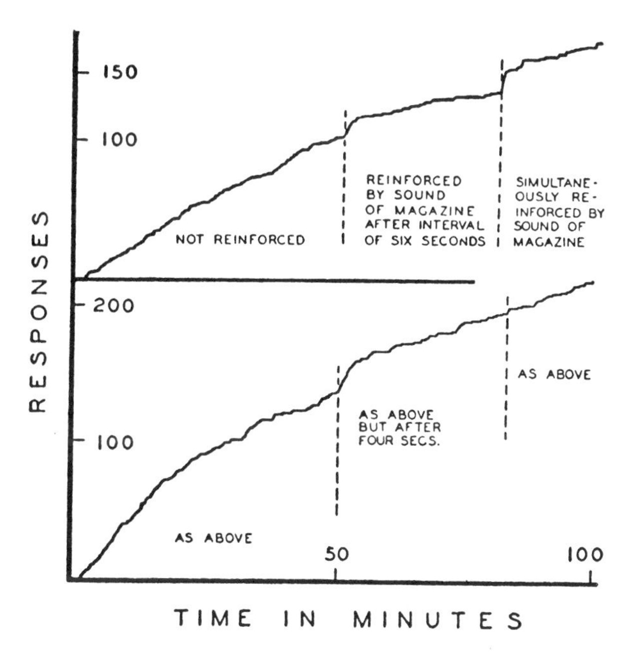

FIGURE 27 (18)

EXTINCTION OF THE MEMBERS OF THE CHAIN WITH DELAYED REINFORCEMENT

The response is first not reinforced, and extinction occurs. The response is then followed by the sound of the empty magazine after different delays as marked. Later the sound of the magazine is made simultaneous with the response.

is first introduced (at the first vertical line). In this case the interval was six seconds. In the lower figure no effect is felt upon changing to simultaneous reinforcement, although the introduction of delayed reinforcement (after an interval of four seconds) gave a significant curve. There is unusual variability in the effect of the delay, and in the following chapter this will be found to be characteristic of experiments involving delayed reinforcement.

The sound of the magazine, following a response after an interval, reconditions the response to some extent. The resulting gain in strength is lost through further extinction. It might be expected that when the discriminatory response to the sound had not been reinforced with food, it would no longer recondition the response to the lever, even when it again followed it immediately. But in the majority of cases some reconditioning occurs. The explanation probably is that the later member (the differentiated response to the sound of the magazine) does not become fully extinguished during the extinction with delayed reinforcement, because the magazine does not sound often enough to produce complete extinction. Consequently the sound again reconditions the response to the lever when it follows immediately. I do not regard this as a wholly satisfactory explanation of the result. Several records fail to show the effect, and the experiment should obviously be repeated on rats with a shorter experimental history.

The Possibility of Negative Conditioning

One kind of reinforcing stimulus in Type R apparently produces a decrease in the strength of the operant. If pressing the lever is correlated with a strong shock, for example, it will eventually not be elicited at all. The result is comparable with that of adaptation or extinction, but there is little excuse for confusing these processes. The distinction between extinction and a decline in strength with 'negative' reinforcement rests upon the presence or absence of the reinforcement and should be easily made.

The effect of such a reinforcing stimulus as a shock in decreasing the strength may be brought about either by a direct reduction in the size of the reserve or by a modification of the relation between the reserve and the strength. Only in the former case should we speak of negative conditioning. The process would then be the opposite of positive conditioning and could be described as a reduction in reserve not requiring the actual expenditure of responses as in the case of extinction. It is not clear, however, that a reduction of this sort actually occurs, at least when the change begins after previous positive conditioning rather than at the original unconditioned strength.

The alternative case of a modification between the strength and the reserve comes under the heading of emotion as defined later. The emotional reaction to the shock is conditioned according to Type S in such a way that the lever or incipient movements of pressing the lever become a conditioned stimulus capable of eliciting it. The effect of the emotional state is to reduce the strength of the response. Responses are not made when the lever is presented, not because there are no responses in the reserve, but because the lever sets up an emotional state in which the strength is depressed. The resulting failure to respond is obviously related to the phenomenon of repression.

The second alternative makes it difficult to demonstrate the first in the case of simple conditioning. Some experiments on the general subject of negative conditioning which utilize the technique of periodic reconditioning will be described in the following chapter.

Comparison of the Two Types of Conditioning

In view of the belief expressed by Pavlov and many others that there is only one fundamental type of conditioning, which is applicable to all 'acquired' behavior, it may be advisable to list some of the differences between the two types defined and described above.

(1) The fundamental distinction rests upon the term with which the reinforcing stimulus (S^1) is correlated. In Type S it is the stimulus (S^0), in Type R the response (R^0).

(2) Type S is possible only in respondent behavior because the necessary S^0 is lacking in operant behavior. The term originally correlated with S^0 must be irrelevant, because otherwise the correlation would be with R^0 as well (even though the temporal conditions of the correlation differed). Type R is possible only in operant behavior, because S^1 would otherwise be correlated with a stimulus as well.

(3) In Type S the change in strength is in a positive direction only and $[S^0 . R^1]$ may begin at zero. In Type R $s . R^0$ must have some original unconditioned strength, and the change may possibly be negative.

(4) In Type S a new reflex is formed. There need be no correlation between S^0 and R^1 to begin with. In Type R the topography of the operant does not change except by way of a selective in-

tensification. A new form is established in the sense that a strong response emerges having the unique set of properties determined by the conditions of correlation of S^1.

(5) A conditioned reflex of Type R is apparently always phasic, perhaps because S^1 follows the response and terminates it. As an operant its strength is measured in terms of a rate, and the reserve has the dimension of a number of responses. Conditioned reflexes of Type S are chiefly non-phasic. As respondents their strength is measured in terms of static properties, and the reserve has the dimensions of a total amount of elicitable activity.

(6) In both types the reserve may be built up through repeated reinforcement, and the subsequent extinction seems to have the same qualitative properties: the strength declines when reinforcement is withheld; the total activity during extinction depends upon the amount of previous reinforcement; successive extinction curves exhibit a smaller total activity (although the intervening amounts of conditioning and the momentary strengths at which the processes begin may be the same); and the reflex recovers to some extent when allowed to remain inactive. Pavlov has noted that 'conditioned reflexes of Type S spontaneously recover their *full* strength after a longer or shorter interval of time [(64), p. 58],' but this applies to strength, not to reserve, and is equally true of Type R. The initial rate in a curve showing recovery is frequently as great as in original extinction (see Figure 9). Pavlov has also reported rhythmic fluctuations during extinction of Type S. The interpretation advanced above for the fluctuations during extinction of an operant will not hold in the case of a respondent. It requires a reduction in the rate of responding, during which a cumulative emotional effect is allowed to pass off, but in the respondent the rate of elicitation is held constant through control of the stimulus. The reality of a genuine cyclic fluctuation in the extinction of a respondent may, however, be questioned. I have plotted all the data for extinction given by Pavlov and fail to find anything more than a considerable irregularity.

(7) Because of the control exercised through S^0 in Type S, extinction may pass through zero in that type. If Pavlov's data are valid, a negative reserve may be built up. Thus, Pavlov's measure of extinction below zero is the extra reinforcement required for reconditioning. In other words, a certain number of responses must be contributed to the reserve before any effect is felt upon the

strength. A negative reserve is impossible in Type R because further elicitations without reinforcement are not available when the strength has reached zero.

An analysis of differences between the two types has been made by Hilgard (44), who points out that both types usually occur together and that 'reinforcement' is essentially the same process in both. The present distinctions are, however, not questioned.[1]

The two types may be characterized somewhat more generally as follows. The essence of Type S is the substitution of one stimulus for another, or, as Pavlov has put it, signalization. It *prepares* the organism by obtaining the elicitation of a response before the original stimulus has begun to act, and it does this by letting any stimulus that has *incidentally* accompanied or anticipated the original stimulus act in its stead. In Type R there is no substitution of stimuli and consequently no signalization. The type acts in another way: the organism selects from a large repertory of unconditioned movements those of which the repetition is important with respect to the production of certain stimuli. The conditioned response of Type R does not prepare for the reinforcing stimulus, it produces it. The process is very probably that referred to in Thorndike's Law of Effect (70).

Type R plays the more important rôle. When an organism comes accidentally (that is to say, as the result of weak investigatory reflexes) upon a new kind of food, which it seizes and eats, both kinds of conditioning presumably occur. When the visible radiation from the food next stimulates the organism, salivation is evoked according to Type S. This secretion remains useless until the food is actually seized and eaten. But seizing and eating will depend upon the same accidental factors as before unless conditioning of Type R has also occurred—that is, unless the strength of sS^D*: food . R: seizing* has increased. Thus, while a reflex of Type S prepares the organism, a reflex of Type R obtains the food for which the preparation is made. And this is in general a fair characterization of the relative importance of the two types. Thus,

[1] The problem has been discussed in several other recent papers. Schlosberg (Psychological Review, 1938, *44*, 379–394) agrees with the distinction here drawn and insists that the differences are too great to justify the use of the term conditioning in the case of Type R. Mowrer (Psychological Review, 1938, *45*, 62–91) holds out for the possibility that the two processes may eventually be reduced to a single formula. These papers, as well as that of Hilgard, should be consulted for the relation of the problem to the traditional field of learning.

Pavlov has said that conditioned stimuli are important in providing saliva before food is received, but 'even greater is their importance when they evoke the motor component of the complex reflex of nutrition, *i.e.*, when they act as stimuli to the reflex of seeking food [(64), p. 13].' Although 'the reflex of seeking food' is an unfortunate expression, it refers clearly enough to behavior characteristic of Type R.

The distinction between Types R and S arising from their confinement to operant and respondent behavior respectively implies a rough topographical separation. Reflexes of Type S, as respondents, are confined to such behavior as is originally elicited by specific stimuli. The effectors controlled by the autonomic nervous system are the best examples, one of which was used almost exclusively by Pavlov in his classical studies. This subdivision of behavior is a very small part of the whole field as defined here, and much of it is perhaps excluded if the definition is interpreted strictly. To it may perhaps be added a few scattered skeletal responses—flexion of a limb to noxious stimulation, winking, the knee-jerk, and so on. Most of the experiments upon skeletal behavior which have been offered as paralleling Pavlov's work are capable of interpretation as discriminated operants of Type R, as will be shown in Chapter Six. It is quite possible on the existing evidence that a strict topographical separation of types following the skeletal-autonomic distinction may be made.

Any given skeletal respondent may be duplicated with operants and hence may also be conditioned according to Type R. Whether this is also true of the autonomic part is questionable. Konorski and Miller have asserted that it is not (57). There is little reason to expect conditioning of Type R in an autonomic response, since it does not as a rule naturally act upon the environment in any way that will produce a reinforcement, but it may be made to do so through instrumental means. In collaboration with Dr. E. B. Delabarre I have attempted to condition vasoconstriction of the arm in human subjects by making a positive reinforcement depend upon constriction. The experiments have so far yielded no conclusive result, but there are many clinical observations that seem to indicate conditioning of this sort. The operant field corresponds closely with what has traditionally been called 'voluntary' behavior, and the 'voluntary' control of some autonomic activities is well established. The child that has learned to cry 'real tears' to produce a reinforcing stimulus has apparently acquired a conditioned autonomic operant.

The mere existence of the 'voluntary' control of a respondent does not prove that the response has been conditioned according to Type R. The respondent may be chained to an operant according to the formula:

$$s\,.\,R^{II} \rightarrow S^{I}\,.\,R^{I},$$

in such a way that the 'control' is exercised through $s\,.\,R^{II}$. Four cases arise from the possibilities that S^I may be either conditioned or unconditioned and either exteroceptive or proprioceptive. Examples are as follows.

(1) Unconditioned and exteroceptive: $s\,.\,R^{II}$: *sticking pin in one's own arm* $\rightarrow S^I$: *prick* . R^I: *increase in blood pressure.* This would not ordinarily be called the voluntary control of blood pressure, but it does represent one kind of operant control. If the rise in blood pressure is correlated with a reinforcement (through instrumental means), $s\,.\,R^{II}$ will increase, provided the reinforcement is great enough to overcome the negative reinforcement from the prick. Thus, if a hungry man were given food whenever his blood pressure rose, he might resort to using a pin to produce that effect.

(2) Unconditioned and proprioceptive: $s\,.\,R^{II}$: *rapid muscular activity* $\rightarrow S^I$: *proprioceptive stimulation* . R^I: *rise in blood pressure.* This case is very similar to the preceding. If the change in blood pressure were again correlated with a positive reinforcement, $s\,.\,R^{II}$ would increase. Our hungry man might resort to this device also.

In both cases the correlation of reinforcement with R^{II} is equivalent to correlating it with R^I because the connections between R^I and S^{II} and between S^{II} and R^{II} are practically invariable. What is done in such a case is essentially to base the reinforcement upon an operant defined as any movement producing a change in blood pressure. The single condition of reinforcement specified would not necessarily produce the response of pricking the skin in the first case and rapid movement in the second. *Any* response which occurred and produced the effect would be conditioned. But it is still true that in both cases some sort of 'voluntary' control over the blood pressure has been acquired.

(3) Conditioned and exteroceptive. $s\,.\,R^{II}$: *looking at picture* $\rightarrow$

S^I*: picture* . R^I*: emotional effect.* In this case instrumental means need not be resorted to, as some examples of R^I are in themselves reinforcing (*i.e.*, 'pleasurable'). The case applies to the very common behavior of reading exciting books, looking at or painting exciting pictures, playing exciting music, returning to exciting scenes, and so on. The result is again not ordinarily called the 'voluntary' control of the emotional response, but it represents one kind of operant control.

(4) Conditioned and proprioceptive. *s* . R^{II}*: subvocal recitation of poetry* → S^I*: proprioceptive effect* . R^I*: emotional response.* Here again R^I may be reinforcing in itself and therefore may not require instrumental connection with an external reinforcing stimulus.

The well-known experiment by Hudgins (46) falls within the last two categories. A stimulus of Type S (the word 'contract') is conditioned to elicit contraction of the pupil by presenting it simultaneously with a strong light. Eventually the subject may produce the stimulus himself (operant behavior), and contraction of the pupil will follow. When the stimulus is produced aloud, the case is (3); when sub-audibly, it is (4). It should be clear from the foregoing analysis that the experiment does not demonstrate contraction conditioned according to Type R and hence does not resemble a true conditioned operant, such as contracting the lips. Delabarre and I have easily confirmed Hudgins' experiment in the case of vasoconstriction. The subject said 'contract' and a gun was fired to produce strong vasoconstriction. Eventually constriction followed the unreinforced saying of 'contract.' The sub-audible case was not clearly demonstrated.

In view of the possible chaining of operant and respondent it is difficult to tell whether an autonomic response can in any case actually be conditioned according to Type R. The report of the organism itself as to how it exerts 'voluntary' control is not trustworthy. In the experiments on vasoconstriction we found that in the case of an apparently successful result the subject was changing the volume of the arm by changing the amount of residual air in the lungs. The depth of breathing was in this case conditioned according to Type R because of the reinforcement of its effect on the volume of the arm. The 'successful' result was obtained many times before the intermediate step was discovered by the subject.

Aside from the question of whether all responses may be conditioned according to Type R, it may definitely be stated that a large part of behavior cannot be conditioned according to Type S. Following the classical investigation of conditioning of Type S it was quite generally held that the type was universally applicable. But most of the original work had been based upon the responses of glands and smooth muscles, and the extension of the principle to striped muscle met with indifferent success. It was assumed that the formula would apply with the substitution of a skeletal for a glandular response, as in the following typical example:

Here the tone was to be followed by flexion in the absence of the shock, as in the Pavlovian experiment it was followed by salivation in the absence of food. The results were not satisfying. The conditioned skeletal respondent was usually found to develop slowly and to be relatively unstable. A low rate of conditioning was not a disturbing result in the case of responses of glands and smooth muscle, which are characteristically slow, but the laboratory result for skeletal responses did not give much promise of accounting for the ordinary highly mobile properties of skeletal behavior. In the light of the operant-respondent distinction this is easily understood. Most (if not all) skeletal behavior is operant and conditioned according to Type R. The formula for Type S is applicable, if at all, only within a limited field. Much of the plausibility given to the extension of Type S has come from a confusion with Type R, which arises from the fact that most of the stimuli which elicit skeletal respondents are also reinforcing stimuli for Type R. It is difficult to set up conditions for Type S in a skeletal response which are not also the conditions for Type R. Some of the resulting mixed cases are considered later in Chapter Six.

Chapter Four

PERIODIC RECONDITIONING

Intermediate States of Conditioning

An operant may be strengthened or weakened through reinforcement or the lack of it, but the phenomena of acquisition and loss of strength are only part of the field defined by reinforcement as an operation. In the traditional field of 'learning' there is frequently an implication of an all-or-none 'knowing' or 'not-knowing.' In the present system we have also to consider sustained and relatively stable intermediate states of strength due to the operation of reinforcement. Outside the laboratory very few reinforcements are unfailing. Perhaps least so are the tactual reinforcements correlated with visual discriminative stimuli in the external world. It is almost invariably true that in the presence of certain kinds of visual stimuli certain movements of my arm result in the tactual stimulation of my hand. Only under illusory conditions (as in mirrors) or where the discriminative stimuli are so vague as to be ambiguous (as in a dimly lighted room) are the necessary mechanical connections between visual and tactual sources of stimulation lacking. Of other reinforcing stimuli the contingency is uncertain. This is particularly true in the verbal field, which may be defined as that part of behavior which is reinforced only through the mediation of another organism. The contingency of water as a source of reinforcing stimuli upon the spoken response 'Water!' is obviously of an entirely different order of magnitude from that of touch upon sight. This is also true for other kinds of purely mechanical reinforcement. My pipe is not always in my pocket, and a match does not always light. In reaching for my pipe and striking a match, my behavior is marked to some extent by the effect of previous failure of reinforcement. The strengths of these reflexes are not wholly dependent upon the state of the drive, of my emotional state, and so on, but upon the degree of the conditioning as well.

In general the states of strength of the conditioned reflexes of an organism are submaximal with respect to the operation of reinforcement. This important property of behavior can be dealt with adequately only by going beyond the notion of momentary strength. Special properties of conditioned reflexes arise under periodic reconditioning which have no counterpart in the original conditioning and extinction of a reflex. They are properties of the reflex reserve and of the relation of the reserve to the rate of elicitation. We may approach the subject by examining the effect of periodic reconditioning upon the state of our representative operant. The periodicity of the reinforcement will be held constant, even though it is by no means so outside the laboratory.

Periodic Reconditioning

In Chapter Three records of extinction after small amounts of reconditioning were described. The reconditioning and extinguishing process may be repeated at will. But if a second reconditioning follows closely upon the first, the remainder of the first curve for extinction appears to sum with the second. If the reconditioning is periodically repeated at some interval shorter than the average effectual length of the extinction curve for the amount of reconditioning employed, the successive curves continue to sum, until eventually a complete fusion takes place and the strength of the reflex remains at a constant value so long as the periodic reconditioning is maintained.

The development of a fused state is shown in Figure 28. The records are for two rats in which the response had been well extinguished on the preceding day. Some loss of extinction is shown in both cases (at A and A′), but the rate of elicitation quickly reaches a very low value, which is maintained for approximately 25 and 35 minutes respectively. At B and B′ a schedule of periodic reconditioning was instituted, in which the response to the lever was followed by the delivery of a pellet of food every 5 minutes and the intervening responses were allowed to go unreinforced. The reinforced responses have been indicated with vertical bars. The extinction curves at B and B′ have the usual form, if we allow for the difficulty of approximating such a curve with as few as 10 or 12 discreet steps. The curves at C and C′, however, show much longer rising limbs, which apparently contain the responses remaining to the preceding curves. The third curves have still longer rising limbs, and, as the other reconditionings ensue, the character of each curve is finally lost and the records assume a constant slope. Thereafter the elicitation proceeds at a steady

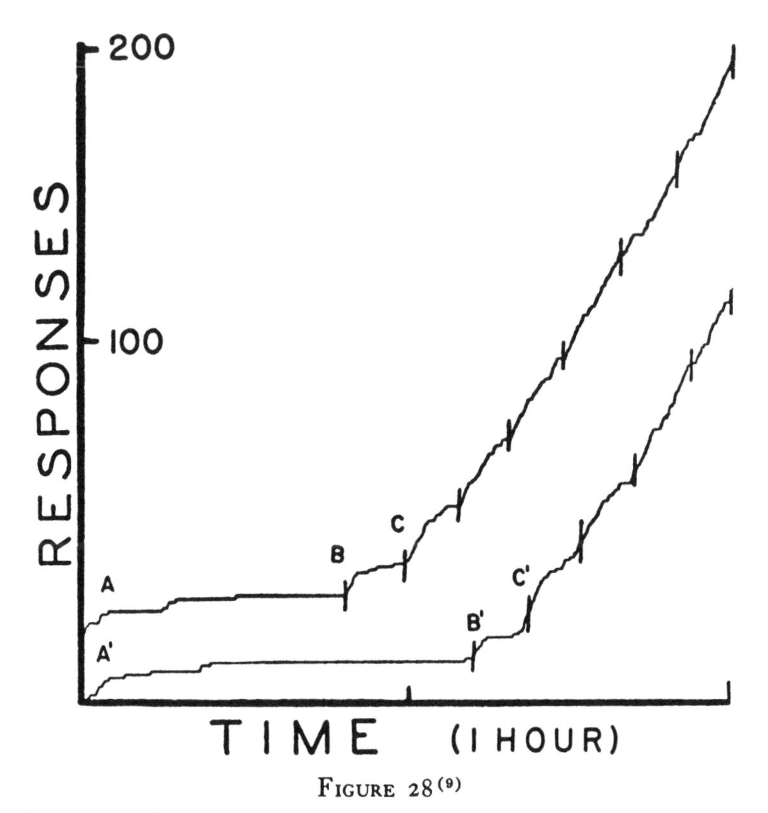

FIGURE 28 (9)

FUSION OF SUCCESSIVE EXTINCTION CURVES FOLLOWING THE REINFORCEMENT OF SINGLE RESPONSES (AT B, C, ETC.)

rate, and it is not possible to tell from the recorded behavior where the successive reinforcements occur.

The constant strength assumed by the operant under periodic reconditioning is highly stable. In an experiment which tested the strength at four different rates of reconditioning, essentially constant values were maintained for twenty-four experimental hours. Intervals between successive reinforcements of 3, 6, 9, and 12 minutes were assigned to four male rats, P7, P9, P8, and P10, respectively. The rats were 106 days old at the beginning of the experiment. Since it was not possible to obtain the elicitation of a response immediately after the magazine had been turned on, the intervals could not be exactly determined. The delay in making a response was added to the interval, and the resulting average intervals for the series were 3 minutes 6 seconds, 6 minutes 27 seconds, 9 minutes 47 seconds, and 12 minutes 59 seconds, respectively. In the other experiments to be described the delay was deducted from the succeeding interval, so that the average was exactly of the scheduled length.

Prior to this series of observations the response to the lever had been conditioned and extinguished. Each animal was placed in its box at the same hour daily (9:00 A. M.) and approximately two minutes later was released into the main part of the box, where the lever was accessible. The first response to the lever was reinforced, and thereafter single responses were reinforced according to schedule. All responses to the lever, whether reinforced or not, were recorded. A characteristic constant rate of elicitation was attained in each case. The animals were removed from the boxes at the end of one hour, and extra food was given to them in their living-cages for about two hours. No other food was given until the following (non-experimental) day, when a full ration was given at 9:00 A. M. This procedure was repeated for 24 experimental days.

The results are contained in four sets of 24 kymograph records, which describe the behavior of the rats with respect to the lever for a total of 96 experimental hours. The complete experiment is represented in Figure 29 (page 120), constructed by plotting the total number of responses attained by the end of each day against the number of days. The approximate course of each daily record was drawn (free-hand) between these separate points. The details of the original records are lost in the unavoidable reduction in size, but typical records from each set are reproduced in Figures 30 and 31, and their positions in Figure 29 are indicated with brackets in the latter figure, the order of occurrence corresponding in the two cases. Thus, the curve for P7 in Figure 29 represents a series of 24 kymograph records pieced together to form a continuous graph, of which Figure 30 shows, reading from left to right, the fourth, sixth, and seventeenth records.

The most obvious conclusion to be drawn from Figure 29 is that

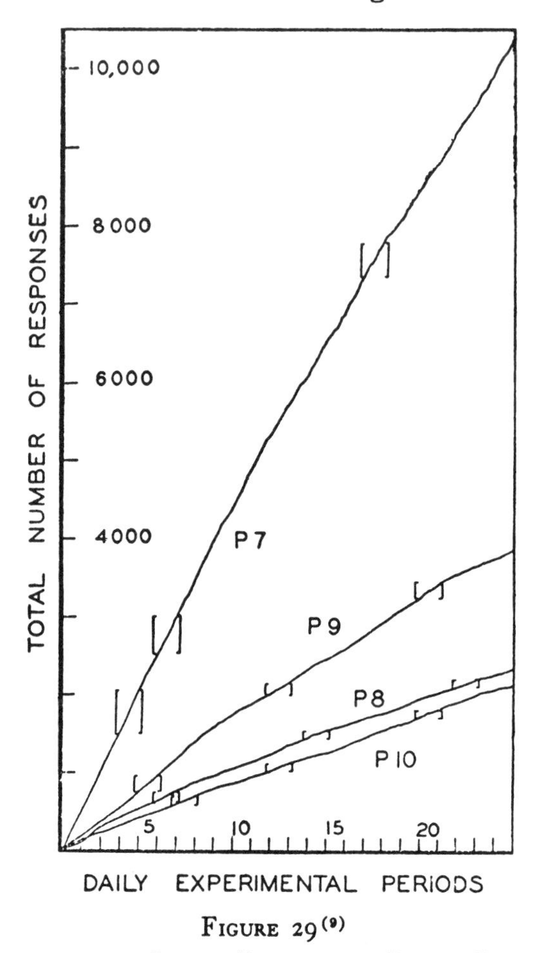

FIGURE 29(9)

RESPONSES TO THE LEVER DURING 24 DAILY EXPERIMENTAL PERIODS OF ONE HOUR EACH

Responses were reinforced as follows: P7 every three minutes, P9 every six minutes, P8 every nine minutes, and P10 every twelve minutes.

the value of the assumed rate is a function of the interval between successive reconditionings. The shorter the interval, the steeper the slope of the graph. I shall return to this relationship. It is also apparent that no very significant change in rate is to be observed at any given interval during the period of the experiment. So far as each daily record is concerned, there is, in fact, no consistent

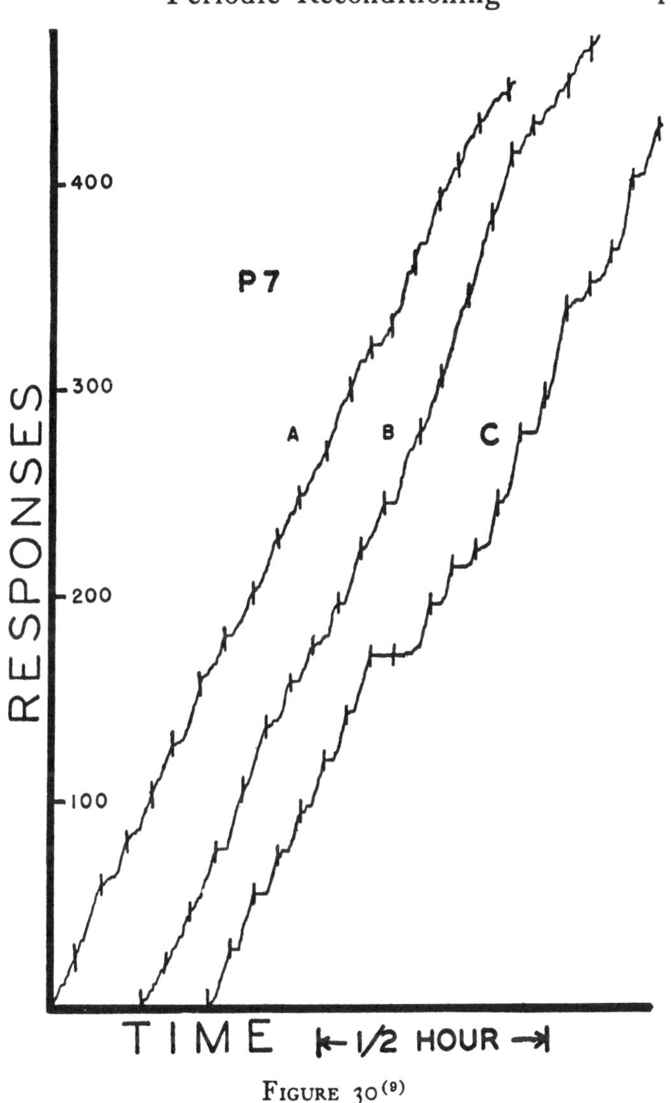

FIGURE 30[9]

THREE DAILY RECORDS FROM THE SERIES FOR P7 IN FIGURE 29

The positions in Figure 29 are indicated with brackets.

deviation from a straight line, although all records show local irregularities. The curvature observed in some few instances (see P9, Figure 31) follows no consistent rule from day to day. In the

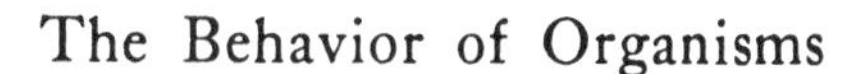

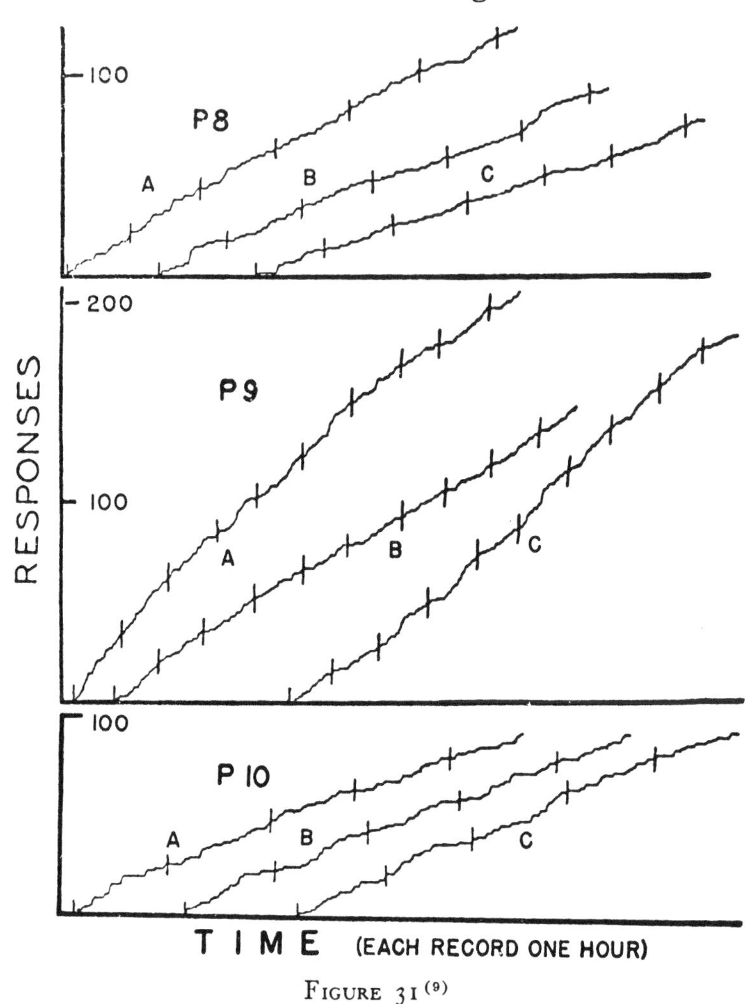

FIGURE 31 [9]

THREE RECORDS EACH FROM THE SERIES FOR P9, P8, AND P10 IN FIGURE 29

The positions in Figure 29 are indicated with brackets.

complete series, on the other hand, the rate of elicitation is not strictly constant, but shows a significant though slight decline. It is impossible at present to correlate so slight an effect with any specific condition of the experiment. The curves in Figure 29 cover a period of 47 days, and there are many progressive changes to which an effect of this magnitude could be attributed. For any short section of the graph the curve may be regarded as essentially rectilinear.

Local deviations from a straight line peculiar to these records are of four clearly distinguishable orders. The first is unimportant. Ignoring the curvature of the graphs in Figure 29, we must still take account of a variation of the following sort. The rate of elicitation first shows a slight but continuous increase, which may extend over a period of several days and which may be obscured by the progressive decline in rate that occurs simultaneously. Then, within a day (or at most two days) the rate falls to a value that is minimal for the curve in question. From this point there is a slow recovery, which may persist for several days. When the rate has again reached a relatively high value, another rapid drop occurs, and the recovery is subsequently repeated. The range over which the rate varies is small, but it is sufficient to give the records a slight 'scalloped' effect. This characteristic is exceptionally prominent in the curve for P9, Figure 29, but it can be detected in the other curves by strongly foreshortening them. It has appeared in other experiments of the same kind. The cause of the variation is unknown.

Deviations of a second order are much more striking and are definitely correlated with the special conditions of the experiment. They are found only at the higher rates of elicitation (induced with more frequent reconditioning) and are not present during the first days of the experiment. There is no reliable sign of the effect in Figure 30 A, for the fourth day of the series, although deviations of this type have clearly appeared by the sixth day (Record B). They resemble those of the first order, with the important exception that the duration is of an entirely different order of magnitude. In Figure 30 B it may be observed that the average rate of elicitation gradually increases until the sixth reinforcement. It is then suddenly depressed. There is a subsequent steady recovery, leading to a second maximum at the fourth reinforcement from the end. Here another depression appears, the recovery from which is interrupted by the termination of the period. The record is thus divided into three segments, each of positive acceleration, which meet at points showing definite discontinuity. The effect has become more clearly marked (although in this case somewhat more irregular) by the seventeenth day, Record C.

In spite of these second-order deviations the average basic rate of elicitation is unchanged; the three records in Figure 30, for example, are approximately parallel. The effect has, therefore, two discrete phases, one marked by an increase above the basic rate, the other by a depression below it, which may combine to give the basic rate as a resultant. The two phases may be observed in Figure 30 and in the figure about to be described (Figure 32). When

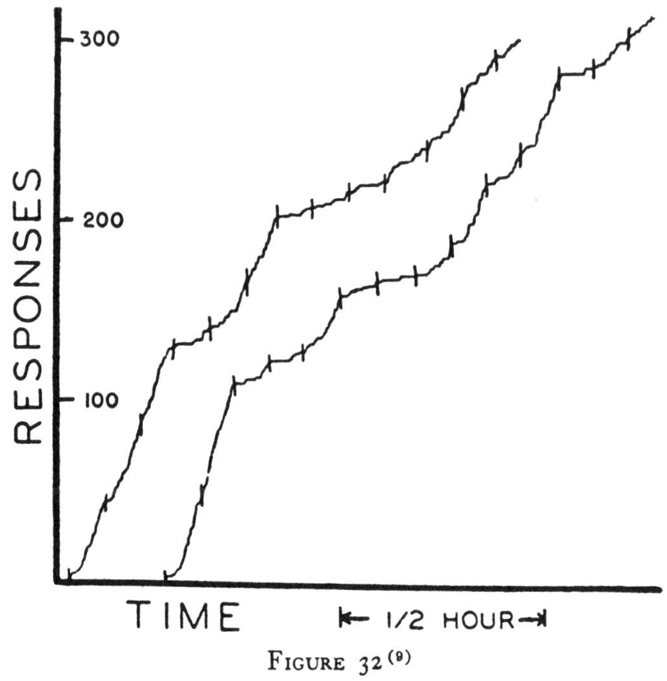

FIGURE 32 (9)

UNUSUAL DEVELOPMENT OF DEVIATIONS OF THE SECOND ORDER

a restriction is placed upon one phase, there is a corresponding adjustment of the other. In Figure 30 the basic rate is already so high that any increase is probably faced with a practical limit. (The apparatus is capable of handling a rate at least 50 per cent above the highest here observed.) Accordingly, the depressed phase has no extensive compensation to make and is of fairly short duration. The two records in Figure 32, which may be compared with Figure 30, were obtained on successive days at the end of a long series of observations upon another rat where the interval of periodic reconditioning was five minutes. The basic rate (given approximately by a straight line drawn through the end points of each record) is moderate. A very large increase is therefore possible and is actually observed—notably in the initial segments of the two curves. The periods of depressed rate are correspondingly more extensive. Figure 32 is a much better example of the second-order effect, since it demonstrates a fuller development of the augmented phase. It also shows clearly that the fundamental deviation is the increase in rate and that the depressed phase enters only as a compensating factor. The nature of deviations of the second order will be taken up again in Chapter Seven.

Deviations of a third order appear as depressions in the rate of elicitation after the periodic reconditioning of the reflex. They are followed typically by compensatory increases, so that the total rate is unchanged. With a relatively short period of reconditioning the depressions begin to appear within a few days. They are already present to some extent in Figure 30, Record A, and have become quite definite in Record B. They give a step-like character to Record C, on the seventeenth day of the series. The same effect is apparent, though to a lesser degree, in the records obtained with reconditioning at longer intervals (see Figure 31). Here their development is considerably retarded. The third-order deviation will be shown in Chapter Seven to be a temporal discrimination, which cannot be avoided in periodic reconditioning. The reflex in response to the lever immediately after receipt of a pellet is weakened because it is never reinforced at that time.

Deviations of a fourth order are not peculiar to the procedure of periodic reconditioning. They have already been discussed in connection with extinction and are characteristic of the general behavior of the organism. They are the expression, on the one hand, of the tendency of responses to occur in groups—a form of the facilitation described in Chapter One—and on the other, of the occasional prepotency of stimuli not successfully eliminated from the experimental situation. Here again there seems to be an adequate and probably complete compensation for deviations in either direction. Examples of the two effects may easily be located in Figure 31 and other figures.

These four orders are clearly definable in terms of their observed properties. Taken together, they include all the deviations from a straight line found in the present records. The only possibility of confusion is between the third and fourth, which are of approximately the same magnitude: when a deviation of the fourth order occurs immediately after the reinforcement of a response (and this is likely when it is due to the prepotency of other stimuli), it is indistinguishable from one of the third. For this reason it is difficult to detect the first appearance of the third-order effect, although the difficulty could be surmounted through a statistical treatment.

At the lower rates of responding which accompany lower frequencies of reinforcement a difficulty arises in interpreting the constant rate, which cannot be accounted for simply in terms of the superposition of extinction curves, since it is less than the rate observed in the initial limb of the extinction curve. In Figure 28 the initial rate is approached because the final rate is high, but in cases such as those of P8 and P10 in Figure 31 the constant rate is far below the values originally observed in extinction. The constant rate is attained only through a *reduction* in the rate immediately following reinforcement. A reduction cannot come about through a superposition of extinction curves but must involve other factors. In Chapter Seven the production of a constant rate will be discussed in some detail. As in the case of third-order deviations a temporal discrimination is involved.

In summary, the periodic reconditioning of a reflex establishes a constant stable strength which may persist without essential change for as long as twenty-four experimental hours covering forty-seven days. The rate is due to a fusion of successive extinction curves, but the character of each curve is lost because of a temporal discrimination. This stability is of considerable importance in the normal behavior of the organism and will be seen to be of great value experimentally when we come to the study of discrimination, which necessarily involves alternate reinforcement and extinction.

The Extinction Ratio

Having examined the extent to which our experimental curves are to be regarded as rectilinear, we may turn to the relation between a given rate of elicitation and the period of reconditioning at which it is attained by the rat. That the rate and the interval between reinforcement vary inversely was apparent in Figure 29, but the relationship is not reliable because individual differences are probably present. These may be allowed for either by using a large number of animals at each slope or by rotating a small number through a series of slopes. The former method has not been tried, but an experiment with the latter yielded the following result.

Four rats (P3, P4, P5, and P6), of the same strain and approximately the same age, were used. They had been conditioned 50 days before. Meanwhile, on alternate days, several aspects of the response had been studied, most of the experiments requiring periodic reconditioning at 15-minute intervals for periods of 1 hour. The four rats were tested at intervals chosen to cover the greater

TABLE 1

INTERVALS BETWEEN REINFORCEMENTS

Days	P3	P4	P5	P6
1–2	3	5	7	9
3–4	5	7	9	3
5–6	7	9	3	5
7–8	9	3	5	7
9–10	3	5	7	9

part of the change in slope in Figure 29, *viz.*, 3, 5, 7, and 9 minutes, according to the schedule in Table 1. Each rat remained at each interval two days and returned on its last two days to the interval at which it began. The mean rate in responses per hour for each rat at each interval is given in Table 2. Each entry is the

TABLE 2

RESPONSES PER HOUR AT DIFFERENT INTERVALS OF REINFORCEMENT

Interval	P3	P4	P5	P6	*Mean*
3 min.	315	197	458	305	319
5 min.	256	196	391	221	266
7 min.	226	158	333	164	220
9 min.	170	144	230	131	169
Mean	242	174	353	206	

average of two observations, except for the interval at which each rat began, where four observations were taken. That there are large individual differences is shown by the mean rate for the series regardless of length of interval, which range from 174 to 353 responses per hour. The mean at each interval is free of the effect of this individual variation and is the result required. The possible effect upon the means of the extra weight of the four-observation averages may for the present purposes be ignored.

The result is represented graphically in Figure 33. The means shown in the heavy line give an approximately linear relation. The

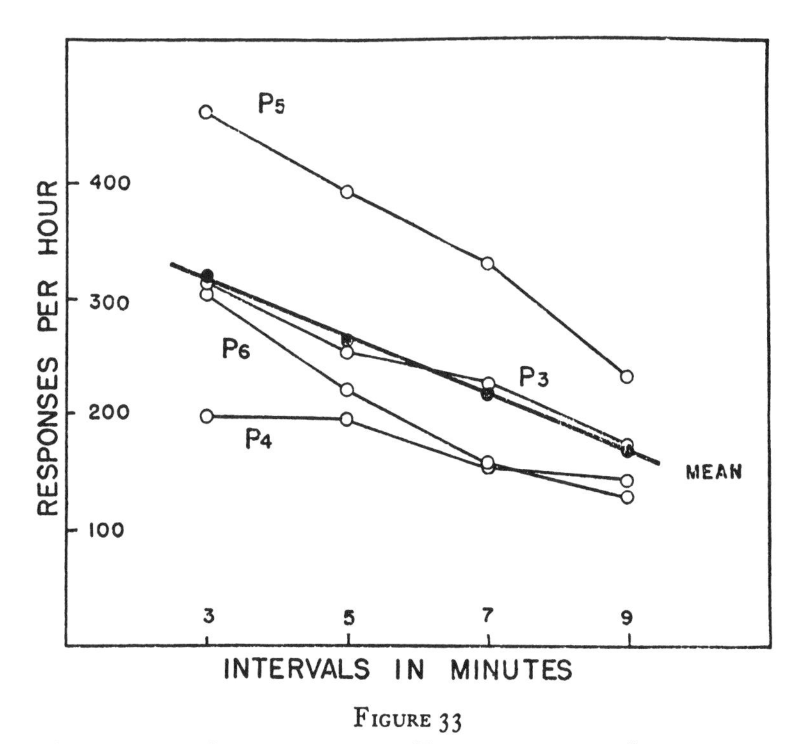

FIGURE 33

THE RATE OF RESPONDING AS A FUNCTION OF THE INTERVAL BETWEEN PERIODIC REINFORCEMENTS

lighter lines for individual rats differ somewhat in slope but are also roughly linear. This relationship is to some extent fortuitous, since it is impossible in so short a series of observations to correct for the effect of 'carry-over' from one experimental day to the next. If the rat has been responding to the lever at some constant rate, that rate will obtain at the beginning of a new experimental period, even though the frequency of reconditioning in the new period is going to justify either a lower or a higher rate. Adjustment to the new frequency is quickly made, but the average for the full period necessarily shows the effect of the anomalous initial rate. Accordingly, since the intervals of 5, 7, and 9 minutes in the present schedule follow *shorter* intervals in each case, we should expect the rates observed upon the first days at those intervals to

FIGURE 34

ACCELERATION IN RATE FOLLOWING CHANGE FROM NINE- TO THREE-MINUTE REINFORCEMENT

The slopes prevailing under nine-minute reinforcement are represented by the straight lines.

be to some extent too high. The three-minute intervals, on the other hand, follow longer intervals in every case, and the observed rates are therefore probably too low. Of the two cases the transfer from a lower to a higher rate takes place the more slowly and is the more clearly indicated in the records. In Figure 34 two typical curves are reproduced to show the change from an interval of reinforcement of nine minutes to one of three. The straight lines give the slopes prevailing on the preceding day at the greater interval. The experimental curves show a uniform acceleration from this value to the maximal values prevailing at the new frequency. The maximal value is reached at about the middle of the hour. Some allowance could have been made for our present purposes by estimating the slope eventually attained at each interval, rather than using the total number of responses per hour. The result given in Figure 33 is not intended to be independent of the particular conditions of the experiment. A relationship between slope and interval, free of the effect of any significant individual difference, is demonstrated, but its precise nature is not shown.

One way of expressing the constancy of the rate obtaining under periodic reconditioning is to say that the effect of the reinforcement of a single response is constant. As we have already noted, the number of elicitations observed during the first interval in a series (as recorded, for example, in Figure 28) is obviously less than the number due to the initial reconditioning, since the extinction curve is clearly interrupted; and this is also true for a few succeeding intervals. The rate of elicitation is rapidly accelerated, however, until a given constant rate is attained. If we attribute the acceleration to the eventual appearance of the responses remaining in the interrupted curves, then the attainment of a constant rate must be taken to mean that the number of responses now observed in a single interval is precisely the number due to the reconditioning of one response. The output is just balancing the input, and there are no responses left over to cause further acceleration.

I shall represent this relation with the ratio N_e / N_c, where N_e is the number of responses not reinforced and N_c the number reconditioned. When $N_c = 1$, I shall refer to it as the *extinction ratio*. In the present case its value is constant during an extended experimental series. Dividing the total number of responses in each record in Figure 29 (10,700, 3980, 2420, and 2200) by the number of minutes in the series (1440), we obtain average rates of 7.10, 2.51, 1.57, and 1.45 responses *per minute* for rats P7, P9, P8, and P10 respectively. If we now multiply by the number of minutes in the average interval at which each was observed, we obtain 22.8, 16.1, 15.2, and 18.8 as the number of responses *per reinforcement*. In order to take account of uncompleted intervals at the end of each experimental period, we may assume a complete compensation on the following day and obtain another set of values by dividing the total number of responses in each series by the total number of reinforced responses. The resulting values are approximately the same: 22.8, 17.4, 15.1, and 18.4. From the first method a mean of 18.2 is obtained; from the second, 18.4. Similarly, in the experiment represented in Table 2 we obtain 319, 266, 220, and 169 responses *per hour* for the four intervals of 3, 5, 7, and 9 minutes respectively, and dividing by the number of times each interval is contained in 60 minutes, we obtain values of 16.0, 22.2, 25.6, and 25.4 responses *per reinforcement.* It has been noted that the value for the three-minute interval is probably too low and the others too high, and this is in agreement with these means. Within rather wide limits, therefore, the extinction ratio is roughly independent of the frequency of reconditioning.

It may be concluded that for every response reinforced, about 18 will be observed without reinforcement. This value of the extinction ratio depends, however, upon many factors, such as the drive and the conditions of reinforcement, some of which will be examined later. It does not hold when N_c is greater than one.[1] The relation between a number of responses *continuously* reinforced and the number obtained in subsequent extinction is by no means of the same sort, as was seen in Chapter Three. In periodic reconditioning little or no change is observed when two or three responses are reinforced together periodically. In a minor experiment on this point the reinforcement was produced every eight minutes and the number of responses reinforced at each time was either one, two, or three. Thirty-five records one hour long and about equally divided between the three cases were obtained from a group of four rats, each of which contributed to each group. The average rates expressed in responses per hour were:

1 response reinforced every eight minutes	212
2 responses " " " "	221
3 " " " " "	213

The experiment shows no significant increase in rate with an increase in the number of responses reinforced. This can be explained

[1] Youtz (*J. Exper. Psychol.*, in press) has obtained ratios of only 1.2:1 when 40 reinforcements are followed by extinction and 1.7:1 with 10 reinforcements.

as a discrimination (see the following chapter) where the discriminative stimulus for the second or third reinforcement is supplied by the preceding reinforcement. A discriminated response contributes little or nothing to the reserve.

It has already been noted that to attribute the constant rate assumed under periodic reconditioning merely to the superposition of

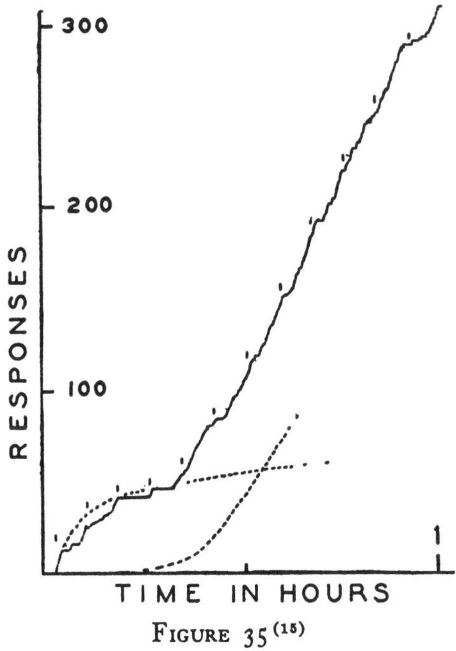

FIGURE 35 [15]

A COMPOSITE CURVE OBTAINED EXPERIMENTALLY

Original extinction is combined with the acceleration obtained under periodic reinforcement.

extinction curves is inaccurate because the final rate under periodic reconditioning may be far below the initial rate in extinction. A second reason may be noted here. If the periodic reconditioning is begun when the rate of responding is already high, it does not immediately maintain its eventual rate. An extinction curve first appears, during which the original rate falls, and from which the periodic rate finally emerges. Figure 35 is a typical record showing the original extinction of a reflex during which the response was periodically reinforced (as marked by dashes above the record). The first part of the curve cannot be distinguished from simple extinction. The periodic slope emerges at the fourth or fifth reinforcement. The dotted lines are theoretical curves for the processes of extinction and of the development of the constant slope, the sum of which gives the experimental curve. It will be seen that the positive acceleration due to the periodic reinforcement cannot be said to begin before the third or fourth reinforcement. The same explanation is available here as in the case of deviations of the third order. The reserve which underlies the constant rate observed during periodic reconditioning involves a temporal discrimination. The present effect will be discussed again when the problem of discrimination is taken up.

An average of four curves showing the same superposition is given in Figure 36 on page 134.

Extinction after Periodic Reconditioning

The effect of periodic reconditioning upon the reserve, which is reflected in the stable strength assumed by the reflex, is also felt in the extinction curve which ensues when reinforcement is omitted altogether. Examples of extinction curves after periodic reconditioning have already appeared in the preceding chapter (Figures 22, 24, and 26) where they were used in place of curves for original extinction because of the absence of cyclic deviations. The difference is presumably due to the adaptation of the emotional effect following failure to reinforce, ample opportunity for the adaptation being provided by the periodic procedure.

The resulting smoothness is only one of the distinguishing properties of the post-periodic curve. Another is a reduction in the rate at which the rate declines. This is noticeable after a very few periodic reinforcements. In Figure 36 the average for the four original extinction curves from Figure 7 is given at A, and the area enclosed during one hour has been shaded. The same figure gives the averaged curve for another group of four rats, where extinction followed a brief periodic reconditioning. The two groups had undergone comparable amounts of conditioning prior to the experiment. In neither case had any preceding extinction taken place, but for the group at B the original extinction occurred at the beginning of the record. The first part of the curve represents a combination of extinction and positive acceleration due to periodic reconditioning similar to that described above (see Figure 35). A constant slope had been clearly established by the seventh reinforcement, and at that point the reinforcement was discontinued. The base-line for the extinction curve which followed has been added to the

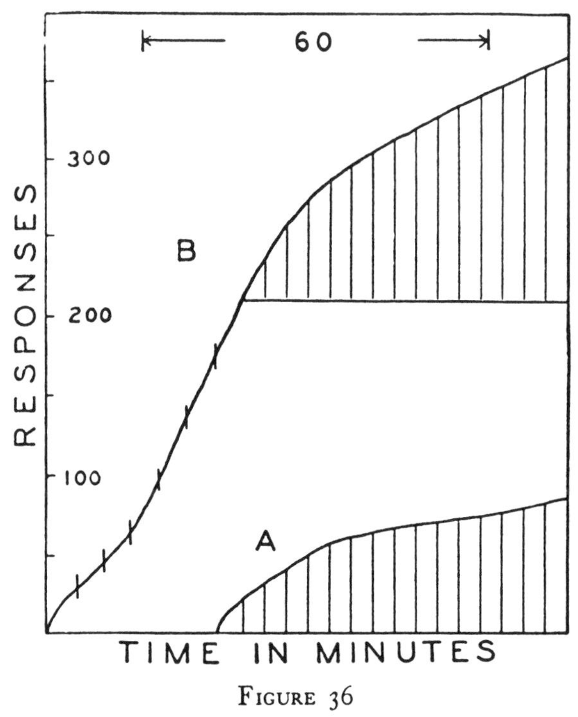

FIGURE 36

ORIGINAL EXTINCTION COMPARED WITH EXTINCTION AFTER PERIODIC RECONDITIONING

A: average of curves in Figure 7. B: a similar group periodically reinforced seven times without previous extinction (*cf.* Figure 35) and then extinguished.

figure and the area enclosed during fifty-five minutes shaded. For a reason to be given in Chapter Seven the extinction begins at the first *omission* of a reinforcement. The area is considerably greater than that of the curve for original extinction at A. A much greater difference would be observed if the periodic reconditioning had been carried out for several daily hours.

The use of curves arbitrarily brought to an end in this fashion is, of course, dangerous. The increase in area is not the basic result of the procedure, and the increase shown in Figure 36 is conditional upon the relatively high frequency of periodic reinforcement. Where a low rate of elicitation has been established by infrequent reinforcement, the area may actually be less in the case of post-periodic curves if they are not carried too far. In Figure 37 the extinction curves for P9 and P10 taken immediately after the prolonged periodic reconditioning shown in Figure 29 are compared

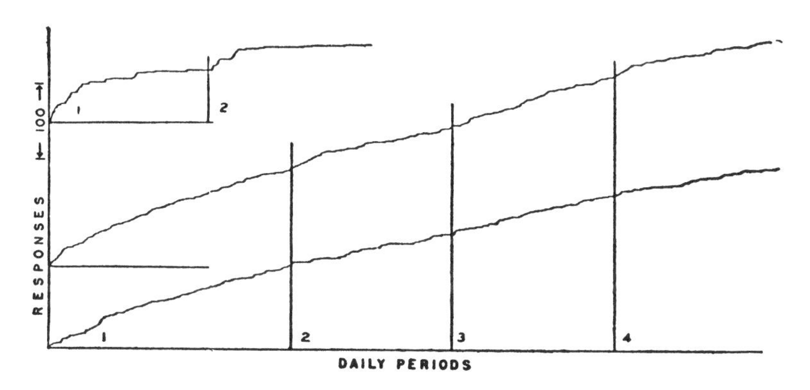

FIGURE 37

EXTINCTION AFTER PERIODIC REINFORCEMENT AT A LOW RATE COMPARED WITH ORIGINAL EXTINCTION

with the original extinction curve for P9 from Figure 9. The area for the first hour is greater for original extinction, because the curves after periodic reconditioning begin at low rates. The *retardation* has been greatly reduced by the procedure, however, and the post-periodic curves eventually reach a much greater height. A second day only is available for original extinction, but the result is clear. The record was taken after an intervening day of no responding, as noted above, in order to make the opportunity for 'spontaneous recovery' comparable in the two cases. The recovery is unusually great, but in spite of it the average slope for the two days is already less than that of either of the other records.

Representative records of extinction after considerable periodic reconditioning are reproduced in Figures 38 and 39. Figure 38 was constructed in the same way as Figure 29. The first two days give the control slopes under periodic reconditioning at intervals of five minutes. (The record for the second day for P6 was destroyed through a fault in the apparatus. The slope given was estimated from the previous performance.) The individual variation in rate follows the order already given for these rats in the fifth line of

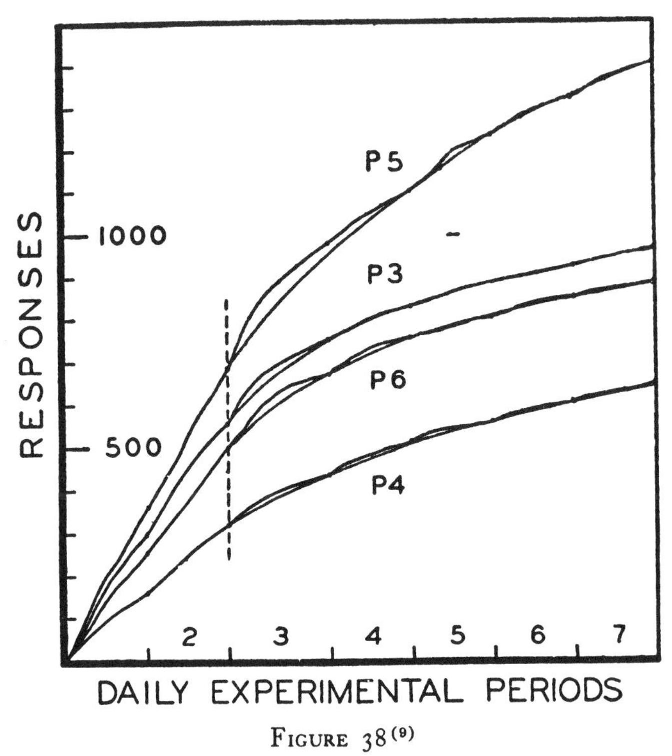

FIGURE 38[9]

EXTINCTION AFTER PERIODIC REINFORCEMENT

The first two days show the slopes obtaining under periodic reinforcement. No responses were reinforced after the vertical line.

Table 2—that is, P5 > P3 > P6 > P4. The extinction extends from the third to the seventh day. The lines drawn through the records are theoretical curves to be used in a later chapter. Figure 39 gives the first four daily records for Rat P3. It shows the absence of cyclic deviations characteristic of this kind of extinction. In order to reveal the character of the change more effectively the first period was extended to 1¼ hours.

The principal effects of periodic reconditioning upon extinction may be summarized as follows. Two effects are independent of the frequency or amount of reconditioning. The first is the absence of the cyclic deviations that are characteristically observed in original extinction. The second is the change in retardation: given the same initial rate, the curve for post-periodic extinction will fall off much more slowly. The third effect depends upon the frequency of reinforcement and concerns the initial rate, which is reduced by periodic reconditioning, except when the frequency is very high. A fourth is a function of the amount of reconditioning that has taken place, and is related to the change in the ratio N_e / N_c when N_c is increased above 1. We cannot compare the heights of extinction curves (at some arbitrary point) without taking the preceding reinforcement into account. The effect of periodic reconditioning upon the height may be stated in this way: the most efficient means of building a reserve with a given number of reinforcements is to administer them periodically. The heights observed in Figure 38 would not have been reached if all the reinforcements in the preceding periodic reconditioning had been grouped together solidly. A more

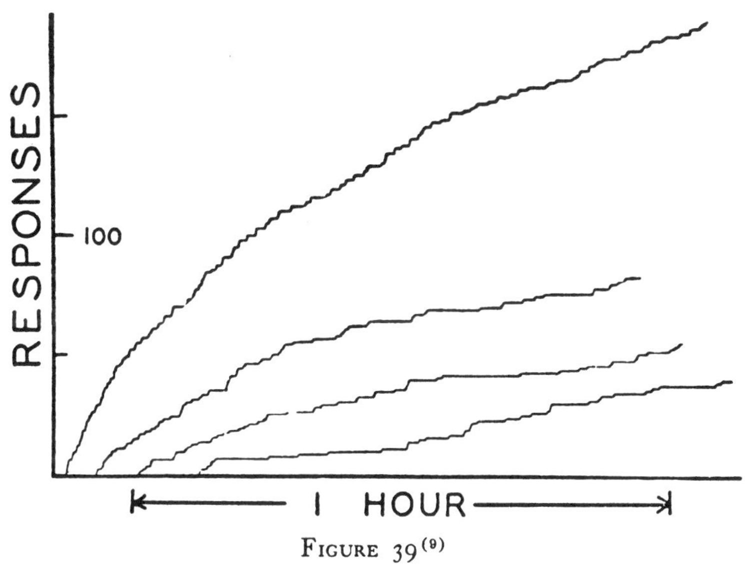

FIGURE 39[9]

FIRST FOUR DAILY RECORDS FOR P3 FROM THE SERIES BEGINNING AT THE VERTICAL LINE IN FIGURE 38

extreme statement is perhaps justified. In my experience no amount of continuous reconditioning will yield an extinction curve of the height obtained through even small amounts of periodic reconditioning.

Two curves showing reconditioning after extinction after periodic reconditioning for this group of rats were given in Figure 10.

The stability of reflex strength under periodic reconditioning and the prolongation of the extinction curve following it are important properties of normal behavior. They are responsible for a measure of equanimity in a world in which the contingency of reinforcing stimuli is necessarily uncertain. Behavior would be clumsy and inefficient if the strength of an operant were to oscillate from one extreme to another with the presence or absence of its reinforcement. Sudden changes in this variable are undesirable because other parameters which must be brought into action if the strength is low cannot be quickly withdrawn when they are no longer needed. The degree of drive needed to elicit a response when the conditioning is weak would be embarrassing if the conditioning were suddenly to become strong. When a reinforcement fails, it would be disadvantageous if all the available responses in the reserve were immediately expended as in original extinction.

Still more important for our present purpose is the constancy of the extinction ratio. As a means of studying the effect of various properties of reinforcement, it has many advantages over original conditioning.

The Extinction Ratio in the Study of Conditioning

In dealing with original conditioning in the preceding chapter, it was found to be difficult to obtain quantitative uniformity between cases because of the problems of topography and chaining. Consequently it was impossible to deal satisfactorily with the effect of any one condition of reinforcement upon either the rate or the reserve. The constancy of the extinction ratio provides a method for the investigation of these problems. The fact that the ratio is approximately the same for intervals as far apart as three and nine minutes indicates that it is a relatively precise measure of the effect of the reinforcement. By varying the kind or condition of reinforcement we should be able to obtain a direct measure of the result in terms of the rate of responding, provided the frequency of reinforcement is held constant. There are adequate controls for the maintained constancy of the rate (as for example the curves in Figure 29) if the experiment is not too long. The local deviations which appear between reinforcements are of no significance, since they do not affect the final average rate.

The Effect of an Interval of Time between Response and Reinforcement

I have only one experiment to offer which utilizes the extinction ratio in this way, but it is a fairly adequate demonstration of the general usefulness of the method. It is concerned with the effect upon the degree of conditioning of an interval of time elapsing between the response and the reinforcement. It is generally recognized that the temporal relation is of some importance in the process of conditioning, at least if the delay is long. However, it is often assumed that a short delay is insignificant, particularly in psychological systems which suppose something corresponding to a 'perception of the relation between the act and the result.' Systems of that sort usually have no way of measuring intermediate states of strength, and it is perhaps unfair to criticize them for that reason.

The plan of the experiment was simple. Various intervals of time were introduced between the elicitation of the response and its reinforcement, and the effect was observed as a modification of the rate of elicitation under periodic reconditioning. The intervals were obtained with the apparatus already mentioned in Chapter Three, which was introduced into the circuit between the lever and the magazine. When the rat responded, the apparatus was put into operation, and a given number of seconds later a pellet was discharged from the magazine. The device has this important property: if a second response is made during the interval, the timing begins again, so that a full interval must again elapse before reinforcement occurs.

This arrangement encounters the following difficulties. (a) No provision is made against the possible coincidence of a second response with a delayed reinforcement. In such a case a response is reinforced simultaneously. Examples occurred only infrequently in the experiments that follow. Their effect would be in the direction of making the average interval shorter and it is assumed that this is irrelevant in the present degree of approximation. (b) A second difficulty is inherent in the problem. An interval may be measured either from the beginning of a response to the lever or from the end —that is, from either the initial pressing from or the release of the lever by the rat. In the first case the result is that the rat may still be holding the lever down at the time of reinforcement, even after a considerable interval. But in the second case the reinforcement may follow the initial pressing by a considerably longer interval than is intended. This is more than a mere technical difficulty. If the rat presses the lever down and holds it there for, say, three seconds, at what *moment,* if any, can the response be said to occur? In the present experiments it has been assumed that the important part of the response is the initial pressing, and the intervals have been measured from that point on. Occasionally, especially at the shorter intervals, the lever is therefore still being held down at reinforcement. A possible effect upon the result in the direction of reducing the interval must be allowed for, although it is not expressly treated herein.

In one experiment a group of twelve rats, approximately 150 days old, were conditioned in the usual way. After one hour of extinction the response to the lever was periodically reconditioned. During the first daily hour the interval of reconditioning was four minutes; during the rest of the experiment it was five. All twelve rats assumed an approximately constant rate of responding. After three days of periodic reconditioning intervals were introduced before the periodic reinforcements as described above, three rats being assigned to each interval. This procedure of delayed reinforcement was continued for three days. On the two following days the reinforcement was again simultaneous.

Since the frequency of reconditioning remains constant, the result could be expressed as a reduction in the extinction ratio, but I shall here speak simply of the rate. As recorded, the effect is a reduction in the slope of the summation curve. The change is indicated in Figure 40, where the rate, expressed as responses per hour, is followed throughout the experiment. The averages for all rats without respect to the length of interval are given in the heavier line. The rates during the second and third days of periodic reconditioning (Days 1 and 2 in the graph) were 190 and 193 responses per hour respectively. On the first day of delayed reinforcement (average length of interval five seconds) the rate fell to

120 responses per hour and during the next two days remained close to this value (121 and 118). Upon returning to simultaneous reinforcement the rate rose immediately to 173 responses per hour and on the following day to 175 responses per hour. The latter values indicate a reduction in the original rate of about 10 per cent, which is roughly of the order for the spontaneous decline of the extinction ratio (see Figure 29). The effect of the introduction of the intervals is a reduction in rate of about 37 per cent.

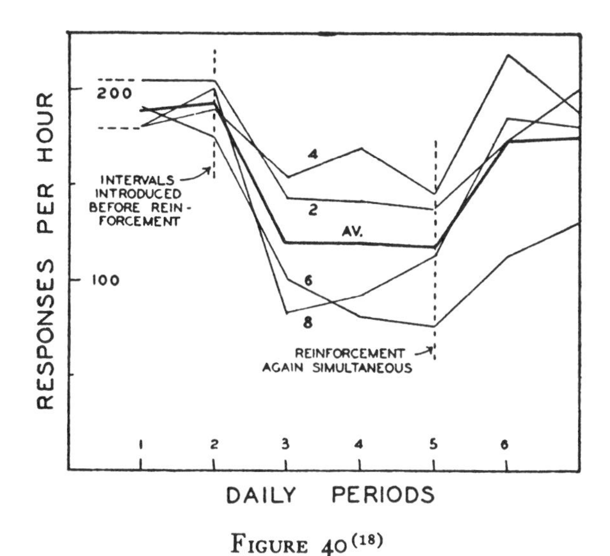

FIGURE 40[18]

DECLINE IN RATE WITH THE INTRODUCTION OF INTERVALS BETWEEN RESPONSE AND PERIODIC REINFORCEMENT

The intervals (2, 4, 6, and 8 seconds) are marked.

The curves for the separate groups are less smooth, partly because of the small number of animals in each group (three); but each set shows the same result as the average. In general, the longer the interval the more marked its effect, but there is an inversion in the case of the two- and four-second groups. This is partly explained by the fact that four seconds is an optimal interval for producing coincidences between the reinforcement and a succeeding response, at least for the animals and the degree of drive used in these experiments; but this is probably not an adequate explanation of the whole difference. The inversion is even greater than it appears to be in the figure, when a correction is made for a difference in the original rates (see Days 1 and 2). The eight-second group is more significantly below the six-second group when a similar correction has been made.

Taking the average for the three days as the basic rate for each interval, we may express the relative effect of the interval as the percentage of the original rate lost. The figures in Table 3 are calculated from the uncorrected data:

TABLE 3

Interval	*Percentage decline in rate*
2	32
4	18
6	51
8	52

In another experiment with four groups of three rats each, aged approximately 130 days, the intervals were introduced before reinforcement on the first two days of periodic reconditioning. On the following three days reinforcement was simultaneous; on the following three days it was again delayed with the same interval, and on the following three days again simultaneous. This was continued for 23 days. Various exploratory experiments were tried during this time, such as changing the procedure during the hour (see Figure 42, page 146), introducing a day of complete extinction, and so on, but it is possible to obtain from the records the following data on the present point:

1. In 25 cases three successive daily records were obtained with delayed reinforcement following two or more days with simultaneous reinforcement.

2. In 20 cases three successive records were obtained with simultaneous reinforcement following two or more days with delayed reinforcement.

The first of these sets is composed of seven cases at two, seven cases at four, five cases at six, and six cases at eight seconds. The rates for the last simultaneous day and for the three days of delayed reinforcement are given in Figure 41 at A. Here the records have been plotted in their original form (number-of-responses *vs.* time) rather than as rate *vs.* time as in Figure 40. The curves represent only the end-points of the actual records. No attempt is made to reproduce the changes taking place during each hour. The original differences in periodic slope have been corrected for by assigning to each group a correction-factor, the values of which were 0.93, 1.00, 1.22, and 0.91 for the groups of 2-, 4-, 6-, and 8-second inter-

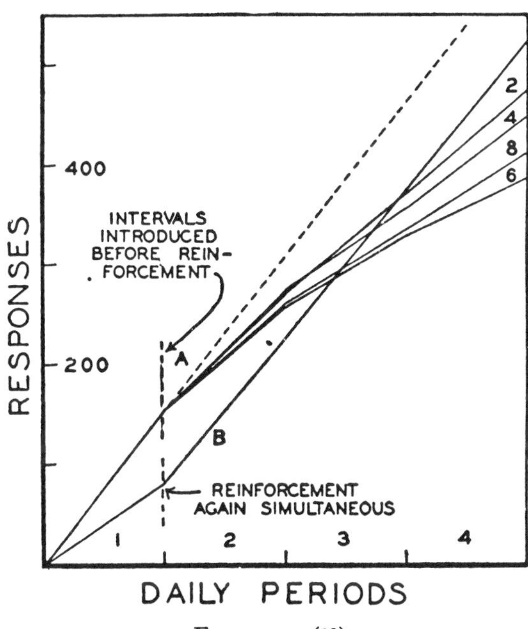

FIGURE 41 (18)

DECLINE IN RATE UNDER DELAYED REINFORCEMENT

A second set of data similar to those in Figure 40 are here plotted in the recorded form. A: the average slope of the daily records under periodic reconditioning declines when intervals are introduced before reinforcement as marked (2, 4, 6, and 8 seconds). B: the average slope for the four groups at A returns to its original value when the reinforcement is again simultaneous.

vals respectively. When a series is multiplied throughout by its factor, the rate for the first day (simultaneous reinforcement) is converted to the average for the four groups. The succeeding records may then be compared directly. In Figure 42, which will be discussed shortly, a typical set of records is reproduced from a single experiment.

In Figure 41 the group at A shows the effect of the various intervals. The single curve on the first day gives the average slope obtaining under periodic reconditioning on the day before the change to delayed reinforcement. The four curves coincide here because of the correction. The introduction of intervals produces a drop in this slope, which becomes even greater on the following days. Here again it will be seen that in general the longer the interval the greater the effect but that again an inversion appears, this time between six and eight seconds. Expressed as percentage decline in the original periodic rate calculated from the uncorrected data (only the values for the last two days were averaged, since the rate had not stabilized itself on the first day), the effect of the intervals is given in Table 4:

Table 4

Interval	*Percentage decline in rate*
2	33
4	42
6	57
8	51

The record at B, Figure 41, shows the return to the original slope under simultaneous reinforcement. The curve on the first day is the average for the preceding days irrespective of the length of interval. On the following three days the reinforcement was simultaneous. Here the full value of the original slope is reached (or very nearly) because the data have been selected at random from the whole series and a progressively spontaneous decline will consequently not be appreciable.

In neither experiment is the effect of the interval shown to be a simple function of its length. In a certain sense, however, the two experiments cancel each other so far as deviations from a simple *order* are concerned. At least it can be said that no consistent discontinuity has been shown. The irregularities are of the sort that should be expected from a failure to exclude extraneous factors but not from the complex or discontinuous nature of the function. We have already noted two factors to which irregularities may be assigned. One of these (coincidental simultaneous reinforcement) could be eliminated by allowing the experimenter to watch the rat.

The other needs to be investigated directly. We need to find the point in the sequence of events called 'the response' from which measured intervals show the greatest simplicity in their effect. Such an experiment would call for a large number of animals and would be of an entirely different order of rigor. A third sort of irregularity is described below.

The present result may be expressed in this way: the effect of an interval, if not orderly, is at least great. The average percentage declines in rate for both groups are given in Table 5. In spite of

TABLE 5

Interval	*Percentage decline*
2	33
4	30
6	54
8	52

the two inversions the effect is roughly proportional to the length and is of large magnitude. An interval as short as two seconds reduces the effect of the reinforcement by one-third.

It is important to make clear that we are not studying the effect of the reinforcement upon the *rate* of conditioning. Presumably the rate at which the effect takes hold is not involved or is of a different order of magnitude. The experiments are therefore not comparable with investigations of the effect of 'delayed reward' upon a learning curve. It is the number of responses contributed to the reserve by a single act of conditioning that is affected. The constant rate represents a balance between input and output; if the input is affected by reducing the efficiency of the reinforcement, the output must fall. I have already considered the extreme case in which the input ceases altogether. The result is that the rate falls off along an extinction curve to an ultimate value of zero. In the present case the ultimate value is determined by the efficiency of the remaining reinforcement. In order to conform to the preceding interpretation there should be a gradual, rather than an abrupt, change from the higher to the lower rate. On the first day of delayed reinforcement the apparent ratio of the responses not reinforced to those reinforced should be greater than the value warranted by the new reinforcement, for some of these responses are due to the previous simultaneous reinforcement, the effect of which is now being extinguished. The rate should drop from one value to the other, not abruptly, but along an extinction curve. That this is actually the case is apparent in Figure 41, even without detailed information about the daily records. The slope on the first day of delayed reinforcement is intermediate. The second day finds it at approximately its final value for each interval; in the case of the six-second interval only is there any further drop, and this may not be significant. On returning to simultaneous recon-

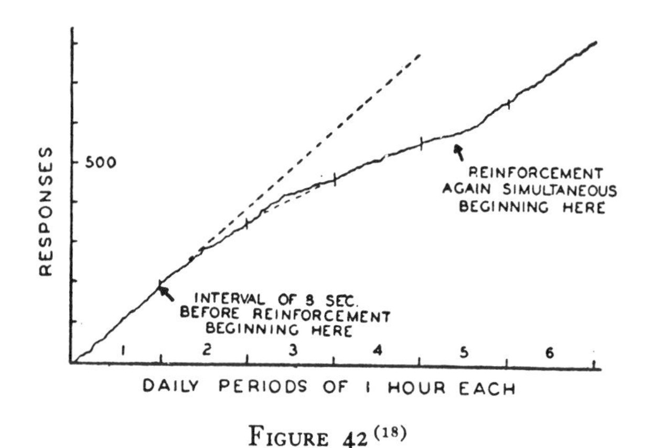

FIGURE 42 [18]

SET OF ORIGINAL RECORDS FOR ONE RAT IN FIGURE 41

ditioning we should find a rapid acceleration toward the old maximal rate similar to that observed after extinction or discrimination. It will be seen by sighting along Curve B in Figure 41 that the change to the new rate is not quite abrupt.

In Figure 40 practically the full extent of the change in rate has been accomplished at the end of the first day. But these averages are misleading, and it is necessary to turn to the individual records to follow the change accurately. These records are of two kinds, one showing the extinction curve clearly (where the rate drops from one value to another along a smooth curve), the other showing an effect which has not appeared elsewhere in exactly the same form in experiments with this method. Figure 42 gives a representative case of the normal change. The curve is composed of six consecutive daily records for one of the rats in Figure 41.

The first day is with simultaneous reinforcement. The probable course of the curve, if simultaneous reinforcement had been continued, is indicated by projecting the slope for this day in a broken line. On Days 2, 3, 4, and for 20 minutes on Day 5 an interval of eight seconds was introduced before reinforcement. The resulting change is of the sort we should expect. The curve judged by inspection is closely similar to those previously reported for extinction after periodic reconditioning and for discrimination (see Chapter Five). On Day 3 some slight 'overshooting' occurs (the initial rate is too high), but it is compensated for by subsequent retardation.

On the fifth day the conditions of reinforcement were changed during the hour in order to observe the result without interference from spontaneous recovery. Although the record is severely reduced in the figure, a smooth positive acceleration is apparent beginning at least by the second simultaneous reinforcement. The rate reaches a maximum within five or six reinforcements. In many cases involving a return to a higher slope, in this and other experiments, the rate rises for a short time above its later value, so that on extrapolating backward from the final stable curve one often hits the point at which the reinforcement was changed. This is the case in Figure 42. It is as if the change in rate should have been instantaneous and as if the failure of the rate to rise properly during the first three or four intervals were compensated for during the fifth and sixth. The sample records of the return to periodic reconditioning after extinction in Figure 34 (page 129) should be compared.

In the other kind of daily record obtained under these circumstances the orderliness of the change is destroyed by sudden depressions in rate. An unusually severe case is shown in Figure 43, which is for one of the rats in Figure 40, where the interval was eight seconds. The second day in the graph is the first day on which the reinforcement is delayed. As will be seen from the figure, the rate begins to decline as in Figure 42, but shortly after the middle of the hour it drops suddenly to zero and remains there for the rest of the period. On the following day the rate begins at a low value but shows some acceleration during the hour. On the third day further acceleration occurs.

In this figure a curved broken line has been drawn to suggest the probable course of the record in the absence of any sudden depression in rate. There is no reason, of course, to assume that the compensatory acceleration of the fourth day is now over and that henceforth the curve will follow the extrapolation of the curved broken line. Unfortunately the plan of the experiment did not permit following this change far enough to determine the extrapolation more precisely. On the next day the reinforcement was again simultaneous, and the rat reproduced quite accurately its former rate, having already given that rate at the end of the fourth day. The example in Figure 43 is exceptionally severe, so far as

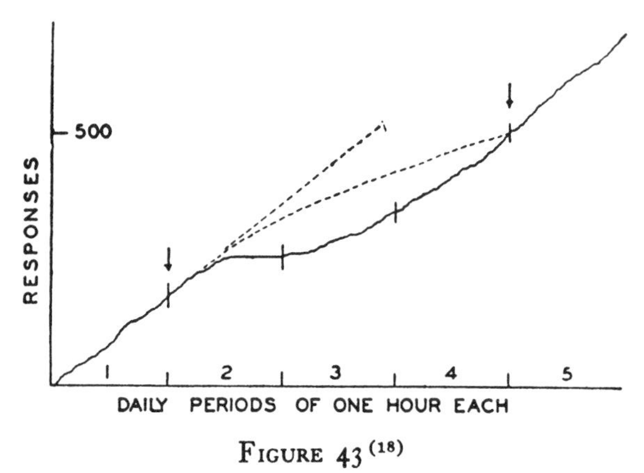

FIGURE 43 [18]

SET OF ORIGINAL RECORDS FOR ONE RAT IN FIGURE 40

At the first arrow an interval of eight seconds was introduced between response and reinforcement. Thirty minutes later the rate dropped suddenly to zero and recovered only gradually during the next two daily periods. At the second arrow the reinforcement again became simultaneous.

its duration is concerned. A sudden reduction to a zero rate is quite common, but the effect seldom lasts so long. A subsequent recovery is usually obvious.

The nature of the effect is not clear. It seems definitely to be associated with the procedure of delayed reinforcement and is more likely to occur at the longer intervals. Its presence makes any group average of little value in determining the course of the change from one rate to another. It tends especially to mask the orderly decline due to extinction, since it takes responses from records early in the series and transfers them to later parts. If the records for Days 2, 3, and 4 in Figures 42 and 43 were averaged, the result would be nearly a staight line, which would indicate nothing of either kind of change. In Figure 40 the points on the third day of the graph are lower and those of the two subsequent days higher than would be the case in the absence of deviations. The apparently abrupt drop to a new rate is thus illusory.

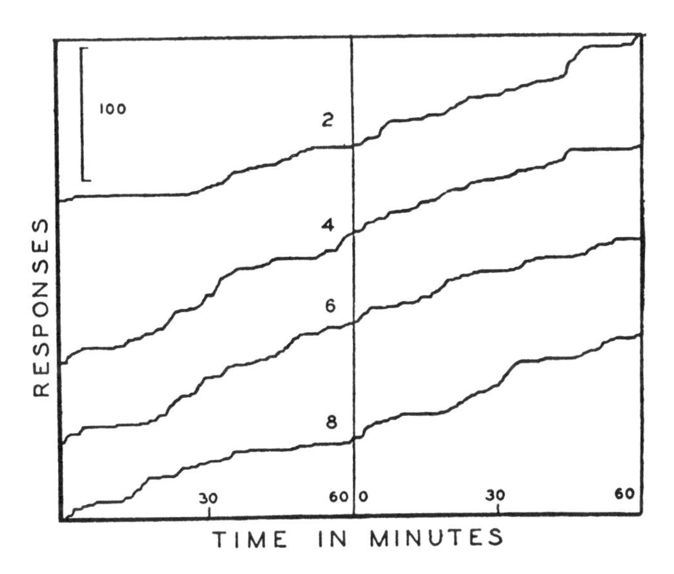

FIGURE 44 [18]

FIRST TWO DAYS OF PERIODIC REINFORCEMENT DELAYED AS MARKED

The rates begin low because of previous extinction. No very high rate is developed, even with the shorter delays, and there is considerable irregularity. Compare Figure 28 (page 118) for the case of the non-delayed reinforcement.

In the experiment represented in Figure 41 the averages for the group show the result of extinction, because the disturbing effect is lacking. This might be due to the fact that, since the effect apparently occurs only once in the case of each rat, and since the curves in Figure 41 are the averages of repeated tests, an occasional deviation would be fairly well concealed. But there are no similar deviations to be concealed. With this group, as I have said, the intervals were introduced on the first two days of periodic reconditioning. Here we should not expect a deviation exactly comparable with that described above, although we might expect something similar. There is, in fact, a noticeable disturbance, as may be seen in Figure 44, which gives records for these two days for the first four rats in the group. These curves should be compared with Figure 28 (page 118), beginning at *B* and *B'*, in order to observe the effect of the interval. The rate obtaining under the interval is reached fairly quickly, but it is not equal to the rate observed upon returning to the interval after simultaneous reinforcement. Moreover, the "grain" of the record is rough. The rat oscillates between periods of slow and rapid responding, and the average rate itself is seldom realized. All four records show this characteristic. Partly as the result of their subnormal values the four slopes are of the same order, in spite of the difference in intervals. All four rats showed an immediate acceleration on the following day when the reinforcement was for the first time simultaneous.

As the result of this initial delayed reinforcement the sudden deviations found in the other group do not appear when the reinforcement is again delayed. Consequently the individual curves are similar to that in Figure 42, and the negative acceleration is clearly shown in the average for the group (Figure 41).

The character of these anomalous deviations makes it possible to assign them to factors lying outside the system that we are immediately investigating. Setting them aside, we may say that the strength of the reflex passes from one value to another in a way that is in harmony with previous descriptions of its state and of the factors determining its state.

In view of the presence of responses due to extinction in the reduced rates observed with this method, it must be supposed that the percentage decline given for each interval is somewhat less than a full expression of the effect of the interval. The effect must be at least as great as indicated, and it is probably much greater. The failure to develop a higher rate when the interval is introduced at the beginning of the periodic procedure (Figure 44) is strong evidence in support of this conclusion. It is probable that the effect of a delay of eight seconds is a reduction in efficiency of nearer 100 than 50 per cent.

Negative Conditioning and Periodic Reinforcement

The procedure of periodic reconditioning is also valuable in studying the hypothetical case of 'negative conditioning' discussed in the preceding chapter. In experiments now to be described the form of negative reinforcement used was a sharp slap to the foot or feet used in pressing the lever, delivered by the lever itself in the course of being depressed. The apparatus consisted of an electrically operated double hammer striking upward against the two shafts of the lever behind the panel. The slap could be administered or omitted at will. Since a rat presses the lever with nearly the same force each time, the effect of the slap given by the sudden upward movement of the lever was relatively constant and was therefore to be preferred to an electric shock, which is the commonest form of negative reinforcement. The only stimulation in addition to the slap arising from the apparatus was a fairly loud click.

The first experiment concerns the effect of negative reinforcement upon extinction. Extinction curves after periodic reinforcement with food were obtained from four rats. On the third day the slapper was connected for the first time at the end of twenty minutes, and all responses made during the rest of the hour and on the following day were negatively reinforced. On the fifth day the slapper had been disconnected, and on the sixth and seventh the response was again periodically reinforced with food.

The results are shown in Figure 45 (page 152). The first effect was an immediate strengthening of the reflex. A second quick response followed the first slap, and for the next two or three minutes a relatively rapid responding in the face of sustained negative reinforcement was observed. A second quick response was found to be the rule in some exploratory experiments in which a shock strong enough to cause a violent jump was administered, and in general an initial strengthening of the reflex is clearly indicated. This phase was followed by practically complete suppression. On the following day only a few scattered responses occurred and were negatively reinforced with the slap. On the next day (Day 6 in the figure) no responses were negatively reinforced, but *a very low rate was nevertheless maintained*. When the response was later reconditioned, the rate rose very much as in original periodic reconditioning. A set of individual records from this experiment is given in Figure 46. The numbers refer to the days in Figure 45.

Figure 45 suggests that a reduction in the size of the reserve of the reflex was brought about by the slap, such as might be expected if there were a process of negative conditioning exactly

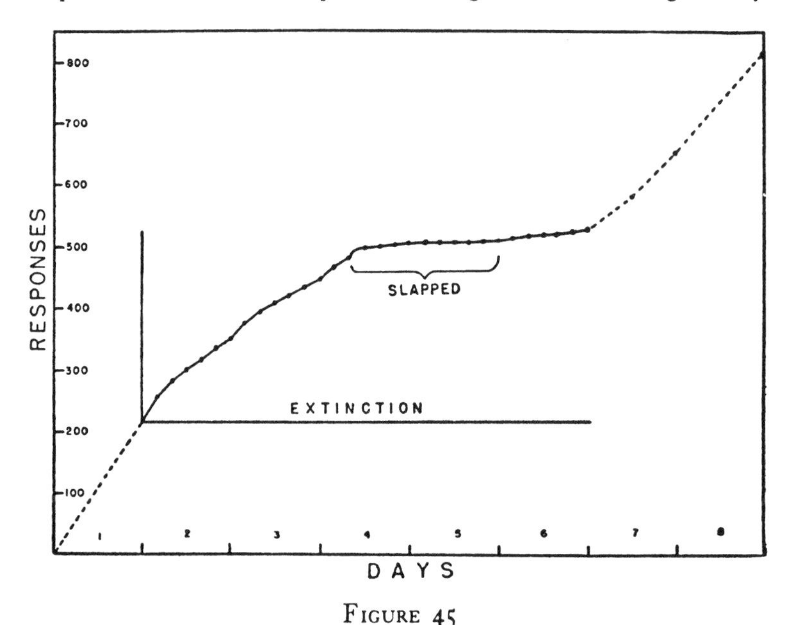

FIGURE 45

EFFECT OF NEGATIVE REINFORCEMENT UPON THE EXTINCTION CURVE

The extinction curve is the solid line. The part during which all responses were slapped is indicated. There is little or no recovery on the last day of extinction without slaps. The dotted lines give the slopes under periodic reinforcement.

opposed to that of positive conditioning, in which each negative reinforcement *subtracted* a number of responses from the reserve. According to this view the low rate on the last day of extinction, when there was no negative reinforcement, was due to the emptiness of the reserve and was comparable with the rate at a much later stage of ordinary extinction. It will be seen in the experiments that follow that this conclusion is unjustified and that a conditioned emotional state is probably responsible for the suppression of activity on the last day of extinction in this experiment. The present interpretation is inconclusive for two reasons: (1) the use of the slapper was so prolonged that the emotional state

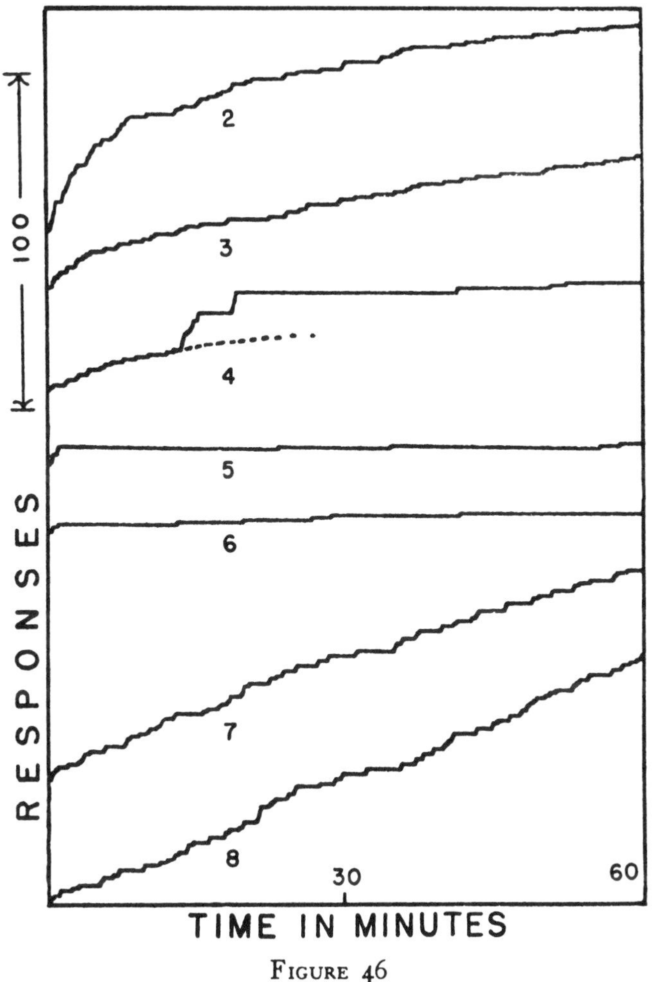

FIGURE 46

SET OF RECORDS FOR ONE RAT IN FIGURE 45

could become almost maximally conditioned, and (2) the responses occurring on the last day without the slap were too few to permit an adequate extinction of this effect.

In an experiment in which the negative reinforcement was brief and the chance for extinction of an emotional effect increased, no reduction in reserve was discovered. Two groups of four rats each with no previous experience with the slapper were periodically reconditioned for three days. The reflex was then extinguished in both groups for two hours on each of two successive days. In one group all responses were slapped during the first ten minutes of the first day.

The result is shown in Figure 47. The effect of the slap in de-

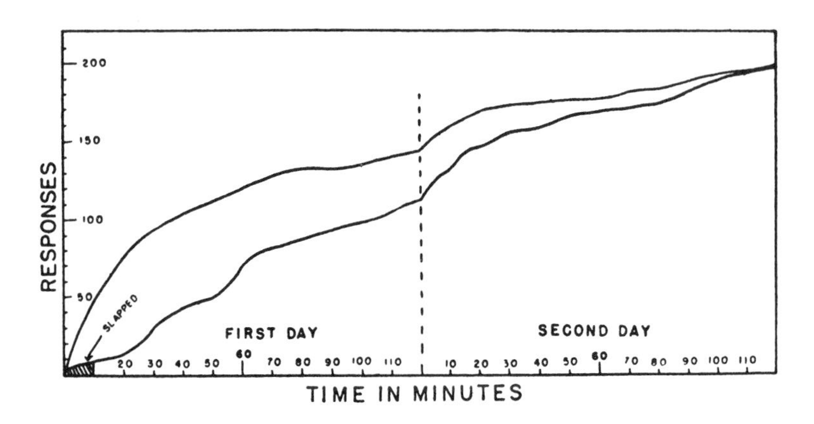

FIGURE 47

EFFECT OF NEGATIVE REINFORCEMENT UPON EXTINCTION

The two curves are from groups of four rats each, with the same experimental history. All responses made by one group during the first ten minutes of extinction were slapped. The rate is depressed for some time but eventually complete recovery is made.

pressing the rate is obvious, but a full recovery when the negative reinforcement is withheld is also plain. At the end of the second day the two groups had emitted practically the same mean number of responses. In comparing groups in this way account must be taken of their extinction ratios, but no explanation of the present result is forthcoming from the ratios demonstrated on the days preceding extinction. The ratio of the group that was slapped was about twenty-five per cent *lower* than the group not slapped. In the absence of any negative reinforcement the group that was slapped should accordingly have given a *lower* extinction curve. The fact that the same height was attained is therefore all the more significant; indeed, it might be argued that the effect of the slapping was to *increase* the reserve.

Figure 47 must lead us to revise the conclusion based upon the preceding figure. It is true that there is a temporary suppression of responses, but all responses originally in the reserve eventually emerge without further positive reinforcement. Such an effect is, by definition, emotional. It is an effect upon the relation between the reserve and the rate, not upon the reserve itself. In this experiment there is no evidence whatsoever for a process of negative conditioning directly the opposite of positive conditioning. The behavior of the rat, on the other hand, is quite in accord with the assumption that the slap establishes an emotional state of such a sort that any behavior associated with feeding is temporarily suppressed and that eventually the lever itself and incipient movements of pressing the lever become conditioned stimuli capable of evoking the same state. The effect was not clearly shown in Figure 45 for reasons already given.

A second way to test for a negative reinforcing action that would reduce the size of the reserve is to interpolate negative reinforcements during positive periodic reconditioning. For example, let the response be reinforced with food every four minutes and the value of the extinction ratio obtained. Then let negative and positive reinforcements be administered alternately every two minutes, so that the frequency of positive reinforcements remains the same as before while an equal number of negative reinforcements is also given. If the negative reinforcement has an effect upon the reserve, the difference between the resulting rate and that prevailing under positive reinforcement alone should permit the calculation of the number of responses subtracted by a single negative reinforcement.

Three experiments, each upon four rats, are represented in Figure 48 A (page 156). The mean result (heavy line) is representative of the individual cases. The points for the first two days give the rates per hour under positive periodic reinforcement at four-minute intervals. On the four following days slaps were alternated with pellets every two minutes (pellets thus being received every four minutes). There is a significant drop in rate only on the second day of negative reinforcement. On the seventh day of the experiment no responses were slapped, but the rate did not change appreciably.

If there is a real negative reinforcing effect revealed in this experiment, it is not permanent. The only apparent reduction is on the second day. It might be argued that a similar reduction on the first day was masked by the temporary strengthening that is the first result of negative reinforcement, and that a reduction in

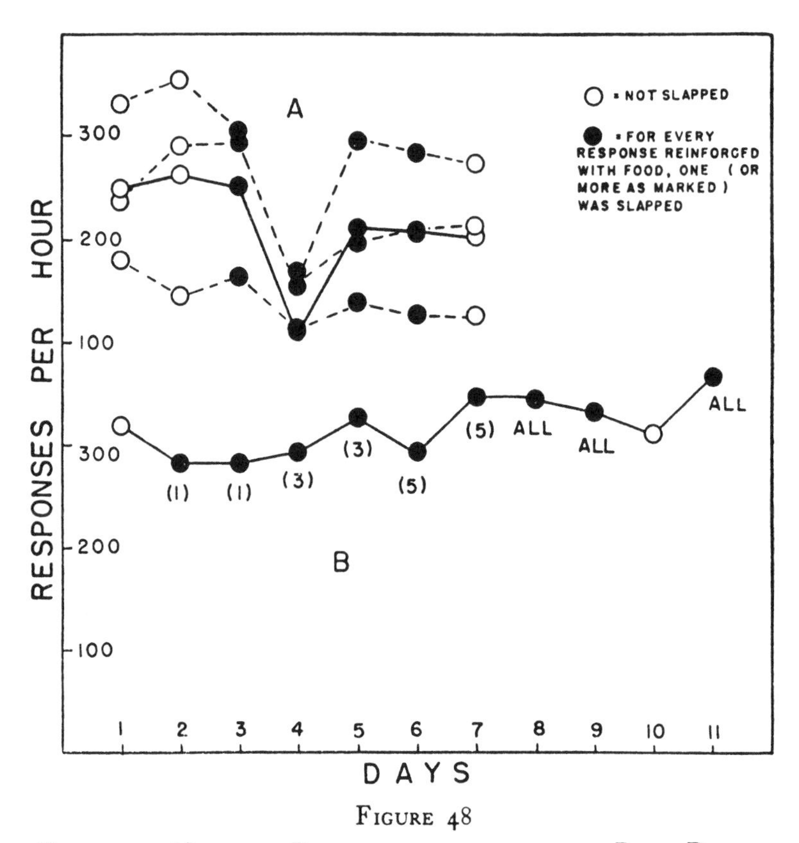

FIGURE 48

EFFECT OF NEGATIVE REINFORCEMENT UPON THE RATE DURING PERIODIC REINFORCEMENT WITH FOOD

reserve obtained for two days. But a sudden reduction in rate under periodic reconditioning is not the result to be expected from a reduction in reserve. The decline in rate should be gradual, especially since the negative reinforcements were spaced out. On the other hand the result fits the hypothesis of an emotional effect which characteristically adapts out as the experiment progresses. The slap is not a highly noxious stimulus and adaptation is to be expected. Why there is no compensation for the reduced output on the second day is hard to say, unless we assume that compensation occurs only within a reasonably short space of time and cannot hold over for twenty-four hours. Another explanation may be derived from the case of a reduced rate due to lowered drive. As will be pointed out in Chapter Eleven emotion and drive are closely related phenomena, but it will be shown in Chapter Ten that a reduced rate due to lowered drive is not compensated for subsequently.

Further proof that, eventually at least, no reduction in the extinction ratio is felt as the result of negative reinforcement is given in Curve B, Figure 48. Two of the groups in Curve A were extinguished after the last day in the figure and subsequently periodically reinforced with food. The second day of this reconditioning is shown as Day 1 in Curve B. When negative reinforcements were again interpolated, little or no effect was felt (Days 2 and 3). When the slaps were increased in number to three every five minutes, there was a slight increase in rate. Later five negative reinforcements were administered every four minutes, and eventually all responses were slapped, food still being received periodically as before. The rate shows no significant effect. When the slapping was omitted altogether, there was a *reduction* in rate; when all responses were again slapped, the rate increased rapidly.

To sum up, the experiments on periodic negative conditioning show that any true reduction in reserve is at best temporary and that the emotional effect to be expected of such stimulation can adequately account for the temporary weakening of the reflex actually observed.

It could be argued that, even though the size of the reserve is not affected by 'negatively reinforcing' stimuli, the accessibility of the responses it contains might be modified. Any such change in accessibility should be revealed in an extinction curve following negative reinforcement. To test this possibility the reflex was extinguished after the days shown in Figure 48 A. It will be noted that on the day preceding the extinction no responses were slapped, nor were any slapped during the extinction. The curve was recorded for two hours. The mean heights of the eight curves measured every ten minutes are given in Figure 49. The curve is for the equation

$$N = 108 \log \frac{t}{10} + 35 + 0.5\,t,$$

which is the same type of equation used to describe simple extinction and discrimination. Here the equation is applied to the behavior of the rats during a single experimental period. Although the fit is not perfect, we are perhaps justified in concluding that

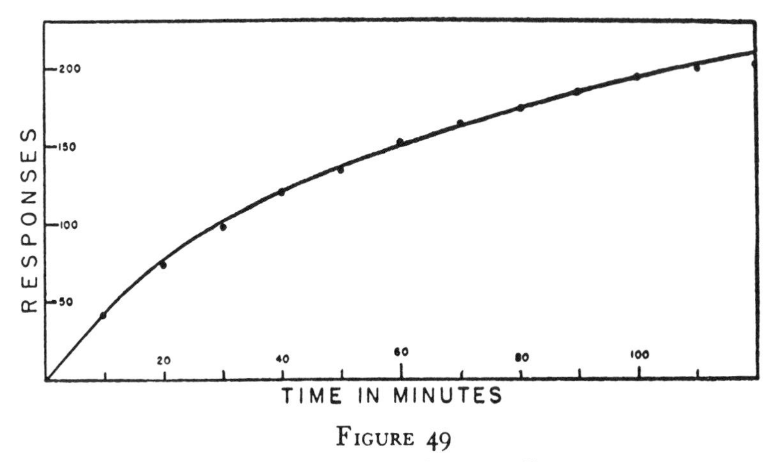

FIGURE 49

EXTINCTION AFTER PERIODIC NEGATIVE REINFORCEMENT

The extinction began on the day after the last day in Figure 48 A.

in spite of the preceding negative reinforcement the reserve presents itself for emission in essentially the same fashion.

Another opportunity to use the extinction curve in testing for a reduction in reserve arises when the negatively reinforcing stimulus has been adapted to. In the following experiment it will be seen that the process of extinction goes on undisturbed even when all responses are negatively reinforced. In Figure 50 extinction curves after periodic reconditioning are given for two groups of four rats each, having the same history of adaptation to being slapped. During the first four and a half daily periods all responses made by one group were negatively reinforced and all by the other were not. The slapped group shows a somewhat slower decline in rate, which is probably attributable not to the slapping but to the greater mean extinction ratio for this group. Shortly before the middle of the fifth period the slapping was reversed, and a sixth day under the same reversed conditions was recorded. The only significant result of the change was a slight increase in rate for the group originally not slapped. Some survival of the initial emotional phase of increased strength could account for this increase, and in general the conclusion seems justified that the

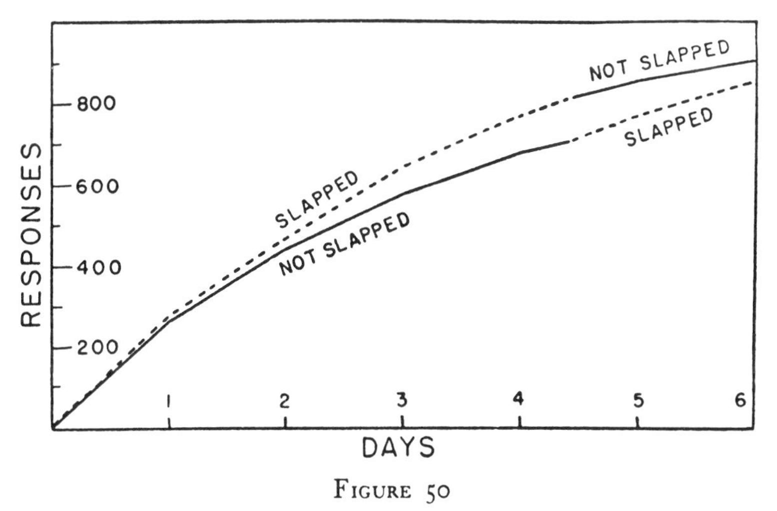

FIGURE 50

EFFECT OF NEGATIVE REINFORCEMENT UPON EXTINCTION AFTER ADAPTATION TO THE REINFORCING STIMULUS

The curves are for two groups with the same experimental history. All responses were slapped as marked, with very little effect upon the curve.

sustained negative reinforcement of the response had no effect upon the reserve.

In interpreting all the preceding results it should be remembered that the negatively reinforcing stimulus was relatively weak. That it had a definite effect is clear from Figures 45 and 47, but it should not be regarded as necessarily comparable with such a stimulus as a strong shock. Whether a stronger stimulus would actually bring about a reduction in reserve to be ascribed to a genuine negative conditioning is questionable. It would be strange if a mild negative reinforcement did not show some sign of reducing the reserve in so delicate a test as that provided by extinction or periodic reconditioning if a stronger stimulus were to do so. The stronger stimulus, however, might conceal the effect by generating a stronger and more lasting emotional effect, with little probability of adaptation with time.

The use of a negative reinforcement provides another case of a reduction in reflex strength in which the term 'inhibition' is often invoked. The failure to execute a strong response because of a previous negative reinforcement is perhaps the commonest case to which the term 'inhibition' is popularly applied. If the preceding interpretation is correct, the effect should be classified as a conditioned emotional reaction, comparable with the suppression of eating behavior by a 'frightening' stimulus, except that in the present case the stimulus which arouses the emotional state happens to be also the external discriminative stimulus upon which the execution of the response depends. The nature of such an emotional reaction and its relation to inhibition as defined in Chapter One will be discussed in Chapter Eleven.

The distinction between the weakening of a reflex through the exhaustion of a reserve and weakening through an emotional modification of the relation between reserve and strength is obviously the distinction between mere 'forgetting' or 'loss of interest' and an active 'repression.'

The emotional effect of a 'negatively reinforcing stimulus' in Type R provides another explanation of the occasional failure to obtain instantaneous conditioning discussed in the preceding chapter. If the tactual and auditory stimulation arising from the downward movement of the lever happens to be negatively reinforcing in this sense, two effects follow the first response—the positive reinforcement of the food tending to increase the strength of the response and the negative effect of the movement tending to decrease it. The net result may be only a moderate positive effect or even no effect at all. Later compensation and the adaptation of the negative stimulus allow for an eventual increase in strength.

The Negative Correlation of Response and Reinforcement

We return to the effect of an interval of time between response and reinforcement. With an increase in the length of the interval the effect of the reinforcement should at some point disappear, and when the length has been carried to its practical extreme it is possible that an opposite effect should be observed. There would then be a negative correlation between the response and the reinforcement. The possibility that 'not-responding' may be reinforced deserves consideration. The notion of conditioning an organism not-to-respond is not to be confused with negative conditioning (where a response is followed by a negative reinforcement) nor with that kind of temporal discrimination in which a response is reinforced if it has not been preceded by another response for a certain length of time. Conditioning not-to-respond is possibly a separate and important way of reducing the reserve. A critical experiment should not be difficult to design, although this has not yet been done. It would involve comparing two sets of extinction curves (obtained preferably after periodic reconditioning) in one of which intervals of no responding were reinforced. Every interval might be reinforced as soon as it reached a required length, or, if this threatened to involve a change in drive by providing too much food, intervals might be reinforced as nearly periodically as possible. The existence of 'conditioning-not-to-respond' would be proved by a reduction in the height of the extinction curves for the group receiving the reinforcement, which would indicate that the reflex reserve of this group had been reduced.

One experiment in which the interval approached, if it did not reach, a value yielding conditioning-not-to-respond, is represented in Figure 51. The record was taken after prolonged periodic reconditioning. On the day of the experiment pellets of food were delivered to the rat periodically but never within fifteen seconds of the last response. The average interval was much greater than fifteen seconds, particularly toward the end of the experiment. It will be seen that the negative correlation results in a rapid decline in rate—from about twelve responses per minute at the beginning to about one per minute at the end of an hour and a quarter. There was considerable recovery on the following day. In spite of the shortness of the interval, this case is very near the practical limit. It is impossible to tell when a pause is to come to an end. We cannot reinforce just as the rat starts to respond, because that would reinforce the starting movement; hence we cannot take advantage of occasional long intervals. Some rule must be set up, like that in the present case: reinforce periodically, provided the rat has not responded within a certain length of time, or wait until an interval of that length has elapsed. If the rat is responding at a fairly high rate, however, the desired frequency of periodic rein-

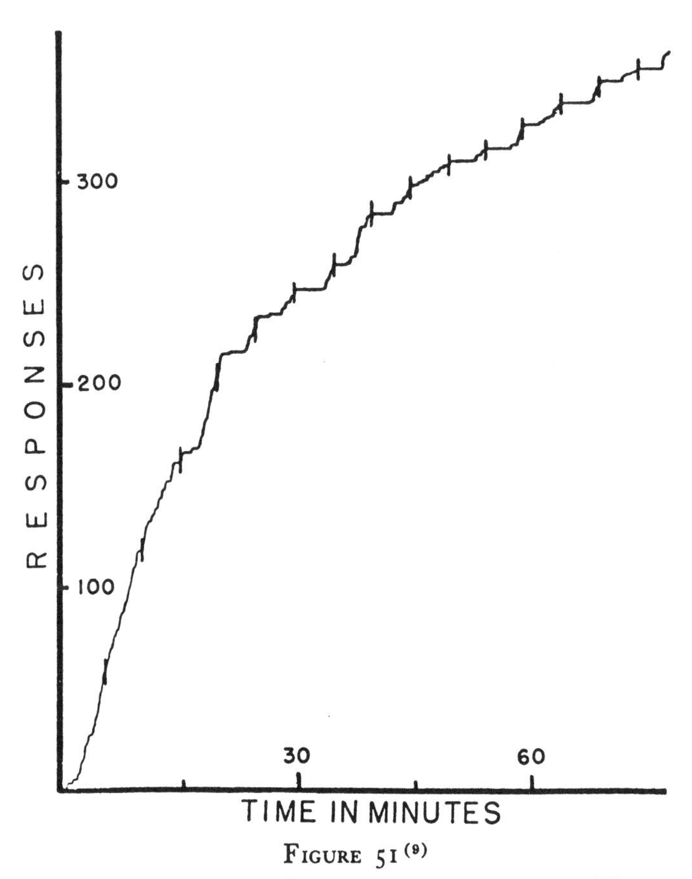

FIGURE 51 (9)

CHANGE IN RATE UNDER PERIODIC REINFORCEMENT WHEN THE DELIVERY OF PELLETS BEGINS TO BE CORRELATED NEGATIVELY WITH RESPONSES TO THE LEVER

forcement cannot be achieved with an interval much longer than fifteen seconds.

This experiment differs fundamentally from merely introducing a long interval between responses and reinforcement, where the reinforcement is contingent upon the interval *but also upon the preceding response.* In the present case the responses might be withheld altogether and the magazine would nevertheless periodically deliver pellets of food. The case of the interval approaches that of negative correlation when the experiment is performed upon rats already responding under periodic reconditioning. The difference would be more strongly marked if both procedures were adopted in original conditioning. No positive conditioning would take place under the conditions of negative correlation. If the rat were not already responding to the lever with some unconditioned strength, it would never come to do so because of this procedure. On the other hand, where original conditioning depends upon the elicitation of a response plus a pause, conditioning takes place provided the pause is not too great (see Chapter Three).

No Correlation between Response and Reinforcement

In the preceding experiment where the essential condition is the negative correlation of response and reinforcement, the rat might be said (in the vernacular) to learn not to press the lever because the periodic delivery of food is found to depend upon an absence of responses. A somewhat similar case should be considered here, in which there is no correlation, either positive or negative, between response and reinforcement. The rat responds according to any previously established strength, but the periodic delivery of pellets is timed by a clock, and there is only an accidental temporal relation between a given reinforcement and a response or the absence of a response. Since single reinforcements are effective, the resulting rate should fluctuate to some extent at random. But the average effect is no correlation whatsoever and ultimately the behavior of the rat should reflect that fact. In the vernacular, the rat should learn that its responding has no connection with the periodic delivery of food.

An experiment to discover the effect of passing from periodic reinforcement to the independent periodic administration of food was performed upon four rats 120 days old. The operant was conditioned as usual, and periodic reconditioning began the following day. An interval of five minutes was found to yield an unusually high rate with this group, and a change was made to six minutes on the second and all following days. After three days of periodic reconditioning the magazine was operated at the same frequency but wholly independently of the behavior of the rat at the time. The pellet of food either reinforced a response or did not according to the momentary behavior of the rat. This procedure was maintained for nine days, and on the tenth the reinforcement was again correlated with a periodic response.

The way in which the rat 'learns that the delivery of food bears no relation to its responding' is shown in Figure 52, which gives the average rates for the group from the third day to the end of the experiment. The initial rate under periodic reconditioning was

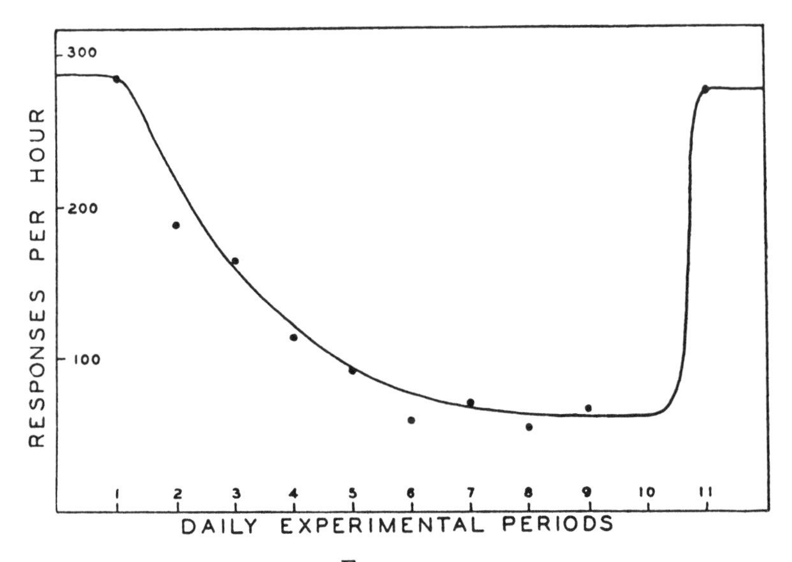

FIGURE 52

DECLINE IN RATE WHEN THERE IS NO RELATION BETWEEN THE PERIODIC PRESENTATION OF FOOD AND RESPONSES TO THE LEVER

slightly less than 300 responses per hour. Immediately upon abolishing the correlation between reinforcement and response (the second day of the graph) the rate fell. It was reduced by fully one-third on the first day. This result could be predicted from the preceding experiment upon the introduction of an interval of time between response and reinforcement. The rate continued to fall for two or three days but then stabilized itself at about one-quarter of its original value. There is no indication during the period of the experiment that the rat would eventually cease responding, even though all responses were actually useless. When the correlation was reestablished (on the eleventh day of the graph), the rate immediately rose to very near its original value, the usual slight downward drift being evident.

The explanation of the failure to cease responding is probably as follows. The average interval between a response and a reinforcement is determined by the rate of responding prevailing at the time, as is also, therefore, its average effect in reinforcing responding or possibly not-responding. When the rate is nearly zero, as it is prior to conditioning, the average interval is long, and an effect is either lacking or is on the side of reinforcing not-responding. (We cannot assume that the presentation of a reinforcing stimulus has no effect.) The rat will not be conditioned to press the lever under this procedure of periodic administration of a reinforcing stimulus if it has only this low unconditioned rate to begin with. But when a high rate has been previously established, as in the present experiment, the average interval is of a different order, and the completely independent delivery of food must be supposed to have a considerable, if accidental, reinforcing effect. The stabilization at a rate of from one to two responses per minute observed in this experiment is possibly due to the fact that the occasional proximity of a response to the administration of food for which that rate is responsible just suffices to maintain the rate. With an extinction ratio of 20:1 three coincidences per hour would suffice, although what would actually obtain is the equivalent of this in a larger number of less perfect coincidences. Whether we must appeal to an input large enough to allow for subtraction due to conditioning not-to-respond must remain an open question until the latter phenomenon has been investigated.

The experiment shows, then, that if the rate begins at a high value, it does not reach zero because it cannot pass through a point at which accidental reinforcement is sufficient to maintain it. The rat can never learn 'that there is no connection *whatsoever* between its responses and the delivery of food.' In making this statement I am not allowing for the random character of the experiment. Given a favorable period lacking in coincidences, the rate might fall to a value at which coincidences would become less and less likely and eventually reach zero.

An experiment of this sort cannot successfully be performed upon rats which have begun to show the temporal discrimination of the third-order deviation described above (page 125). The resulting concentration of responses toward the end of each interval significantly raises the probability of coincidental reinforcement and vitiates the result.

Periodic Reconditioning of a Respondent

Although this chapter has been written entirely with reference to operant behavior, a respondent may also be periodically reconditioned and usually is so reinforced in nature. Its strength will presumably behave in approximately the same way, although the two cases are not exactly parallel. In a respondent there is a modification of the intensity of the response with a change in strength. The notion of a rate of responding is invalid, and the notion of a number of responses and of an extinction ratio is therefore meaningless. But given a constant rate of presentation of the conditioned stimulus (S^0) with a periodic correlation of the reinforcing stimulus (S^1), a similar stable state might be reached. If at intervals of one minute stimuli were presented as follows: $S^0 + S^1$, S^0, S^0, S^0, $S^0 + S^1$, S^0, S^0, S^0, $S^0 + S^1$ . . . , the magnitudes of the responses for all single presentations of S^0 should at first show a tendency toward extinction between presentations of S^1 but on the present analogy should eventually assume an intermediate stable value. A constant value could be used in the same way as an extinction ratio in studying the effect of a given condition of reinforcement (for example, the effect of a delay). It would be interesting to know whether the extinction curve (which is usually irregular in this type) would show a similar improvement in smoothness after periodic reconditioning.

The negative correlation of S^0 and S^1 in a respondent would have the form of the presentation of a reinforcing stimulus in the absence of S^0 and the absence of presentation in its presence. A different result is here to be expected, at least in the extreme case in which S^0 is presented more or less continuously for short periods of time. A positive response should then appear upon cessation of S^0 (Pavlov). The case of no correlation whatsoever would have the form of random presentation of both S^0 and S^1, any coincidence being accidental. Again unlike the case of the operant, a conditioned reflex would presumably arise, even if the procedure were instituted at the beginning, because S^0 would become part of the experimental situation to which conditioned value always becomes attached in conditioning of this type.

Chapter Five

THE DISCRIMINATION OF A STIMULUS

The Nature of the Problem

The different states of strength and changes in strength described in the preceding chapters were due to the different ways in which a reinforcing stimulus could be correlated with a response or with another stimulus. The cases were neither few nor simple, and they have not exhausted the subject of reinforcement as an operation or of its effect upon behavior. Another kind of correlation raises an entirely fresh problem. It is a correlation of a reinforcement with a stimulus or response *possessing some specific property*. Such a correlation cannot be fully represented by one reinforced elicitation or even by the repetition of a single reinforcement. For example, presentation of food as a reinforcing stimulus may be correlated with a tone of a given pitch but not with other tones, but repeated presentation of the tone together with food will not establish a conditioned reflex showing this restricted correlation because neighboring tones also acquire the property of eliciting salivation through 'induction' (Chapter One). The behavior may come to follow the actual relation more or less precisely because a correlation with a tone of a given pitch implies that tones of other pitches sometimes occur, but responses made to them will be extinguished under the terms of the reinforcing correlation. Eventually, the organism responds to the selected tone but not to others, within certain limits. The process through which this is brought about is called Discrimination. The example given is one of the three possible types to which the following three chapters will be devoted.

The process of induction which gives rise to the problem of discrimination was described in Chapter One in the following law: A change in the strength of a reflex may be accompanied by a similar but not so extensive change in a related reflex, where the relation is due to the possession of common properties of stimulus or response. The dynamic changes to which induction applies are those in which the operation involves the elicitation of the reflex—namely, reflex fatigue and conditioning and extinction. These are also the operations affecting the reserve, and this fact will later be found to be important. The present chapter is concerned with induction due to similarity of stimuli. The general problem is as follows. In stating the correlation of a specific stimulus (*e.g.*, a shock of a given strength applied to a given spot on a leg) with a response (*e.g.*, the flexion of the leg in a given direction), it cannot be said that an isolated unit has been set up. The unit will obey the laws which apply to it as experimentally treated, but it is not necessarily totally unrelated to the rest of the behavior of the organism. We may set up a number of flexion reflexes differing in their loci of stimulation and in their direction of flexion which, so long as we are concerned simply with correlation, we may regard as separate units. Nevertheless, in examining such a dynamic change as fatigue it will be found that an operation performed upon one of them affects the others also, presumably according to the proximity of the stimuli (Sherrington). This was seen in Chapter One to be closely related to the problem of the definition of a unit. The problem arises because stimuli may usually be arranged in a continuous order, such as the spatial continuum in the case of the flexion reflex, where adjacent members differ only slightly. In unconditioned respondents we control the stimuli, and the phenomena of induction therefore arise as simple interactions and may be described as such. In accordance with this view I shall speak of each experimentally isolable correlation as *a* reflex and treat any group of reflexes showing inductive interaction simply as a group.

Induction raises a special problem in conditioned reflexes where the reinforcement is correlated with a stimulus or a response exhibiting a special property. It is impossible, in view of induction, that behavior should reflect such a correlation precisely without recourse to extinction. If the presentation of food is correlated only with the presentation of a tone of a special pitch, the reinforcement will condition responses to tones of other pitches with which there is actually no correlation of food. When these tones are later presented, extinction necessarily follows. In this way different strengths come to be assigned to closely related reflexes contrary to the natural effect of induction. This process of discrimination arises from the restricted correlation of the reinforcing stimulus. The organism 'generalizes' the effect of the reinforcement through induction, but the external conditions of the correlation fail to support it.

In a discrimination of a stimulus we have in the simplest case two reflexes differing with respect to a property of their stimuli. Let l represent the property or selected value of a property with which the reinforcement is correlated, and λ, either the absence of this property or some other value of it on a continuum.[1] The object of the discrimination is to give $Sl . R$ a significant strength while holding $S\lambda . R$ at a lower, or preferably at zero, strength. We first reinforce $Sl . R$, but [$S\lambda$. R] also increases. (The process may begin with both reflexes at equal strengths if both have been previously reinforced. In such a case the discrimination begins when the reinforcement is first withheld from $S\lambda . R$ after having previously been accorded to it.) The next move is to extinguish $S\lambda . R$, but $Sl . R$ also decreases in strength. If we then recondition $Sl . R$, $S\lambda . R$ also rises in strength. And so on. The process is not futile because in each case the induced effect is somewhat less than the direct, and the two reflexes draw apart in strength. Each simultaneous movement downward is less extensive for $Sl . R$ than for $S\lambda . R$ and each movement upward less for $S\lambda . R$ than for $Sl . R$. By repeated alternate conditioning and extinction we are able to accumulate the slight differences which occur upon separate occasions.

Expressed in the vernacular a discrimination of the stimulus is a process in which an organism 'tells the difference between two stimuli' or at least 'tells that they are different.' This view has taken a firm hold in psychology because of the importance of discrimination in the study of sensory processes and especially of limens. But it is of little value here. An organism can be said to 'tell that two stimuli are different' if any difference whatsoever can be detected in its behavior with respect to them. From the nature of induction (and especially from its dependence upon proximity) it may be inferred that this is inevitably the case for any supraliminal difference, even though our measure may not be

[1] It is elliptical, but convenient, to speak of the correlation of a reinforcing stimulus with a value of a single property. It does not imply that the property has the status of a stimulus. An organism does not respond to a pitch, but to a tone of a given pitch. The correlation of the reinforcing stimulus is always with a tone, not with a property in isolation. It is not in accord with present usage to speak of single properties (say, 'red' or 'A flat') as 'stimuli.'

delicate enough to detect the effect when the difference is nearly liminal. What 'learning to tell the difference' refers to in this case is the widening of the difference in strength in related reflexes through alternate conditioning and extinction. This is not a form of conditioning. To make a discrimination is to accumulate slight differences which are in themselves properties of the original behavior of the organism. Whether or not it is possible to reduce the amount of induction between two reflexes and so to hasten their separation in strength will be considered later. In such a case the term 'discrimination' might refer to the breakdown of induction, but that is not the process traditionally described by the term.

The present type of discrimination may be stated as a law in the following way:

THE LAW OF THE DISCRIMINATION OF THE STIMULUS IN TYPE S. *A reflex strengthened by induction from the reinforcement of a reflex possessing a similar but not identical stimulus may be separately extinguished if the difference in stimuli is supraliminal for the organism.* This is an incomplete statement, since it ignores the reciprocal effect of the extinction upon the directly conditioned reflex and the need for repeated alternate conditioning and extinction in order to obtain any considerable difference in strength. But since the degree to which the difference is carried is arbitrary and can apparently never be complete, the statement will suffice. The law follows from the Law of Induction, and, as its present statement implies, does not represent a new kind of dynamic process. The changes taking place in discrimination are conditioning and extinction. Nevertheless, the process is interesting in its own right and of considerable importance in the behavior of the organism since by far the greater part of conditioned behavior is discriminative.

Before taking up the two fundamental cases of the discrimination of stimuli, I shall try to clarify the notion of the proximity of stimuli by listing some of the continua along which induction may hold. Some of the properties with respect to which stimuli differ or resemble each other are as follows:

Gross topography. The extreme case of a topographical difference occurs when the stimuli fall within different sense-departments, such as vision and audition. Whether induction takes place across departmental boundaries is a simple experimental problem but one to which there is no clear answer at the present time. Pavlov reports induction of this sort only in long-trace reflexes (see Chapter Seven). It is probable that some effect is usually felt. Stimuli of different kinds have in common at least the property of being sudden changes in the stimulating environment and this may be enough to induce induction. Although the effect is presumably slight, it might possibly be revealed by pooling many cases. Thus, if the same conditioned response were based upon a number of stimuli in all departments but one, an effect might be observed if a stimulus in this excepted department were eventually tested. Only in a rough way can the several sense-departments be spoken of as constituting a continuum along which similarity or dissimilarity may be measured.

Position. Another aspect of topography is the position of the stimulus within a single sensory field. It applies particularly to touch and vision. A classical experiment by Pavlov which shows the effect of the proximity of tactual stimuli will be described below. Stimuli which show the property of position also show the properties of size and shape and may differ from each other in these respects.

Quality. I hesitate to apply this name to perhaps the most familiar property of stimuli, but it will serve. Examples are wave-length in the cases of sounds and lights, and molecular structure in the cases of tastes and odors. A large part of the traditional study of sensory processes has been concerned with finding the least differences with respect to this kind of property that will suffice to establish discriminations.

Intensity. The other great part of the traditional field of sensory processes has dealt with the discrimination of intensity, where the preceding properties are not varied. The intensity of a stimulus is measured in units appropriate to its form—energy for vision and audition, concentration for taste, and so on.

Membership. Within a single sense department two or more tones, spots of light, and so on, may be presented simultaneously, and similar combinations of stimuli are also possible across departments, as in touch and temperature, touch and taste, light and tone, and so on. Combinations will be treated here as single stimuli, although they will be designated as 'composite' and written *SaSb* . . . when this special characteristic needs to be pointed out.

Composite stimuli, otherwise identical, differ if any member differs on any of the continua already listed, and they may also differ simply in membership. Thus a stimulus composed of a tone and a light differs from one composed of the tone alone or of the tone plus an odor. Induction can easily be demonstrated between composite stimuli differing in membership, but, as in the case of gross topography, it is only in a rough sense that a group of composite stimuli may be spoken of as constituting a continuum.

The analysis of sensory continua is scarcely as simple as this list would indicate. I have not mentioned many important problems (such as that of whether all properties are independently manipulable), but the list will suffice to illustrate some of the kinds of induction that enter into discriminations. At the present time the problem of the proximity of stimuli is more important for end-organ physiology than for a science of behavior. Only in a relatively late stage of the study of discrimination should we be interested in the measurement of induction with respect to minimal differences in the properties of stimuli. In beginning the study of discrimination as a process we may profitably use pairs of stimuli that show rather gross differences, because the inductive effect of conditioning and extinction is then considerably below the direct effect and the accumulation of a significant difference is more readily accomplished.

The final measure of inductive proximity is, of course, provided by the organism. Two stimuli which lead to quite different responses in a dog or a scientist may be indistinguishable to a rat. What I am noting here is the independent structure of stimuli and the extent to which it corresponds to the discriminative behavior of the organism. This comparison is possible in the case of the human organism only because of the invention of techniques for revealing differences in stimulating energies and substances that transcend the immediate capacity of the organism. A simple example of such a technique is the use of beats to measure the proximity of two tones beyond the point at which a difference can be detected in a non-simultaneous presentation. A more elaborate example is the use of a spectroscope in distinguishing subliminal differences in color mixtures.

A discrimination of the stimulus always implies the successive presentation of two stimuli, a procedure which is indispensable in

testing a difference in the strength of reflexes. Certain elliptical procedures used with human organisms have led to the supposition that a simultaneous 'comparison' or 'judgment' is possible, as, for example, when the subject is looking at a field divided into two parts differentially illuminated. But such a field is discriminated in one of two ways, both of which involve succession. In the first the subject looks from one to the other half of the field, so that the two stimuli are presented alternately on the same part of the receptive surface. In the second the subject makes a response to a pattern (which will be dealt with here solely as a matter of composition of stimuli), and discriminates between a 'homogeneous field' and a 'divided field,' which are presented in succession. The use of the same elliptical devices (which chiefly involve verbal behavior) have also led to a view that a discrimination is the identification of a property 'by name' or in some similar peculiar way. It is not essential to a discrimination that each property of the stimulus have a corresponding form of response. The basic fact is that the response (whatever it is) is made to one stimulus and not to another.

Discrimination of the Stimulus in Type S

The examples given in the preceding section were of a discrimination of the stimulus based upon a conditioned reflex of Type S. It may be well to add a few further properties of the process, chiefly as reported by Pavlov.[2] Pavlov's data must be accepted with certain reservations. They were presented by him in support of a special conception of 'inhibition'; and when reorganized as a description of the process of discrimination itself, they often leave important points unanswered. They are usually only *instances* of the process being described, as Hull (48) has pointed out. Pavlov does not set up an experiment and proceed to deal with all the resulting data. As a result there is considerable room for unconscious selection of favorable cases. The danger is especially great because the experiments are designed to test hypotheses rather than to provide a simple description. Again, Pavlov is not interested in the course of the changes taking place during a discrimination but only in that more or less final state in which the ex-

[2] Unless otherwise stated all references are to Pavlov's *Conditioned Reflexes* (64).

tinguished member is not elicitable. His usual measure of the rate of discrimination is the time required to reach this point, which is unsatisfactory in a dynamic study of the process.

Pavlov describes an instance of inductive conditioning in Type S and its dependence upon proximity as follows: 'If a tone of 1000 d.v. is established as a conditioned stimulus, many other tones spontaneously acquire similar properties, such properties diminishing proportionally to the intervals of these tones from the one of 1000 d.v.' Examples within other single sense departments are given, as well as the case noted above of induction across a departmental boundary in a long-trace reflex. The inductive effect of extinction is demonstrated in a well-known experiment, the result of which has been confirmed on human subjects by Bass and Hull (26). In Pavlov's experiment five small apparatuses for producing tactual stimulation were arranged along the hind leg of a dog, the first being placed over the paw and the others at increasing distances of 3, 9, 15, and 22 cms. respectively. These five stimuli constituted a fairly homogeneous inductive group; when one of them was conditioned, the others acquired a similar status to very nearly the same extent. A discrimination was established with alternate conditioning and extinction in such a way that all the stimuli remained conditioned with the exception of the stimulus over the paw. When the strengths of the four reinforced responses had been equalized, each was tested after three unreinforced presentations of the extinguished stimulus. The result was that the failure to reinforce the stimulus over the paw reduced the strength of the adjacent point to practically zero and that of the second point to one-half its former value, while the other two points were unaffected or were increased in strength (see below). Pavlov describes this experiment as a demonstration of the irradiation of inhibition. While we are not appealing to a concept of inhibition in the present system, it should be noted that Pavlov's 'irradiation' (either of inhibition or excitation) is very close to what I am here calling induction. In the case of excitation he often uses the term 'generalization.'

Pavlov uses 'induction' in a quite different sense. He appeals to Sherrington for authority, but Sherrington uses the term, as I have said, for both of two opposed processes neither of which corresponds to the present phenomenon. The facts referred to by Pavlov cannot be disregarded. They are described in his chapter on Positive and Negative Induction and seem to show an effect directly opposed to induction. In order to avoid confusion I shall refer to this opposed effect as 'contrast.' The experiments involve pairs of stimuli differing on various continua (touch, sound, and light) which have been used to develop discriminations. The observations may be described independently of the concept of inhibition in the following way. In *Positive Contrast* presentation of the unreinforced stimulus produces a momentary *increase* in the strength of the reinforced member, although a decrease is to be expected from the law of induction. In *Negative Contrast* reinforcement of the reinforced member delays or prevents the reconditioning of the unreinforced. This is also the opposite of the effect to be expected from induction. Until the conditions which determine whether induction or contrast is to occur at a given time have been identified, the observations of contrast stand simply as exceptions to the Law of Induction. Little is at present known except that contrast is usually a temporary phenomenon appearing at only one stage of a discrimination and apparently not sufficing to abolish it in spite of its opposition to induction. Induction is presumably always necessary in order to reach the stage of discrimination at which contrast appears. In the experiment described in the preceding paragraph the effect upon the reflex at the farthest point of stimulation was a slight increase in strength. This suggests that contrast occurs at a certain not too immediate degree of proximity, but other cases in which the nearer point shows contrast and the farther induction are reported by Pavlov, and there may be no relation either way.

It is doubtful whether contrast is a genuine process comparable with induction. In many of the cases cited by Pavlov alternative explanations suggest themselves, although it is difficult to establish their validity on the data given, at least in the non-Russian reports. For example, one procedure would have permitted the development of a conditioned reflex where the stimulus was the change from l to λ as distinguished from either l or λ separately (see page 222), and this reflex being always reinforced and not suffering from inductive extinction would have given the high value observed. It would be futile to take up each case in this way, because the subject needs further investigation. No other occasion for referring to contrast will be met here.

Given the two processes of inductive conditioning and extinction and their relation to the properties of pairs of stimuli, it is not difficult to show that the rate of development of a discrimination will obey the same laws. Many data demonstrating this fact are given by Pavlov, although the course of the process and its relation to proximity are not dealt with quantitatively. In particular it has not been shown whether the process is merely the accumulation of differences between direct and inductive changes in strength or involves a separation of the stimuli on the inductive continuum. In other words it is not shown in Pavlov's work whether the organism ever comes to make a distinction between stimuli that was not already felt in the effect of induction or to broaden such a distinction where it originally exists to a slight extent. This is an important question, which requires a quantitative treatment of the course of the change, and to which I shall devote considerable space in connection with the discrimination of an operant.

Abolishment of the Discrimination

When the response to a discriminated stimulus has been extinguished, it may be restored to its original strength by simple reconditioning. In a typical experiment reported by Pavlov the response to the sound of a tuning fork plus tactile stimulation of the skin had been extinguished while the response to the sound alone had remained conditioned. The combined stimulus was then reinforced at intervals of approximately 15 minutes. The reinforcing stimulus followed the conditioned stimulus after a delay of one minute, during which the conditioned response could be measured. The result is shown in a later figure (59). The strength of the reflex begins at zero but during the course of seven reinforcements increases to the essentially maximal value indicated by 14 drops of saliva. The discrimination may then be said to be abolished.

The Inevitability of a Discrimination in This Type

Conditioning of Type S always involves some amount of induction and eventually some amount of discrimination. The stimulus to be conditioned (S^0) must be presented against a background of stimulation arising from the 'situation' (S^G). In the Pavlovian experiment let S^0 be a tone and S^G the stimulation arising from the experimental stand. Because of induction the presentation of a reinforcing stimulus in the presence of $S^G S^0$ brings about the conditioning of at least three reflexes—namely, $(S^G . R^1)$, $(S^G S^0 . R^1)$, *and* $(S^0 . R^1)$. The power to elicit the response is acquired not only by the precise combination of stimuli present at the moment but also by the component parts. Consequently the dog may salivate when put into the stand again or when a tone is sounded in entirely different surroundings. This triple conditioning cannot wholly be avoided. Theoretically S^G and S^0 could be made to coincide, when a given stimulus would always be present in a given situation and the total stimulating complex would be reinforced. But little if any experimentation would be possible. A more practical solution is to make S^G insignificant with respect to S^0. This can be done in at least two ways: by reducing the value of S^G (as by designing the experimental situation so that a minimum of general stimulation is achieved) and by intensifying S^0 (as by using a strong stimulus or presenting it suddenly just before reinforcement). These devices are common in experimentation on conditioning of this type, but in spite of them some induction to the experimental situation occurs. Since responses to the situation disturb the experiment, they are usually eliminated by extinction. In other words, a discrimination is set up in which responses to $S^G S^0$ are reinforced and the responses to S^G alone extinguished.

The general rule that some amount of discrimination is present in any actual instance of a conditioned reflex of this type obtains as well outside the laboratory. Reinforcing stimuli are practically invariably correlated with stimuli which do not compose all the stimulation affecting the organism at the moment of reinforcement. Consequently the reinforcement has a broader effect than the actual correlation implies and extinction of the extra effect eventually follows.

Discrimination of the Stimulus in Type R: The Correlation of a Discriminative Stimulus with the Reinforcement of an Operant

A connection between an operant and a reinforcing stimulus can be established independently of any specific stimulation acting prior to the response. Upon a given occasion of reinforcement stimulating forces will, of course, be at play, but with constant attention it is possible to reinforce a response (say, a given movement of a leg) under many different sets of stimulating forces and independently of any given set. In nature, however, the contingency of a reinforcement upon a response is not magical; the operant must *operate* upon nature to produce its reinforcement. Although the response is free to come out in a very large number of stimulating situations, it will be effective in producing a reinforcement only in a small part of them. The favorable situation is usually marked in some way, and the organism makes a discrimination of a kind now to be taken up. It comes to respond whenever a stimulus is present which has been present upon the occasion of a previous reinforcement and not to respond otherwise. The prior stimulus does not elicit the response; it merely sets the *occasion* upon which the response will be reinforced.

In a world in which the organism is a detached and roving being, the mechanical necessities of reinforcement require in addition to the correlation of response and reinforcement this further correlation with prior stimulation. Three terms must therefore be considered: a prior discriminative stimulus (S^D), the response (R^0), and the reinforcing stimulus (S^1). Their relation may be stated as follows: only in the presence of S^D is R^0 followed by S^1. A convenient example is the elementary behavior of making contact with specific parts of the stimulating environment. A certain movement of my arm (R^0) is reinforced by tactual stimulation from a pencil lying upon my desk (S^1). The movement is not always reinforced because the pencil is not always there. By virtue of the visual stimulation from the pencil (S^D) I make the required movement only when it will be reinforced. The part played by the visual stimulus is shown by considering the same case in a dark room. At one time I reach and touch a pencil, at another time I reach and do not. There is no way in which I can come to respond only upon the favorable occasions. I can only 'grope'—*i.e.*, reach in the absence of a discriminative stimulus. In neither the light nor the dark does the pencil *elicit* my response (as a shock elicits flexion), but in the light it sets the occasion upon which a response will be reinforced and (through the development of a discrimination) upon which it will occur. Although a conditioned operant is the result of the correlation of the response with a particular reinforcement, a relation between it and a discriminative stimulus acting prior to the response is the almost universal rule.

In dealing with a chain of reflexes rather than with a single member a different kind of discriminative relation must be considered. A discriminative stimulus may be correlated with the eventual reinforcement of a chain but not with that of the member it immediately precedes. A dinner bell is a remote discriminative stimulus, which is correlated with the reinforcement of the final member of what may be a long chain of reflexes. Remote discrimination is much more common than immediate, for the latter is confined chiefly to the behavior of manipulation. The possibility of remote discrimination raises the important problem of the relation of manipulation to eventual reinforcement—that is, of means to end. The formulation is essentially the same in the two cases.

The chain of reflexes studied in the preceding chapters was said in Chapter Two to involve discrimination. The rat comes to make certain responses of progression and postural modification with respect to the food tray, because in the presence of the (principally tactual) stimuli arising from the tray and its surroundings such responses lead to contact with food. Later it responds only after the magazine has sounded because the correlation between the response and food is then restricted in that way. It would have been possible to choose an example that avoided discrimination, as by using a response that did not operate directly upon the environment, but for reasons that I have already considered the case would not have been typical. The use of a sample of behavior containing discrimination is reasonably safe because the latter part of the chain remains essentially constant during the changes in the strength of the initial member that we are studying. If we now make the initial member discriminative and proceed to the study of the process directly, we may lift ourselves by our own boot-straps and return to justify our assumptions about the original sample.

The discrimination to be described here is of the remote kind. A stimulus acting prior to the response to the lever is correlated with the eventual reinforcement of the chain. The stimuli that I have used are a 3 c.p. light, a click, and a buzz. They are all relatively gross and do not raise the problem of a limen. An operant occurring in the presence of a stimulus correlated with reinforcement will be written $sS^D . R$; in the absence of such a stimulus $sS^\Delta . R$. The discriminative property is in this case a matter of membership only.

Induction During Continuous Reinforcement

Before taking up the process of discrimination I shall attempt to estimate the part played by a light in controlling the reserve created by the reinforcement taking place in its presence. Suppose a response to be well conditioned in the presence of the light (in addition to unavoidable general stimulation). If extinction is now carried out in the absence of the light, a measure of the induction from one situation to the other is available, since the curve may be compared with that for extinction in the presence of the light to show the extent to which the stimulus influences the response. Since both curves cannot be obtained with the same organism, an experiment of this sort must be carried out on a fairly large group. It has not yet been done, but a gauge of the difference has been obtained with a slightly different method utilizing the same principle. Toward the end of an extinction curve obtained in the dark after conditioning in the light, the light is introduced and maintained. The responses which are under the control of this special part of the original stimulation appear in an added extinction curve.

In one experiment four rats approximately 100 days of age were conditioned in the usual way, except that the light in the experimental box was on. Fifty responses to the lever were reinforced, and on the following day extinction curves were begun in the absence of the light. After forty-five minutes the light was turned on, and the further course of the extinction was observed for twenty minutes. The average 'latency' of the response to the light was 20.5 seconds—that is to say, the rat first responded twenty seconds after the light was turned on. At the rate then prevailing in the dark the latency due to chance would have been about thirty-six seconds. This positive effect of the light is corroborated by the subsequent change in rate. The average curve for the four rats is given in Figure 53A. It begins in the usual form, and shows a slight deviation below the envelop. At the arrow the light was turned on to restore the exact stimulating complex present at the time of conditioning. There is a significant increase in rate, which is followed by a slight compensatory slackening. The end of the curve is only slightly above the point that would have been reached without the additional stimulation.

It is possible that no greater effect was felt in this experiment because the presentation of the extra stimulus had been too long delayed. In a second experiment, also with four rats, the presenta-

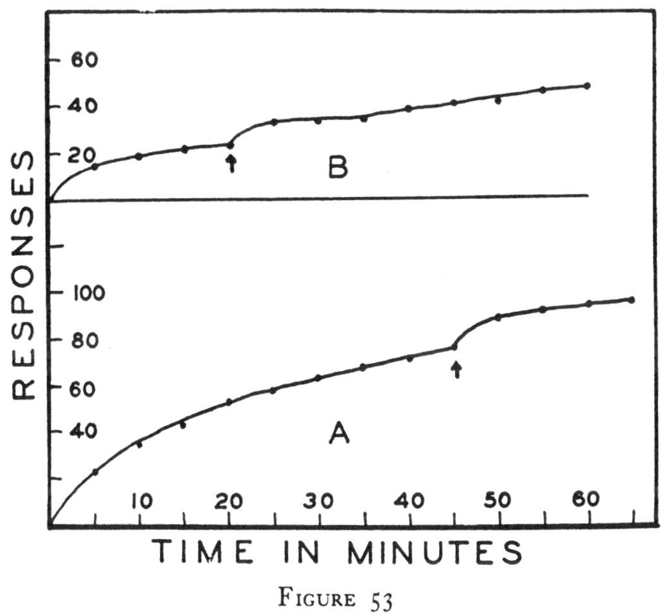

FIGURE 53

INDUCTIVE EFFECTS IN EXTINCTION

The extinction begins in the absence of a stimulus which has always been present during reinforcement. When this stimulus is introduced (at the arrows), the rate increases temporarily.

tion occurred earlier in the course of extinction. The procedure was otherwise the same. The average curve is given in Figure 53 B. The group gave considerably smaller extinction curves, but the result is sufficiently clear. The average 'latency' of the first response to the light was in this case 15 seconds against a chance value of sixty. The rate behaved in the same way as in the first experiment and to relatively the same extent. It is again obvious that the increase in rate momentarily following the introduction of the light is followed by a compensating decrease. The remainder of the curve is not necessarily significantly above the extrapolation of the curve in the absence of the light.

Neither result is clear enough to indicate definitely whether the effect of the light is to produce a second curve (such as that observed in the separate extinction of the members of the chain, p. 104) or merely to provide a temporary increase in rate for which there is later compensation. It is probable from the two experiments that compensation does occur, but it may not be complete. In Figure 53 A the experiment was unfortunately not continued long enough to determine the final position of the curve with respect to an extrapolation of the first part. Whether the curve at B continues significantly above the extrapolation is difficult to say in view of the very short section of the original curve obtained. Much depends upon the weight to be attached to the last point before introduction of the light (the point at the arrow). If this is taken as somewhat too low (cf. the third point from the end of the curve), a satisfactory curve can be drawn with respect to which the increase in rate due to the light is a temporary matter with full compensation. The experiment obviously needs to be repeated with larger samples. Nevertheless, it permits us to say with some assurance that the introduction of the absent member yields a marked increase in rate, followed by some compensation.

As to the reliability of this result, it may be noted that the remarks made in Chapter Three on the averaging of extinction curves apply here. The effect shown in the average is apparent in every individual record with one exception, and some indication of the reliability may be obtained from the fair degree of smoothness of the curves drawn through the points and from the fact that they represent averages of small samples and involve no selection of any sort. Two control experiments in which extinction began under the exact conditions of the preceding reinforcement and continued after a differentiating change was made in the opposite direction gave no increases in rate whatsoever.

It should perhaps be pointed out that the result is not in conflict with a failure to confirm the notion of disinhibition. In the experiments reported in connection with that subject (p. 96) the light produced no effect upon extinction curves except when presented when the curves were below their envelops. In the present case the light has acquired a special power from having been present at the time of the conditioning. How the effect differs in the two cases is what I am here attempting to discover.

A Discrimination after Periodic Reconditioning

A discrimination necessarily involves alternate conditioning and extinction. If we are interested simply in reaching an end-point of more or less complete discrimination, we may turn from one process to the other as the condition of the organism dictates. But if we are interested in the kinetics of the process, the rates of conditioning and extinction must be constant or at least rigidly controlled. Since conditioning takes place much more rapidly than extinction, we shall presumably need a greater amount of the latter. The schedule for establishing a discrimination will therefore be such that $sS^D . R$ is occasionally reinforced and $sS^\Delta . R$ allowed to go unreinforced at all other times. This is a form of periodic reinforcement, except that the reinforced response is always made in the presence of S^D, and the procedure is identical with that of periodic reconditioning except that S^D is presented when the magazine is periodically connected. In the act of throwing switches to both the source of S^D and the magazine at once the fundamental condition of a discrimination of this sort is epitomized.

The result to be expected is the maintenance of $[sS^D . R]$ and the exhaustion of $[sS^\Delta . R]$. If all responses are recorded, the experimental record should give the course of the latter process when the periodically reinforced elicitations of $sS^D . R$ have been subtracted. No measure of the strength of $sS^D . R$ is provided, for its rate is here determined by the rate of presentation of S^D. I shall show later, however, that the state of $sS^D . R$ may be examined in other ways. The effect of the inductive conditioning from S^D to S^Δ should be apparent as a modification of the extinction curve for $sS^\Delta . R$.

The preceding chapter on periodic reconditioning provides a very necessary control and one that is not, I believe, generally made in experiments with other methods. It must be shown that the changes in the rate of extinction of $sS^\Delta . R$ are due to induction from the reinforcement of $sS^D . R$ and are not the result of the alternate conditioning and extinction or the development of a temporal discrimination. The latter possibility may be urged against the method since it holds to a definite schedule of conditioning and extinction. The controls provided by the preceding chapter are adequate. Discrimination is observed as a change in the constant rate obtaining under periodic reconditioning, and for the pairs of stimuli (S^D and S^Δ) used here the process requires a period of time during which this rate may be regarded as essentially constant, provided the constant slope has been fully established prior to the beginning of the discrimination. The temporal discrimination is seen to be taken care of by adequate compensatory changes and in any event is not a serious problem since I shall use a fairly moderate rate of elicitation, induced with a five-minute period of reconditioning, where deviations of the second and third orders are not present to any significant extent. The change observed during the establishment of a discrimination is fortunately of such an order of magnitude that confusion with any one of the types of deviation is unlikely.

A case in which the reflexes draw apart in strength slowly is that in which $sS^\Delta . R$ has previously been periodically reinforced and has a large reserve while $sS^D . R$ has not been previously reinforced, although it has acquired a high strength through induction. An experiment of this sort will be considered first.

Four rats of the same strain, P11, P12, P13, P14, were conditioned to respond to the lever at the age of 100 days. Three days later they were transferred directly to periodic reconditioning, without any intervening extinction, at an interval of five minutes. Figure 35 is from this group. The interval was chosen because it had given the lowest standard deviation in the experiment in Figure 33. It also yields a rate moderate enough to permit significant changes in both directions and practically free of deviations of the second and third orders. The experiments were conducted for periods of one hour on alternate days, and the other details of the procedure outlined in the preceding chapter were followed. Three daily records of periodic reconditioning were obtained, during which characteristic constant slopes were well stabilized. The procedure was then changed to include either the light or the click (click with P11 and P12, light with P13 and P14). In the case of the light the extra stimulus, presented before the reinforced response, was still present when the response was made, but this was not true of the click (since it was not continuous). No significant difference appeared between the results in the two cases, and the records are considered together in the following discussion. I shall refer simply to the discriminative stimulus as a light or as S^D.

The discriminative procedure for a single day was as follows. As the animal was released, both the light and the magazine had been connected, and the first response to the lever, in the presence of the light, was reinforced with a pellet of food. Both light and magazine were then turned off, and the responses during the next five minutes were unreinforced. At the end of that time the light and magazine were turned on, care being taken that this was not synchronous with a response. When a response had been made (and reinforced), the light and magazine were again turned off, and this was repeated for a total of twelve reinforcements. At the end of the last interval no response was reinforced, and the animal was removed from the box. The sound was used in the same way, with the exception already noted. The procedure was repeated for ten days.

The delay in pressing the lever (the 'latency' of the discriminative response) was not added to the interval. This not only simplifies the treatment of the data but is a necessary part of the procedure in a discrimination, since the first significant change in the behavior of the rat is the shortening of the delay. The response to S^D immediately increases in strength from the resultant value given by the constant slope to an approximately maximal value indicated by the latency. I am not here concerned with this change in the strength of the reinforced reflex, but the change in the length of the delay must not be allowed to affect the interval. Otherwise the correlated change in the total slope would enter as an artifact.

The result of the experiment is given in Figure 54, where the records for each rat for two days prior to and for ten days following the change in technique have been pieced together to form continuous graphs (the heavier lines in the figure). The first eight records after the beginning of the discrimination are reproduced from a typical set in Figure 55. The figure is for a rat showing a relatively high basic rate under periodic reconditioning, where the sound was used in establishing the discrimination.

The first two days in Figure 54 establish the slopes assumed under periodic reconditioning immediately prior to the discrimination. As soon as the procedure is converted into that of discrimination by the introduction of a differentiating stimulus, the slope of each curve begins to fall off. The strength of the reflex in the presence of S^D remains high, but the reflex in the presence of the lever alone gradually decreases in strength, and a low value is eventually reached.

The change is not rapid enough to be very clearly revealed at the end of a single hour. Nevertheless, the records for the first day

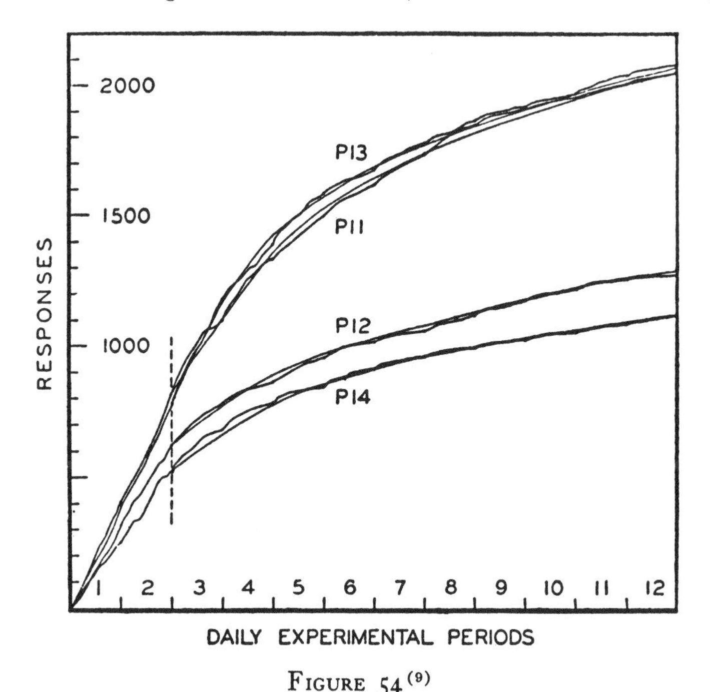

FIGURE 54 [9]

THE DEVELOPMENT OF A DISCRIMINATION

At the vertical broken line a discriminative stimulus was introduced just before each periodic reinforcement. The rates prevailing during simple periodic reinforcement (first two days in the figure) decline along theoretical curves (lighter lines) discussed in the text.

evidently depart from the usual straight line and assume a characteristic curvature, convex upward, which is also assumed on succeeding days. The course may be distorted by deviations of the fourth order. In particular there is a tendency toward overshooting—a period of active responding followed by a compensatory decrease in the rate of elicitation. Six examples have been indicated in Figure 55, where the probable course of each curve has been traced with broken lines. Other deviations are less extensive, and in general the course of each record may be easily inferred. With an occasional exception the daily slope progressively declines during the ten days of the experiment.

No attempt will be made to describe the change taking place during a single hour. It is comparatively slight, and the records are subject to too many incidental variations to make a quantita-

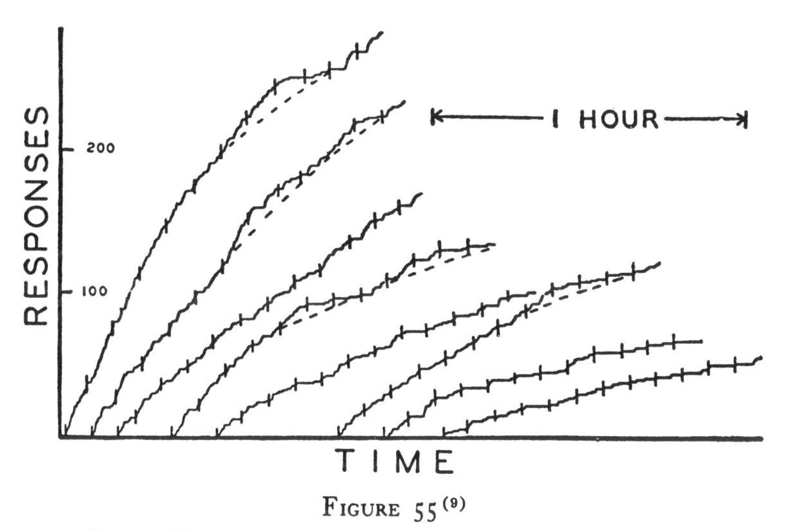

FIGURE 55[9]

SET OF DAILY RECORDS FOR ONE RAT FROM FIGURE 54

Eight records beginning at the vertical line in Figure 54 are given.

tive description either convincing or useful for our present purposes. The greater change during the full course of the discrimination is, however, significant. With one exception the curves given by the end-points of the daily records are closely described with the equation $N = K \log t + C + ct$, when N = number of responses, t = time, and K and C are constants, and where $c = 12$ and is introduced to account for the responses to the lever made in the presence of the light. The exception is for P12, which requires a value of c = about 25. This rat displayed about thirteen extra responses per hour uniformly throughout the process. In Figure 54 theoretical curves have been drawn through the data for this equation where c has been allowed to vary from 12 in order to get the best fit. The values of the constants are given in Table 6. The average curve for the four rats is well fitted when $c = 12$. By subtracting the responses made in the presence of the light and averaging the four sets the data in the upper curve in Figure 56 (page 190) are obtained. The curve is for $N = K \log t + C$, where $K = 612$ and $C = 208$. It will be seen that the experimental point is somewhat high at the end of the first hour but that in general a good fit is obtained.

The validity of this description is restricted. It has been noted that the fit is to the end points of the daily records only. The body

TABLE 6

	P11	P12	P13	P14
K	825	295	765	375
C	265	130	370	105
c	15	25	15	10

of each curve could not be included in so simple an equation. The time represented by t is not continuous, but a succession of discrete periods, and the curve interrupted at the close of one hour does not continue unchanged at the resumption of the experiment 47 hours later. In other words, there is some spontaneous recovery. Lacking a quantitative treatment of the daily change, we must resort to some such treatment as the present, where the records are pieced together to form continuous graphs. In taking the height of the daily curve at an arbitrary point as a convenient measure of the behavior for the day a probable source of error is introduced, since the curve may be at the end of the hour in the course of a deviation. The most serious deviations in the present case are those of 'overshooting.' As a result, the end points of the records for the first days (where the overshooting is most prominent) lie, if not upon, above the theoretical curves.

Another slight irregularity arises both here and in periodic reconditioning because, as will be shown in Chapter Ten, the daily rate depends upon the hunger of the rat. Hunger is maintained as constant as possible, and the results obtained with periodic reconditioning (Figure 29) are an adequate check against any significant variation from day to day, but even with the most careful feeding occasional temporary disturbances in the condition of an animal are certain to appear. The sixth record for P11 in Figure 55 is obviously much too high, and it is reasonable to attribute the abnormality to a condition of increased hunger. It is not clear whether the extra responses in the record are borrowed from later records in the series or whether the total curve is displaced bodily from the anomalous day onward. The curve for P11 in Figure 54 has been fitted as if the increase on the sixth day were compensated for during the two days immediately following. Since a rise in rate in an extinction curve due to increased hunger is compensated for to a considerable extent (see Chapter Ten), the present treatment is correct if the interpretation of discrimination as extinction is correct.

The equation is wholly empirical, and no significance is to be attached to its constants. It might be possible to derive a rational

TABLE 7

	P3	P4	P5	P6
K	215	240	560	335
C	155	70	180	145
c	20	15	25	0

equation for this and the extinction curve from the notion of a reflex reserve but I see no reason to press too eagerly toward this natural conclusion, since all the factors entering into the curve have not by any means been identified.

If the present interpretation of a discrimination is correct, the curves should resemble those obtained during extinction, perhaps with some modification due to induction from the concurrent reinforcement of a related reflex. The interpretation can be tested by applying the same empirical equation to the data previously obtained for extinction after periodic reconditioning. In Figure 38 theoretical curves have been drawn as in Figure 54. In fitting the curves the first period was regarded as 1¼ days, and the other points were taken as 2¼, 3¼ days, and so on. This procedure is based upon the assumption that the important variable is time-in-the-experimental-box. The constants in this case are given in Table 7. The constant c has no significance here comparable to that in discrimination.

The records for the first day of extinction rise considerably above the theoretical curves, although the end-point for the average curve in Figure 56 is no farther off the curve than in the case of discrimination. This overshooting in the case of extinction is to be explained by appeal to the temporal discrimination that comes to be based upon the reception of food during periodic reconditioning. In extinction no food is received and the rate rises because of the discrimination (see Chapter Seven). During discrimination, however, there is a periodic delivery of food and the rat does not go far above that observed during periodic reconditioning.

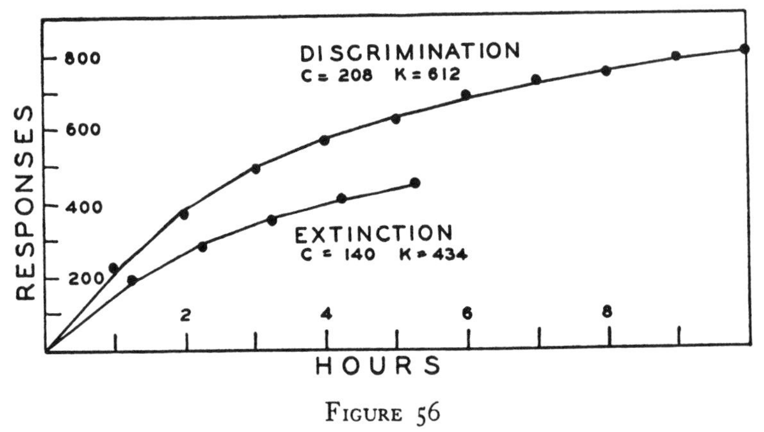

FIGURE 56

EXTINCTION AND DISCRIMINATION COMPARED

The curves are for averages of the data in Figures 54 and 38.

The average data for extinction for the four rats are given in the lower curve in Figure 56 and are fitted reasonably well by the equation $N = K \, log \, t + C$, where $K = 434$ and $C = 140$. The significant difference between the two processes lies in the slope of the curves or in the value of K. But before we may use this as a measure of the effect of induction, we must take account of the following facts. The value of K depends to some extent upon the length of time that periodic reconditioning has been in progress. The histories of the two groups differed considerably in this respect. The rats used in extinction had been periodically conditioned for a long time. This makes the difference in the two curves all the more significant, but the effect is presumably slight, as may be seen by comparing Figure 56 B with Figure 39. A much more serious consideration is that the value of K depends upon the rate of responding during periodic reconditioning prior to extinction or discrimination or, in other words, upon the extinction ratio. This may be seen by comparing the individual records in Figures 38 and 54. There is a relation between the original slope during periodic reconditioning and the slope of the subsequent curve. The rate of responding differs between the two groups in the right direction and to a sufficient extent to account for very nearly all the observed difference. The ratio of the average rates for the two days prior to discrimination and extinction is 347:257 or 1.35:1. The ratio for the values of K is 612:434 or 1.41:1.

In view of the difference in the preceding rates of responding we cannot accept the difference in the value of K as an indication of inductive interference from the periodic reinforcement of $sS^{\Delta} . R$, but must conclude on the contrary that for this value of S^D the process of extinction goes on without significant interference. Expressed in terms of the reflex reserve, the periodic reconditioning of $sS^D . R$ contributes practically nothing to the reserve of $sS^{\Delta} . R$. From the equation for the discrimination we may extrapolate to a point at which $sS^D . R$ is periodically reinforced and $sS^{\Delta} . R$ practically never occurs. Here the periodic reinforcement of $sS^D . R$ obviously does nothing toward strengthening $sS^{\Delta} . R$, and from the curves we have obtained we may conclude that this is practically the case throughout. In addition to revealing the nature of discrimination as extinction, the experiment indicates a sudden change in the induction between $sS^D . R$ and $sS^{\Delta} . R$ when a differential reinforcement has once been based upon S^D.

During a process of this sort we may take the ratio of $[sS^D . R]$ and $[sS^{\Delta} . R]$ as an indication of the extent of the discrimination, and this will be proportional to the rate of occurrence of $sS^{\Delta} . R$. But it will be shown later that the rate also depends upon other variables, such as the drive (see Chapter Ten). By lowering the drive it is possible to reach a point at which we may obtain responses in the presence of S^D but few or none in the presence of S^{Δ}, even when the discrimination is not well advanced. As soon as there is any difference between $[sS^D . R]$ and $[sS^{\Delta} . R]$ it is possible to find a value of the drive such that $[sS^D . R]$ is supraliminal and $[sS^{\Delta} . R]$ subliminal. It is therefore illegitimate to take the point at which a response is made to one stimulus and not to the other as the end-point of the process. The measure of discrimination used by Pavlov and by many others is open to criticism on this score. As formulated in the present system it is obvious that discrimination is a matter of degree. The use of an end-point is based upon the popular all-or-none notion of the development of a capacity to distinguish between two stimuli and has no place here.

Although the strength of the extinguished reflex approaches a point at which no response will ever be made in the absence of the discriminative stimulus, that point is probably never reached. Even when it is reached in practice with a given drive, a response can usually be obtained by increasing the drive. Thus although a verbal response such as 'Water!' depends for its reinforcement upon the presence of another person who has the status of a discriminative stimulus, a response may be elicited in the absence of another person in cases of extreme thirst (see the treatment of drive in Chapter Ten). Similarly, although the response of throwing a ball is reinforced only when the hand is receiving tactual stimulation from a ball (so that we stop the throw if we drop the ball), nevertheless in extreme states of drive we often go through the motion of throwing in the absence of this stimulation, a form of ('irrational') behavior easily observed among ball-players. Certain kinds of tics are operants occurring without appropriate discriminative stimuli.

Abolishment of the Discrimination

As a further check on the state of $sS^{\Delta} . R$ we may examine the rate at which it is reconditioned. Reconditioning may be effected very quickly by successive reinforcements, and the result (like that for reconditioning after extinction) resembles original conditioning very closely. Two typical records are given in Figure 57. At the vertical lines the magazines were permanently connected, and all subsequent responses were reinforced in the presence of what had been S^{Δ}. Some slight delay in accelerating to the maximal rate, typical of many curves involving reconditioning, may be observed. The change is too rapid to permit close analysis, and it is preferable to restore the extinguished reflex not to a maximal strength but to the intermediate strength observed under the original periodic conditioning prior to discrimination. S^D is therefore omitted, and $sS^{\Delta}.R$ periodically reconditioned at the selected interval.

A typical result is shown in Figure 58. Record A is for the seventh day of a discrimination during which responses in the

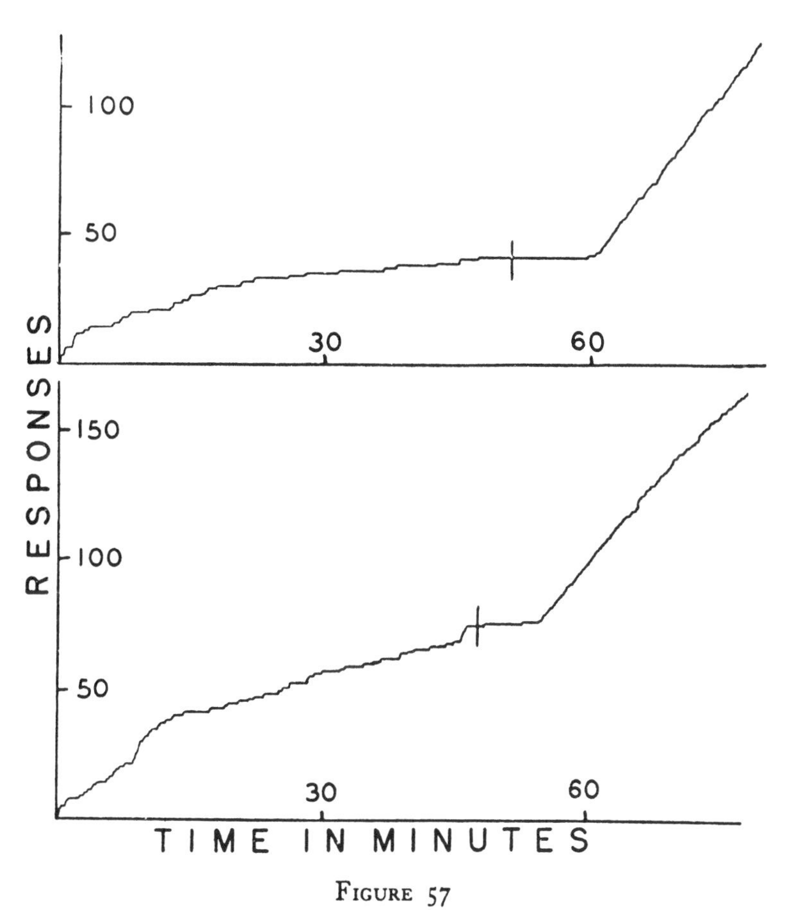

FIGURE 57

RECONDITIONING THE EXTINGUISHED REFLEX AFTER DISCRIMINATION

In the first part of each curve the response was periodically reinforced in the presence of a light; in the absence of the light the rate is low. After the vertical lines all responses in the absence of the light were reinforced. The reconditioning is typical and, like that after extinction (*cf.* Figure 10), closely resembles original conditioning.

presence of S^D were periodically reinforced while intervening responses in the presence of S^Δ were extinguished. Actually S^D was in this case the absence of the light and S^Δ its presence. The case represents a reversal of the foregoing conditions, but, as will be

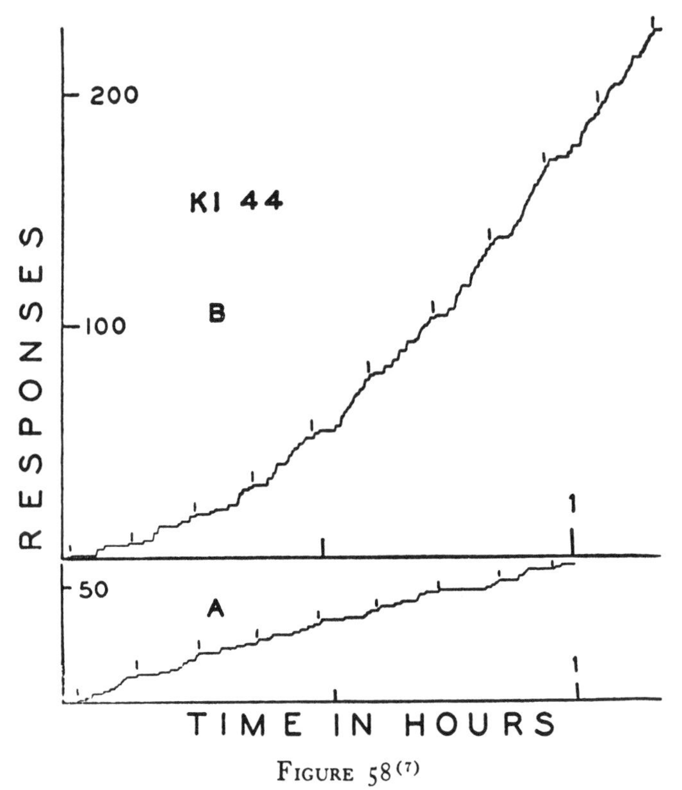

FIGURE 58 (7)

PERIODIC RECONDITIONING AFTER DISCRIMINATION

A: rate in the presence of S^Δ in the seventh hour of a discrimination, with reinforcement in the presence of S^D as marked with dashes. B: increase in rate as responses are periodically reinforced without S^D.

shown later, the two cases are essentially the same. Record B is for the day following Record A. All responses are in the presence of S^Δ, and since these responses have not been reinforced during the discrimination, the rate begins at the low value given in Record A. Periodically, however, the response is now reinforced, and the strength of the reflex gradually increases until it strikes a constant value, which for our present degree of approximation is identical with that originally assumed prior to the discrimination. Except for minor local deviations the curve shows a smooth positive acceleration. It should be compared with Figure 28 which shows the original development of a constant rate under periodic reconditioning.

There is some evidence that a difference between S^{Δ} and S^{D} persists even when $sS^{\Delta} . R$ is reinforced in this way. Since $sS^{D} . R$ should feel the effect of the reinforcement of $sS^{\Delta} . R$ through induction, it may keep a bit ahead in strength. A few experiments on this point gave positive results in every case. A discrimination was established in which S^{D} was a click. The reflex $sS^{\Delta} . R$ was then reinforced continuously until the drive was so low that elicitation was interrupted (see Chapter Nine). When the rat had not responded for several minutes, the click was presented. Except for extremely low degrees of hunger, the click was invariably followed by a response. The response no longer occurred in the presence of S^{Δ} although it had been continuously reinforced, but it immediately occurred in the presence of S^{D}.

In Figure 59 the reinforcement of the extinguished member is compared with that in the case of a respondent. The comparison is arbitrary in several respects. The time elapsing between reinforcements is seven minutes in the case of the operant and fifteen in that of the respondent. The measure of strength is the rate of responding in the case of the operant and the amount of salivation during one minute in that of the respondent. The data for the latter are taken from Pavlov (64). A maximal value of about fifteen drops has been assumed and the beginning of the deceleration noted. The data for the operant are represented as the differential of a smooth curve drawn through Record B in Figure 58. The units on the ordinates (not noted in the figure) have been chosen to bring the maximal value near to that of the respondent. The strength in this case does not begin at zero but at the value prevailing on the previous day as indicated by the horizontal dotted line.

The two curves differ chiefly in the absence of any positive acceleration in the curve for the operant. Each reinforcement raises the rate by a definite uniform amount until the maximal is nearly reached. This is not quite true for the original development of the constant strength under periodic reconditioning (Figure 28) but is in harmony with the notion of a constant extinction ratio when periodic reconditioning has once been established. The respondent shows some delay in acquiring strength which may be due to 'extinction beyond zero'—an effect which is impossible in an operant. In interpreting these curves it should be noted that

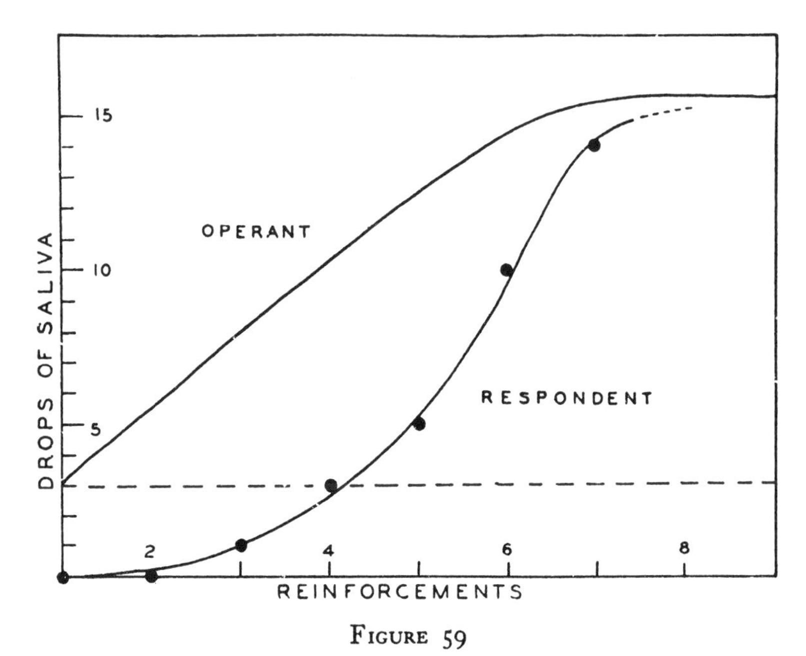

FIGURE 59

ABOLISHMENT OF A DISCRIMINATION THROUGH PERIODIC REINFORCEMENT OF THE EXTINGUISHED REFLEX

The operant curve is the rough derivative of Curve B, Figure 58. The horizontal broken line indicates the rate from which the change begins. The respondent curve is for data from Pavlov (64), p. 81.

they are for single cases, the typicality of which in the case of the respondent is not known.

The orderly return to a constant maximal rate during periodic reconditioning demonstrates that the extinguished reflex is not existing in a state of suppressed excitability (see below) and that toward the end of the discrimination its reserve is, as the rate indicates, nearly exhausted.

The State of the Reinforced Reflex

The rate of elicitation of the reinforced reflex during discrimination depends upon the rate of presentation of the discriminative stimulus and is therefore of no use as a measure of strength. But there are other means of exploring the state of the reflex. In a discriminated operant a certain time elapses between presentation of S^D and the occurrence of the response, which resembles the latency of a respondent and by extension may be called latency. It gives us some measure (though possibly an objectionable one) of the strength of the reinforced reflex during a discrimination. The first change to be observed is a shortening of the latency as the reinforced reflex rises in strength from the intermediate value prevailing during periodic reconditioning to an essentially maximal value. The change takes place quickly but cannot be easily followed. Before the reinforcement in the presence of S^D has had any effect, the 'latency' will be a matter of chance and will depend upon the average rate of responding at the time. Since the responses are not uniformly distributed the actual chance value cannot be calculated from the rate. The element of chance confuses the course of the change in latency on the first day of a discrimination. A group of sixteen rats in an experiment to be described in the following chapter gave as their average latencies for the first nine presentations (not counting the latency at the beginning of the experiment): 13.38, 11.84, 7.69, 8.25, 9.82, 5.76, 4.07, 6.38, and 8.88 seconds. The figures show a considerable scatter in spite of the fairly large sample, and they may be taken as indicating little more than that the better part of the change is accomplished upon the first two reinforcements. The average level may drop slowly during the succeeding days of the experiment, as will be shown later (Figure 70).

There are considerable differences between the latencies of discriminative responses to auditory and visual stimuli at the intensities used in these experiments. In one experiment discriminations were established in a pair of rats with periodic reinforcement in the presence of a light every five minutes. When the latency had stabilized itself, measurements to the nearest fifth of a second were taken for three successive daily periods of one hour. On the following day a buzzer was substituted as the source of the discriminative stimulus. The transfer from one stimulus to the other was readily made, although investigatory responding to the new stimulus produced a few long latencies on the first day. On the second and third days with the buzzer, measurements of the latencies were taken. Two more days with the light were recorded and then two more with the buzzer. Five sets of latencies were thus available for each rat with the light and four with the buzzer. Each set consisted of eleven latencies, since the first latency in each hour was obscured by the release of the rat. The average of all latencies in response to the light was found to be 5.12 seconds; in response to the buzzer, 1.97 seconds. Both rats showed a difference of this magnitude and there was no overlap in the ranges of the daily averages. This result has been confirmed by casual observation in all the other experiments on discrimination; the latency in response to a sound is considerably shorter than that in response to light.

Such a difference may be due to the intensity rather than to qualitative differences in the stimuli. The difference is great enough to suggest further experimentation on the relation between the latency and the intensity as well as on a comparison of the ranges obtained with maximal and minimal intensities in different modalities. Further data on discriminative latencies are given in Chapter Six.

We must now turn from the question of mere strength to that of reserve. During a discrimination $sS^{\Delta} . R$ is extinguished while $sS^{D} . R$ is increased to a nearly maximal strength. But what happens to the reserve of $sS^{D} . R$? Is the principle of an extinction ratio still valid, and are we, by virtue of the procedure of periodic reconditioning, building up a tremendous reserve which can find no expression because of the limited presentation of S^{D}? The question may be answered simply by extinguishing $sS^{D} . R$ after a discrimination has been established.

In an experiment on this point discriminations were set up in the usual way in a group of eight rats. S^{D} was a light, and the interval of periodic presentation and reinforcement was six minutes. On the seventh day $sS^{\Delta} . R$ had reached a low value. Record A in Figure 60 is a typical example of the rate obtaining at that time, where the vertical bars indicate the reinforced responses. Record B was taken on the experimental day immediately following, or on the eighth day of the discrimination. For 48 minutes the usual periodic reconditioning of $sS^D . R$ and the intervening extinction of $sS^{\Delta} . R$ were carried out. The rate shows no appreciable further decline on this day. After the eighth interval S^D was presented continuously and none of the subsequent responses was reinforced.

The extinction curve typically obtained in an experiment of this sort is relatively small. Its area is usually considerably less than

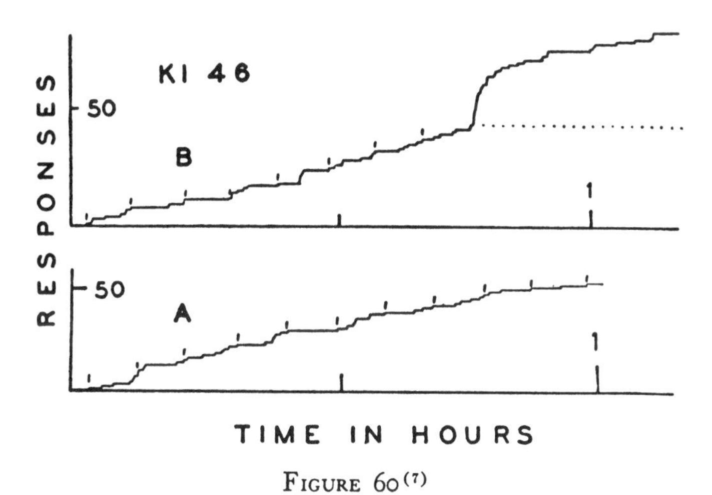

FIGURE 60 (7)

EXTINCTION OF THE REINFORCED REFLEX IN A DISCRIMINATION

A: rate in the presence of S^{Δ} in the seventh hour of a discrimination. B: the eighth hour of the discrimination, in which S^D was presented continuously, beginning at the dotted line, but not reinforced.

that of a curve for original extinction obtained after, say, fifty reinforcements and very much less than that of any curve after periodic reconditioning. It shows a rapid rising limb, similar to a curve for original extinction, and exhibits only slight traces of a cyclic fluctuation. The curve in Figure 60 is typical, except that it may be slightly larger than the average.

The following explanation of the failure to obtain a larger extinction curve might be advanced but is invalid. It will be shown later that the *change* from S^D to S^{Δ} may acquire the properties of a discriminative stimulus where either S^D or S^{Δ} alone is ineffective. It might be argued therefore that in presenting S^D continuously in this experiment we are not in fact maintaining the same discriminative stimulus that was previously correlated with reinforcement. The stimulus might have been the change from S^Δ to S^D rather than S^D itself. It will be shown later, however, that the point at which the change alone becomes effective must be reached through a special procedure. Moreover, the same result can be obtained with a discontinuous stimulus. Figure 61 is a record from such an experiment. In this case the discriminative stimulus was the sound of a click, which did not persist until reinforcement. In the experiment represented in Figure 61 a dis-

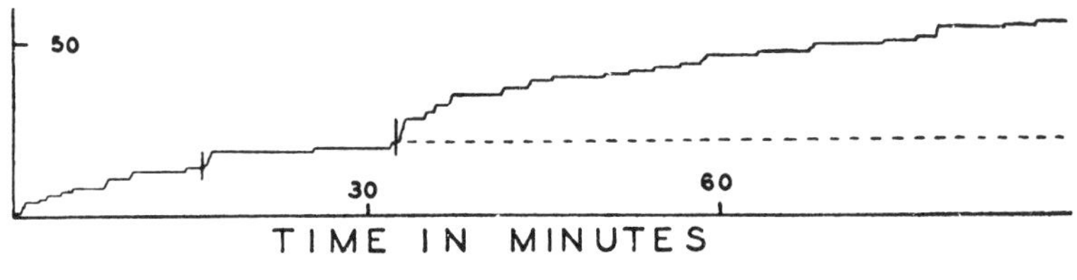

FIGURE 61

EXTINCTION OF THE REINFORCED REFLEX IN A DISCRIMINATION WHERE THE DISCRIMINATIVE STIMULUS IS MOMENTARY

The curve above the broken line is comparable with that in Figure 60, except that S^D (a click) was presented every two minutes without reinforcement.

crimination had been established by reinforcing a discriminated response every fifteen minutes (a relatively long interval). On the day represented in the figure a discriminated response was reinforced at the beginning of the period and after fifteen minutes. The record shows some spontaneous recovery at the beginning of the hour, followed by a low rate of responding. After thirty minutes the click was presented every two minutes without reinforcement. The first presentation led to several responses, the second to a few, and as the experiment progressed, the rate fell off very much as in the preceding experiment. Toward the end of the period the rate was so low that some presentations of the stimulus were not followed by responses. The area of the curve is of an order comparable with that obtained with a continuous stimulus, and this is a typical result.

It is clear that the periodic reinforcement of $sS^D . R$ during a discrimination is not functioning to create a reflex reserve such as we should expect from the notion of the extinction ratio in simple periodic reconditioning. But this is to be expected. It has already been noted that the extinction ratio does not hold when N_c is greater than 1. A few reinforcements distributed periodically will increase the reserve enormously beyond any possible increase from the same amount of continuous reinforcement. In the present experiment $sS^D . R$ is not reinforced periodically in the sense in which that term was used in the preceding chapter. It is characteristic of periodic reconditioning that some responses should go unreinforced; the periodic reinforcement is *re*conditioning. In a discrimination we reinforce *all* responses to S^D and we should not expect a very considerable reserve to be created. On the other hand, we must take account of an inductive effect tending to reduce whatever reserve there may be. The extinction curve obtained for $sS^D . R$ at the end of a discrimination is many times smaller than it would have been if obtained at the beginning of the process. Other examples of inductive extinction will appear shortly.

A Discrimination without Previous Periodic Reconditioning

The 'curve for discrimination' obtained with the foregoing method was seen to be simply the curve for the extinction of $sS^\Delta . R$ and to bear only a doubtful mark of the effect of induction. It is possible to modify the curve by reducing the initial reserve to be exhausted through extinction. One method is to dispense with the preceding periodic reconditioning and to begin with merely the continuous reinforcement of $sS^D . R$. A much smaller curve is obtained, and any induction from S^D to S^Δ should be more clearly revealed.

In an experiment of this sort four rats were conditioned as usual in the presence of a light and fifty responses were reinforced. On the following day the reflex was extinguished both in the light and in the dark. The rats were then put on the discriminative procedure without periodic reconditioning. A typical set of records is reproduced in Figure 62 (page 202). The first record shows some convexity at the beginning, which is due to spontaneous recovery from the extinction of the preceding day. The recovery is great enough to give the curve some of the character of that in Figure 35—namely, an extinction curve followed by a significant acceleration. There is a subsequent decline in rate which continues during the next four days. In the fourth record there is an anomalous burst of activity shortly after the beginning, and the total number of responses for the day is consequently high. The complete sets for all four rats and their average (the heavy line) are given in Figure 63 (page 203). The series from Figure 62 is marked D. By com-

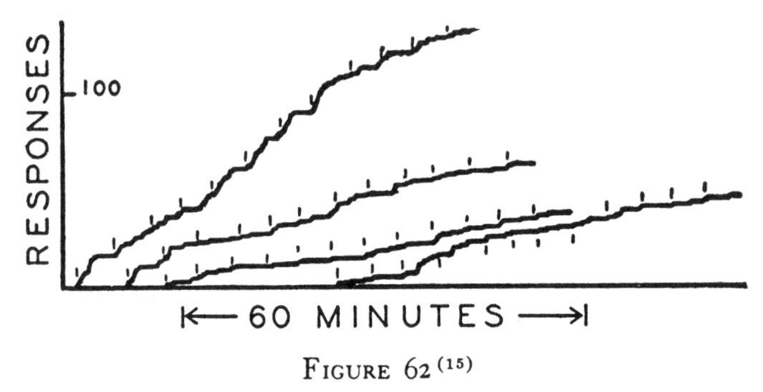

FIGURE 62 (15)

A DISCRIMINATION WITHOUT PREVIOUS PERIODIC REINFORCEMENT
The first four records for one rat.

paring these figures with Figures 54 and 55 it may be seen that a considerable difference in strength between the two reflexes is more quickly approached in the present case. The difference is not in a 'capacity for discrimination' but in the original reserve requiring extinction.

In this experiment there is clear evidence of induction and of a change in its extent during the process. The curves are greater than would be expected from the mere extinction of a previously established reserve, because the reserve was nearly exhausted prior to the beginning of the discrimination. The fairly high rate developed on the first day in the experiment is for $sS^{\Delta} . R$, which has never been reinforced. Its strength and reserve are partly due to induction from the previous continuous reinforcement of $sS^{D} . R$, the effect of which has not been entirely eradicated during the preceding extinction. This much is shown in the spontaneous recovery at the beginning of the records. The positive acceleration on the first day cannot be due to recovery but must be the result of the *concurrent* reinforcement of $sS^{D} . R$. The evidence for concurrent induction was not clear in the experiment following periodic reconditioning but is here unmistakable. As the experiment proceeds, the induction decreases in extent (otherwise the greatest observed slope would be maintained), and the ultimate state is presumably that of no induction and consequently no occurrences

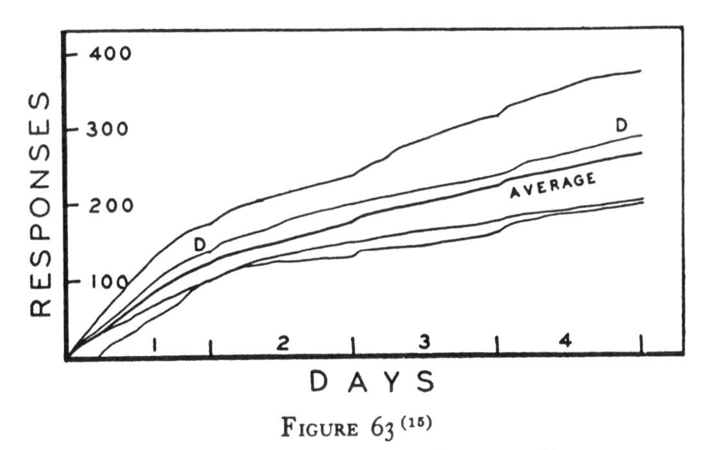

FIGURE 63 [15]

A DISCRIMINATION WITHOUT PREVIOUS PERIODIC REINFORCEMENT
Four sets of records and the average are shown. Curve D is from Figure 62.

of the unreinforced reflex. A similar initial induction and its eventual disappearance are indicated in the following experiment.

A Discrimination without Previous Conditioning

The time required to establish a relatively complete discrimination depends upon the initial reserve of the reflex to be extinguished. The reserve may be reduced still further than in the preceding experiment by beginning the discrimination after the reinforcement of only one response. To do so it is necessary to deal with relatively low strengths, where extraneous factors are likely to introduce irregularities, but the behavior in the present case is satisfactorily uniform.

Eight rats were given the usual training prior to conditioning. I will describe the subsequent procedure in connection with the typical result shown in Figure 64 (page 204), through five successive days for one rat. At the release of the rat on the first day the light was on. A response to the lever occurred quickly and was reinforced. The light and magazine were then disconnected. Some conditioning of $sS^D . R$ very probably occurred. If there were an inductive effect upon $sS^\Delta . R$ an extinction curve should have followed in the absence of the light. Actually, only two re-

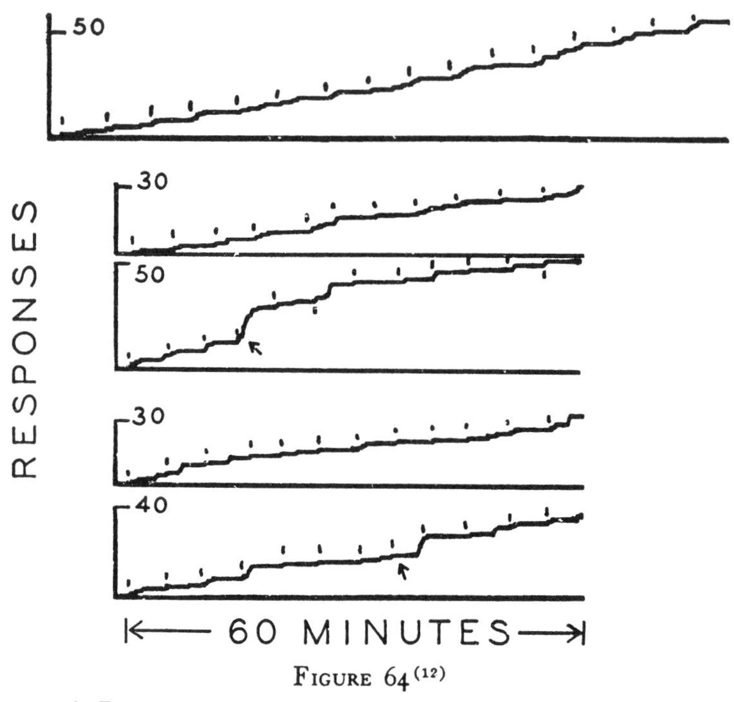

FIGURE 64 (12)

A DISCRIMINATION WITHOUT PREVIOUS CONDITIONING

Records for five successive days beginning at the top. Responses were reinforced at the dots in the presence of a discriminative stimulus. During the remaining time the stimulus was absent and no responses were reinforced. Little or no induction from S^D to S^Δ is to be observed.

sponses were forthcoming during the next five minutes, and it is difficult to say whether they show induction or are similar to the first response. When the light was turned on again, the rat responded after 39 seconds (at the second dot over the first record in the figure). Both light and magazine were then turned off, and two more responses in the dark occurred during the next five minutes. This procedure was repeated for 1½ hours, with the result shown in the figure. By the seventh reinforcement the latency of $sS^D . R$ had reached a more or less stable value of about 20 seconds, which was maintained throughout the rest of the experiment.

The strength of $sS^\Delta . R$ increased slightly through induction during the first 1½ hours to yield a rate of responding of about four responses per reinforcement. On the following day it dropped to an average of about two responses again, and continued to decline as the discrimination became more effective. On the third

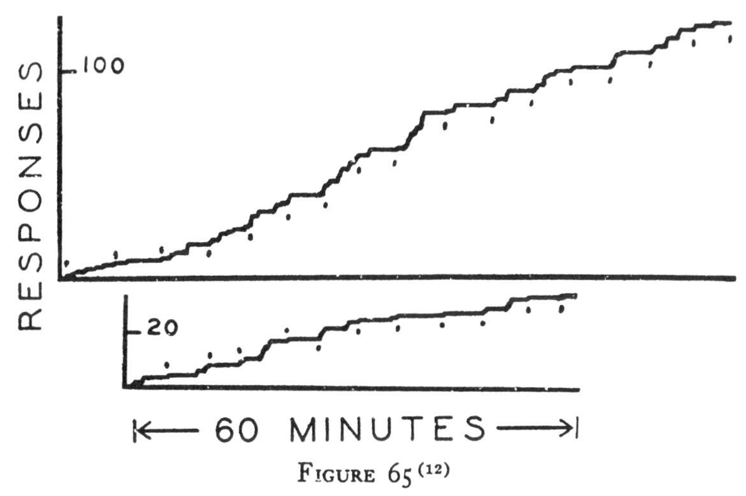

FIGURE 65 (12)

DISCRIMINATION WITHOUT PREVIOUS CONDITIONING

The figure is similar to Figure 64 and shows the greatest amount of induction in eight cases.

day an extinction curve appeared spontaneously after the fourth reinforcement, and a similar effect was observed in two other cases. It is as if the response at this point were made in the absence of the light—as if the reinforcement were applied to $sS^\Delta . R$. This anomalous effect may be accounted for by supposing either that the rat was in the course of a response to S^Δ when the light was turned on, or that it responded to the lever in the presence of the light but in such a way that the light was not an effective part of the total stimulus. In support of this explanation fairly similar curves have been obtained by allowing one reinforcement to take place in the dark. This was done at the arrows in the last records of Figures 64 and 66. In contrast with the spontaneous curves the extinction begins in these two cases only after a delay of approximately three minutes.

Whatever the explanation of this effect may be, it does not seriously disturb the conclusion that if the procedure of discrimination is instituted before either member has been conditioned, the discrimination may be regarded as essentially complete at the beginning. This is a final confirmation of the present interpretation of the process.

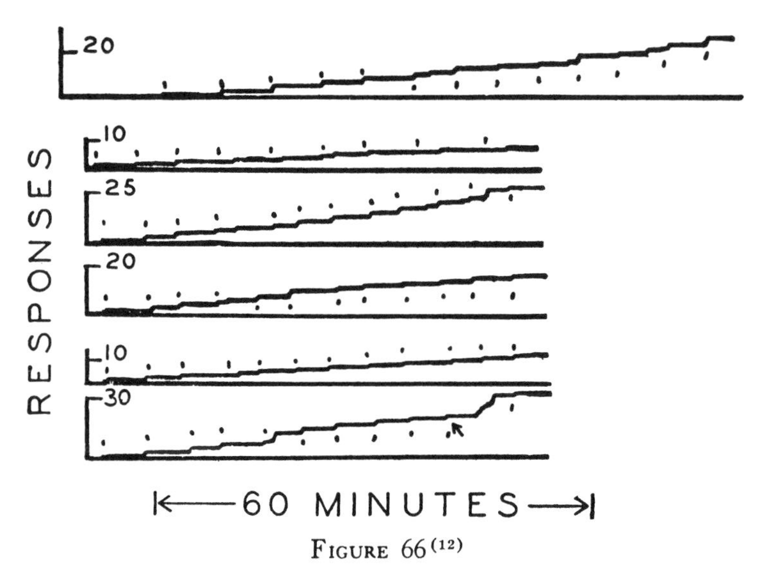

FIGURE 66 (12)

DISCRIMINATION WITHOUT PREVIOUS CONDITIONING

Compare Figures 64 and 65. A reverse induction from S^{Δ} to S^{D} nearly abolishes the reflex on the second day (second record from the top).

The experiment also shows that in some cases the inductive effect may be very slight. Figure 64 is typical of six out of the eight cases. Of the six the greatest induction was observed in the case of Figure 65. (This series was broken on the third day through a technical fault.) The least induction was observed in the two remaining cases, in which an induction in the opposite direction resulted in the early disappearance of all responses. In Figure 66 a similar inductive extinction nearly brought the series to an end in the last part of the second record, where the strength of $sS^D . R$ (as measured from its latency) fell severely. On the following day there was an adequate recovery, although the latency for the whole series was consistently high, averaging 41.0 seconds as against 15.5 for the other five rats. The average latency of 19.7 seconds for all six rats is approximately three times the latency observed with the preceding methods and indicates that with this method the extinction of $sS^\Delta . R$ has in general a more marked effect upon the strength of $sS^D . R$.

To control for any unconditioned difference in the rate of responding in the presence and absence of the light, half of the above cases were actually of the opposite sort. The experiment in Figure 64 was as described, but in Figures 65 and 66 S^D was the absence of the light and S^Δ its presence. The two cases in which the rate dropped to zero under negative induction were of the light-on-at-reinforcement type, and this agrees well with evidence to be considered shortly for a depressive effect of the light. Eight cases are not enough to establish the complete indifference of this condition for the present result, but it is apparent that no very great effect is felt.

An Attempt to Detect Induction at a Late Stage of Discrimination

The effect of induction could presumably be tested by comparing discrimination curves when the rates of periodic reinforcement of $sS^D . R$ were varied. The choice of five or six minutes as an interval between the reinforcements of $sS^D . R$ is purely arbitrary, and an adequate strength of $sS^D . R$ can be maintained with much less frequent reinforcement. This is shown incidentally in the following experiment, which is an attempt to detect the presence of induction at the later stages of a discrimination by varying the relative number of reinforcements. An advanced stage of a discrimination was reached quickly by beginning without previous periodic reconditioning, reinforcing $sS^D . R$ only every fifteen minutes, and increasing the daily experimental period to two hours. The apparatus was arranged so that two successive responses could be reinforced after presentation of S^D (a click). In two of the four cases two responses were reinforced in this way during the early development of the discrimination. In the other two cases one response was reinforced as usual. At the end of six days (twelve hours of experimentation) the average rate for $sS^D . R$ had reached the low value of about thirty responses per hour. The conditions were then reversed so that the first pair now received only a single reinforcement every fifteen minutes and the second received double reinforcement. After five days (ten

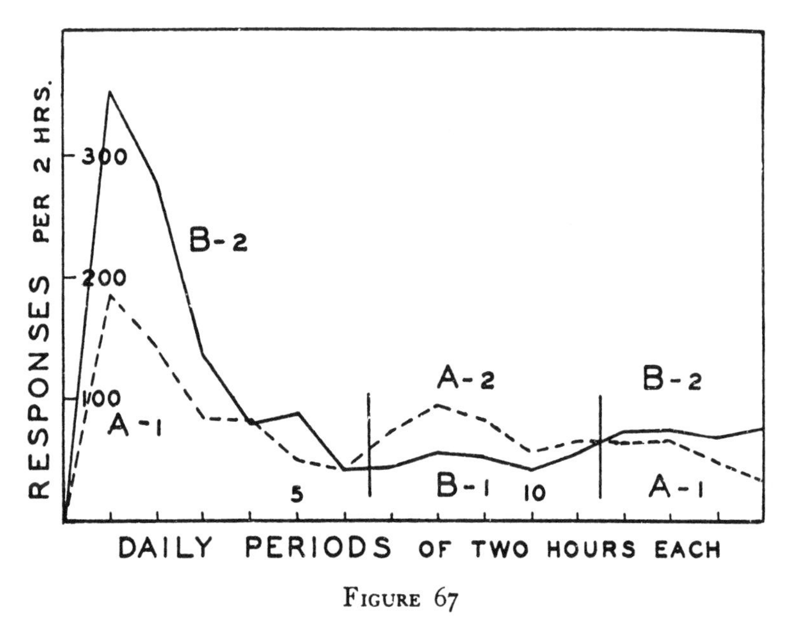

FIGURE 67

INDUCTION IN A DISCRIMINATION WITHOUT PREVIOUS PERIODIC REINFORCEMENT

Reinforcing two responses at each presentation of S^D, rather than one, increases the rate of responding in the presence of S^Δ. The double reinforcement is shown by '–2'.

hours) the original conditions were reinstated for four further days.

The result is shown in Figure 67 where the numbers of responses per two hours for each pair have been averaged. As in the experiment reported above on discrimination without previous periodic reconditioning, a fairly high rate of responding for $sS^\Delta . R$ is reached through the concurrent induction from $sS^D . R$, even at this low rate of reinforcement. The rate is nearly twice as great for the group receiving a double reinforcement. Although this may be partly due to sampling, the subsequent behavior indicates that individual differences cannot account for all of it and that it must be due to greater induction in the case of double reinforcement. Since the S^D did not persist until reinforcement, the reinforced responses of the first day with their long 'latencies' were practically $sS^{\Delta} . R$ and the effect is direct rather than inductive. The rates fall rapidly to approximately the same level at the end of six days, indicating that the curves are for extinction and show little or no persistence of induction. When the conditions are reversed, however, the values for the pair now receiving double reinforcement rise significantly above those for the singly reinforced group, and when the conditions are reversed again a similar change takes place. The averages and standard deviations in responses per hour for the two groups in the last two parts of the experiment are given in Table 8. The numbers in parentheses indicate the number of responses periodically reinforced. Although

TABLE 8

Days	7–11	12–15
Group A	(1) 26 ± 3.3	(2) 47 ± 3.4
Group B	(2) 39 ± 6.5	(1) 28 ± 6.3

the samples are small (two rats in each group) and the standard deviations fairly large, a higher rate of responding for $sS^{\Delta} . R$ is indicated when $sS^D . R$ is reinforced twice rather than once at each interval.

The result is not an unqualified sign of induction in the later stages of a discrimination because it involves another change in the behavior of the rat. The two successive reinforcements raise a problem on their own account, which I have already mentioned in connection with the chained nature of this sample of behavior. In reinforcing two responses after presentation of the click we set up two discriminative stimuli: sS^D*: click* (correlated with the first reinforcement) and sS^D*: click* $S^{D'}$*: reinforcement of preceding response* (correlated with the second). The reflex $sS^DS^{D'} . R$ becomes so strong that it is prepotent over the response to the food tray and is evoked before the rat seizes and eats the first pellet. Now, part of $S^{D'}$ is the mere act of pressing the lever, and this may be responsible for the increased rate in the presence of S^{Δ}. Under the procedure of reinforcing two responses in succession it is necessarily true that the proprioceptive stimulation from a response to the lever is occasionally correlated with the reinforcement of a subsequent response. Hence the occasional responses of the rat to S^{Δ} may supply enough of an effective S^{D} to be responsible for the increase in rate. This fault might be corrected to some extent by using an S^{D} which remained on during reinforcement, as was not the case with the click, but it will always remain to some slight extent an objection to the use of the data for the purpose for which they were here obtained. To vary the rate of reinforcement rather than the number reinforced upon each occasion is probably the better method, although it also is subject to certain disturbing effects. This alternative experiment has not yet been performed.

The 'latencies' for $sS^{D}.R$ in the first day of this experiment are interesting. The first value (for the second reinforcement, since the value at release is meaningless) is 9 minutes 15 seconds. Subsequent values fall with some regularity as follows: 4:31, 4:00, 1:08, 1:52, 0:52, and 0:11. With many methods of measuring discrimination this would be accepted as a discrimination curve. But the first part of the curve is very probably not due to the development of a discrimination at all since very little effect could prevail when the click is followed by a reinforced response only after 9 minutes. The drop in 'latency' is the result of the increase in rate of responding due to the practically direct reinforcement of $sS^{\Delta}.R$ and is simply a change in the probability that a click will be followed in such and such a time by a response. If the value at the end of the two hours (11 seconds) is a true discriminative latency, it must mean that a reinforcement after 52 seconds is capable of establishing the required connection. On the following day the values were too short to be explained as due to the rate of responding and obviously represented true discriminative latencies.

The Reversal of a Discrimination

I shall now consider two further properties of induction and another kind of discriminative stimulus at the same time by examining the way in which a rat acquires a second discrimination in which the conditions of correlation of the reinforcing stimulus are just the reverse of those in the first. For example, if S^D has been a light and S^Δ the absence of the light, the rat is required to reverse its previous behavior by converting the light into S^Δ and its absence into S^D.

In this experiment the superscript designation of the character of the stimulus with respect to its correlation with a reinforcement can no longer be used to imply the presence of a property of the stimulus itself. Consequently I shall specify both the property and the discriminative status. Thus, S^Dl = light correlated with reinforcement; $S^\Delta l$ = light not correlated; $S^D\lambda$ = absence of light correlated with reinforcement; $S^\Delta\lambda$ = absence of light not correlated with reinforcement. The procedure on the day of reversal is as follows. The first response is in the dark and is reinforced; the light is then turned on for five (or in some of the following experiments six) minutes and no response is reinforced. The light is then turned off, the next response reinforced, and the light turned on again; and this procedure is repeated for the rest of the hour.

It should be possible to predict the principal aspects of the behavior of the rat on the day of reversal from what is already known. After the development of a discrimination an extinction curve displaying certain characteristic properties is obtained if the previously reinforced stimulus is presented continuously without reinforcement. Except for the short periods of $S^D\lambda$ which occur every five or six minutes the light is on continuously after reversal, and an extinction curve for the response in the light should appear at the beginning of the day of reversal. The curve need not have so great an area as a curve for original extinction. Emerging from such a curve there should appear a curve of positive acceleration due to induction from the periodic reinforcement of $sS^D\lambda . R$. Curves of these two sorts sum algebraically when the conditions for their development exist simultaneously, as was shown in Figure 35, but the compound curve on the day of reversal should have one modification. The curve of positive acceleration should approach the curve of negative acceleration along which the new discrimination is to develop. In Figure 35 no discrimination was possible, since it was for simple periodic reconditioning, but in the present case it is to be expected. The resulting curve should be composed of at least three separate parts. In spite of its complexity it is actually obtained experimentally.

A group of eight male rats, approximately 110 days old at the beginning of the experiment, was put through the procedure for discrimination with an interval of reconditioning of six minutes. On the eighth day of the discrimination the conditions were re-

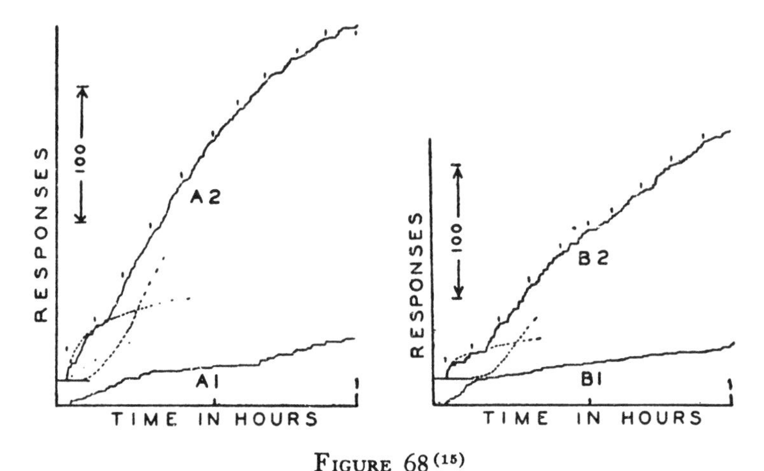

FIGURE 68 (15)

COMPOSITE CURVES OBTAINED DURING THE REVERSAL OF A DISCRIMINATION

The lower curves are for the last day of the original discrimination. The upper curves were obtained when the former S^D was made S^Δ and *vice versa*. Each curve contains (1) the extinction of the response in the presence of the former S^D (cf. Figure 60), (2) the positive acceleration in the presence of S^Δ due to induction (cf. Figure 63), and (3) the negative acceleration of the new discrimination.

versed. Two records for the day of reversal are given in Figure 68. The Records A1 and B1 are for the seventh day of the discrimination. Those marked A2 and B2 are for the day of reversal, and show the predicted characteristics. They begin with extinction curves, but curves of positive acceleration cut through these and lead to negative accelerations as the new discriminations develop. These are typical records. The result may be less clear when the rate fails to return to its full value under periodic reconditioning (two out of eight cases in the present experiment) but from the nature of these exceptions (see below) they should not affect the correspondence with the predicted forms.

Four typical records for the whole process are reproduced in Figure 69. They were obtained by plotting the end points of the daily records and copying the intervening records free-hand. The first three days in the figure are for periodic reconditioning at

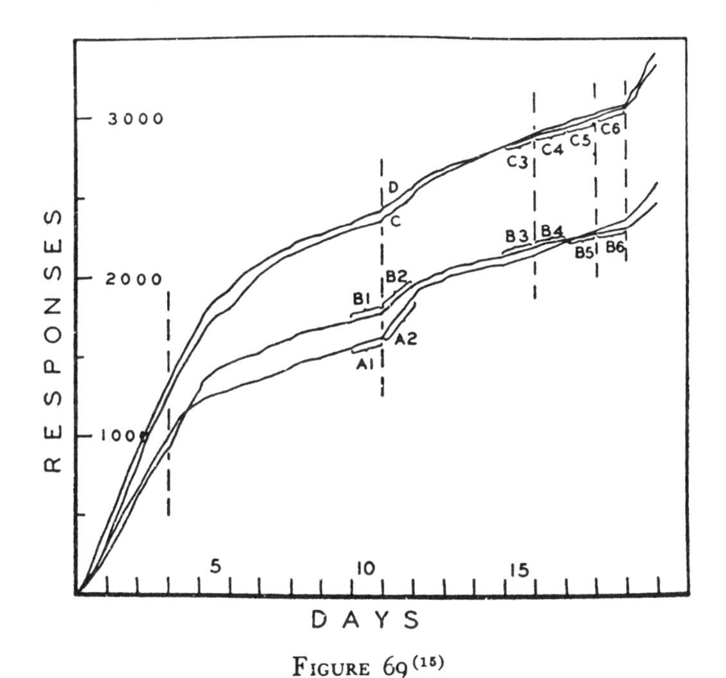

FIGURE 69 (15)

SUCCESSIVE REVERSALS OF A DISCRIMINATION

At each reversal the previous S^D became S^Δ and *vice versa.* Three reversals are shown, at the second, third, and fourth vertical lines. The rates under periodic reinforcement are shown at beginning and end.

six-minute intervals. On the fourth day (at the vertical broken line) the discriminatory procedure was begun, and the next seven days show the development of the discrimination. The minor deviations in these records resemble those described above (p. 186). The daily records A1 and B1 in Figure 68 are shown by brackets in the present figure. At the second broken line the conditions were reversed. The detailed records A2 and B2 in Figure 68 are also bracketed. In the other two curves the return to the periodic slope is slower and never becomes fully developed, but the characteristics of the records for reversal are apparent in spite of the considerable reduction required by the figure. The following five days show the development of the second discrimination, at the end of which the rate of elicitation of the extinguished member has reached a low value. (As I shall note again later, it is not as low as at the end of the first discrimination.)

The curves obtained in the reversed discrimination should have some of the properties of those shown above in Figures 62 and 63. The slight convexity given to the latter was the result of some spontaneous recovery from extinction. Here the convexity is for extinction after a discrimination. In both sets any positive acceleration must be due to concurrent induction, and consequently the maximal slope and the area of the curve will not be as great as that observed after periodic reconditioning. In the present case the areas of the first and second curves should differ by the amount contributed to the first by the periodic reconditioning that preceded them and for which there was no counterpart in the second. The original curve includes the extinction of the effect of three hours of periodic reconditioning; the reversed curve includes, in addition to the extinction of the response in the light, only as much of an inductive effect as can take place before the induction breaks down in the new discrimination. Since the areas are not comparable, the most significant effect of the reversal is the shape of the records for the first day, which reveals clearly enough the nature of the change taking place.

Another important effect of reversal is felt in the state of the reinforced reflex. As in the preceding experiments information as to the state of this reflex may be obtained from its latency, which is measured (in the present case to the nearest second) as the interval elapsing between the introduction of the light and the appearance of the response. The latency reaches a fairly stable value on the first day of the discrimination, and the average for this day is well below the average chance value obtaining under periodic reconditioning. There is a slower daily fall thereafter, which may persist in some slight degree as long as the experiment is carried on (for at least 30 days).

The average latencies for the eight rats in the present experiment (nine latencies per rat per day) are represented by the open circles in Figure 70. The series for the original discrimination begins at the already reduced value of 9.2 seconds. A significant drop is to be observed during the following two or three days. After

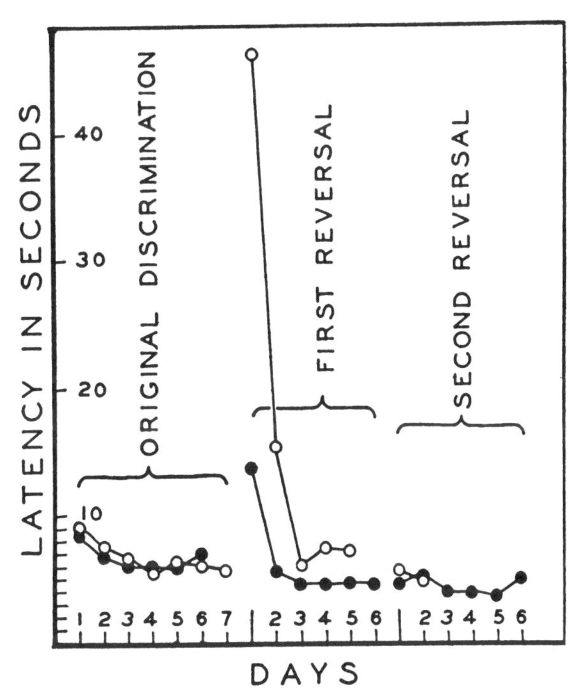

FIGURE 70 [15]

'LATENCIES' OF DISCRIMINATIVE RESPONSES DURING REVERSALS

Two sets of data are shown. The latency fell slightly during the original discrimination. At the first reversal unusually long latencies to S^D (previously S^Δ) were obtained. At the second reversal no lengthening occurred.

the reversal the latency is measured for the response in the dark ($S^D\lambda \,.\, R$) which has previously been extinguished and is therefore at a very low strength. The latency begins at an extremely high value; the average for the first day is 46.0 seconds. The reconditioning is slow, and the effect of the previous extinction is still apparent on the second day where the latency has fallen to only 15.5 seconds. On the third day, however, the strength reaches approximately the value held by the reinforced reflex in the original discrimination and continues approximately at that value thereafter.

If we now reverse the discrimination a second time, we return, of course, to the original set of conditions. There seems to be no reason to expect *a priori* that this will not involve another complex curve of the type already obtained. The response in the dark

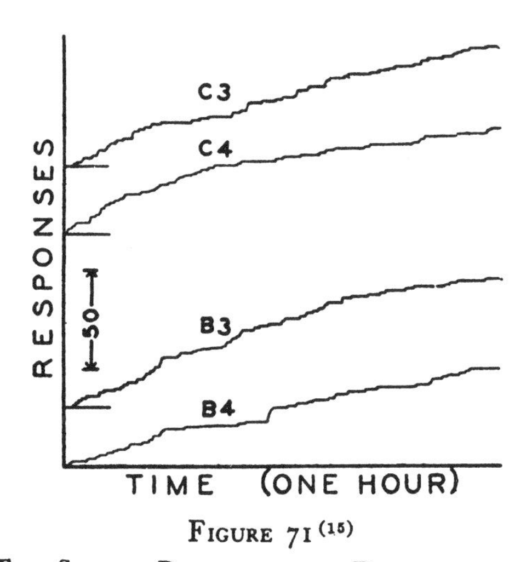

FIGURE 71 (15)

THE SECOND REVERSAL OF A DISCRIMINATION

Individual records from Figure 69 showing no effect at a second reversal comparable with that at the first reversal shown in Figure 68.

has now been reconditioned and should be extinguished, although the curve for extinction might be considerably smaller. The response in the light is at a low strength and should return to essentially its strength under periodic reconditioning. The experimental result, however, which is without exception, shows no such effects of a second reversal. In the present experiment the reversal was made after the new discrimination had been in force for five days. Two pairs of records for the days immediately before and after the second change are given in Figure 71, and their position in Figure 69 is indicated by brackets. It is apparent in comparing these records with Figure 68 that there is no initial increase in rate on the day of reversal that could be regarded as an extinction curve for the previously reinforced reflex, nor is there any positive acceleration with subsequent decline in rate. Some slight effect may be detected by averaging the total number of responses per hour for all eight records. The averages for the three days prior to the second reversal and the two days following it are as follows:

. . . . 96, 92, 88 || 95, 79 . . . .

Toward the end of the discrimination the slope was falling at the rate of about four responses per hour. On the day of reversal there was an increase of seven responses (eleven above the expected number), but on the second day the rate had dropped to approximately what it would have been without reversal. The only significant change, then (if this is to be taken as significant), is a slight increase in rate, apparently distributed evenly throughout the hour.

Similarly there is no change in latency comparable with that observed upon a first reversal. This is apparent in Figure 70 where the latencies are given for two days after the second reversal. (There is an apparent omission of a day at the end of the second series because the graph has been spaced out to accommodate a second set of data to be described shortly. The three series were taken continuously.) It will be seen that the rise in latency at the first reversal has no counterpart at the second.

Upon changing the conditions a third time (returning now to the reversed set) we again obtain no effect, as shown at the fourth vertical line in Figure 69. Two pairs of records are given in Figure 72 (page 218) and their positions are indicated in Figure 69 as before. The number of responses on the day of the third reversal was 76, which follows perfectly in the declining series given above. The latency also undergoes no significant increase, although this is not shown in the figure.

At the end of the experiment a check was made against a possible decline in rate for some unknown reason by returning to simple periodic reconditioning. The recovery of a constant slope is shown on the last day in Figure 69. It is clear that a significantly high strength is quickly developed. The average slope is not, however, equal to that originally observed, and it is in general true that after prolonged discrimination the acceleration toward the normal slope under periodic reconditioning is retarded. I have already noted examples of this in Records C and D in Figure 69, although the effect might there have been due to the reversal. Four other cases, obtained after a long series of discriminations to be described shortly, are given in Figure 73. The typical curve (K) for return to the periodic slope after a short discrimination included on the same coordinates is from Figure 58. The broken lines at the beginning give the slopes reached at the end of the discrimination. Those at the end give the original slopes under periodic reconditioning. It will be seen that only one of the four

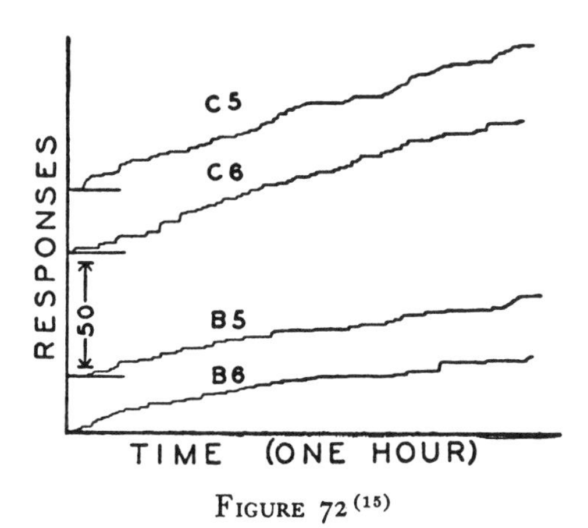

FIGURE 72 (15)

THE THIRD REVERSAL OF A DISCRIMINATION

rats reaches its original slope on the first day. The others show fairly smooth but greatly retarded accelerations, and one at least has not reached the required slope by the end of the sixth day.

The difference between the effects of a first and a second or third reversal demonstrates a new fact about induction. A partial explanation of the difference may be given in terms of previous results. So far as the first limb of the curve (Figure 68, Records A2 and B2) is concerned, it has been shown that extinction after discrimination may yield a smaller curve than original extinction, and it is not unreasonable to suppose that a second curve would average no more than seven responses in one hour and that a third would be negligible. The absence of a positively accelerating limb at the second reversal is the more important result. It might seem to be referable to the retardation in Figure 73 and to the assumption that, if the acceleration can be put off until the new induction is broken down, it will not occur; but since a certain amount of extinction is presumably necessary for the breakdown of induction, this explanation will not hold. The simple fact is that no induction occurs upon the second reversal.

It is a mistake to identify induction with the rate of responding or the breakdown of induction with the decline in rate observed during a discrimination. The separation of the strengths of two reflexes could be accomplished while the induction remained un-

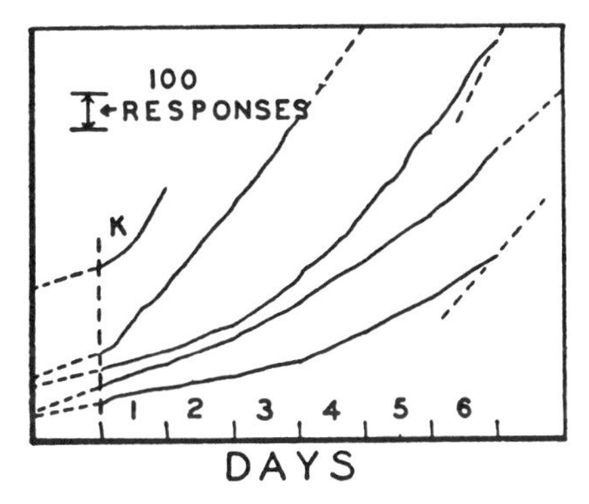

FIGURE 73 (15)

RETURN TO PERIODIC REINFORCEMENT AFTER THREE REVERSALS OF A DISCRIMINATION

The acceleration is, with one exception, much slower than after a single brief discrimination. The curve at K is from Figure 58.

changed. But the degree of induction does vary. For the value of S^D used here it is practically complete during continuous reinforcement, but persists only to a slight extent after differential reinforcement has been begun. The surviving induction disappears quickly and there is no reason to believe that it is effective in any appreciable degree during the last stages of the extinction of $_sS^{\Delta}\,.\,R$ (for the original strength of which it was responsible). In the present experiment the increase in rate at the first reversal must (except for the initial extinction curve) be due to concurrent induction. The increase is identical with that in Figures 62 and 65. At the second and later reversals no similar increase is observed. The facts regarding induction disclosed by the experiment may be stated as follows: (1) The breakdown of induction from Sl to $S\lambda$ (as effected in the original discrimination) does not affect the reciprocal induction from $S\lambda$ to Sl, which must be broken down separately; (2) The separate breakdown of induction from $S\lambda$ to Sl does not restore the induction from Sl to $S\lambda$.

But the fact that the breakdown of induction is irreversible will not wholly account for our present observations. A second reversal should still yield (*a*) the extinction of the previously reinforced member (observed as a small extinction curve on the day of reversal) and (*b*) the reconditioning of the previously unreinforced member (observed as a shortening of its initially long latency). The requirement of a small extinction curve is approximately satisfied by the seven extra responses at the second reversal, but not at later reversals, and the requirement of an initially long latency with subsequent reduction is not satisfied at any reversal after the first. Thus, in Figure 70 no significant increase in latency accompanies a second reversal. This is true not only of the average for the hour but of the first few reinforcements. In another series of experiments (to be described shortly) a reversal was made in the middle of the hour in order to show the simplicity of this change. Four typical records showing three successive days for each rat are given in Figure 74. The upper record in each group shows the last day of a discrimination in which responses in the absence of the light are being reinforced, the middle record shows the first day of (a third) reversal, and the lower record a continuation of the reversed discrimination until the middle of the hour, when the conditions were reversed again. On this last day the light was turned on as usual before the reinforced response at the vertical line. It was then *left on* for the succeeding interval. When it was turned off again a response followed immediately, and it was then turned on. During the rest of the hour $sS^{D}\lambda . R$ was periodically reinforced. The records show no significant change correlated with this reversal. The average latencies for eight rats for the eleven reinforcements during the hour (omitting the reinforcement at release) are as follows:

6.0, 3.4, 5.3, 5.6, 3.0 || 6.4, 4.4, 5.7, 7.0, 5.4, 5.0.

The latency of 6.4 for the first discriminatory response after reversal cannot be taken as significantly above the average for all cases (5.2).

A more crucial experiment on this point was made with the

same eight rats on a later day, when the conditions of the discrimination were changed at every interval during the hour. At release the light was on and the first response reinforced; the light was left on for five minutes, then turned off. A response immediately followed. The light was left off for five minutes, then

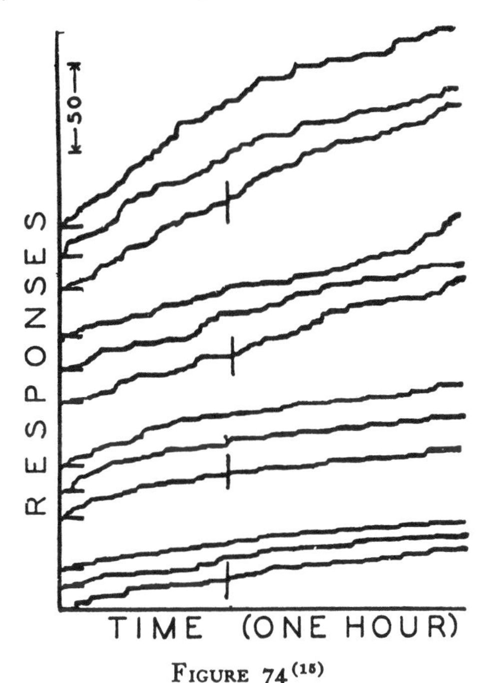

FIGURE 74 [15]

REVERSAL OF A DISCRIMINATION IN THE MIDDLE OF AN EXPERIMENTAL PERIOD

At the vertical mark in each set the properties of S^D and S^Δ were interchanged.

turned on, when another response followed—and so on. In one group of four rats the average latency for an hour of this procedure was 5.1 seconds as compared with 5.0 seconds for the previous day under the usual procedure. In the other group the latency was seven seconds as compared with five seconds on the previous day. The increase in the latter case was due to five especially long latencies (averaging about 25 seconds each) which appeared anomalously in the records of two of the rats. While they are probably significant, they do not seriously affect the conclusion that an increase in latency such as would be required by a theory of rapid reconditioning at each new discrimination is not observed. Similarly the slopes of the eight records show no effect of extinction, such as would also be required; the averages for the two days prior to the day of repeated reversal were 135 and 116 responses respectively. On the day of reversal the average was 110 responses. (These relatively high values are due to the late stage of the experiment. As I have already noted, the slopes approached by later discrimination curves progressively increase.)

This additional property of the behavior following a second reversal may be accounted for by saying that the effective basis for the discriminatory response has become the *change* from one stimulus to the other. The observed fact is that both $S\lambda \rightarrow l$ and $Sl \rightarrow \lambda$ are effective discriminatory stimuli, while Sl and $S\lambda$ are ineffective. This is the condition in which a change is the effective basis for a discrimination. We have then two questions to answer: (1) why is the condition found already established at the second reversal, and (2) why is there no inductive influence of $sS\lambda \rightarrow l \,.\, R$ upon $sSl \,.\, R$ nor of $sSl \rightarrow \lambda \,.\, R$ upon $sS\lambda \,.\, R$—or, in other words, why are the discriminations between $S\lambda \rightarrow l$ and Sl and between $Sl \rightarrow \lambda$ and $S\lambda$ found already in existence after discriminations between Sl and $S\lambda$ have been set up in both directions?

The first point is not difficult. In the original discrimination the reinforced response is really to $S\lambda \rightarrow l$ (if we are now to make this distinction throughout). The extinction, on the other hand, is of the responses to $S\lambda$, not $Sl \rightarrow \lambda$, since the rat is eating for fifteen or twenty seconds after the change back to $S\lambda$ is made. After the first reversal $Sl \rightarrow \lambda \,.\, R$ is periodically reinforced and $sSl \,.\, R$, *not* $sS\lambda \rightarrow l \,.\, R$, is extinguished. The second reversal therefore finds $sS\lambda \rightarrow l \,.\, R$ and $sSl \rightarrow \lambda \,.\, R$ previously reinforced and not subsequently extinguished, and $sS\lambda \,.\, R$ and $sSl \,.\, R$ extinguished and not subsequently reinforced, which is the required condition.

The second point, the lack of induction, is apparently not to be accounted for on any previously established principle. I shall leave it in the form of a descriptive statement as one of the results of the experiment: with the present technique, in which reinforcement occurs very soon after presentation of the stimulus, it is possible to obtain a discrimination based wholly upon the change (as here defined) by exhausting possible discriminations between the stimuli themselves.

[That a stimulus can depend upon the temporal proximity of another stimulus for its effectiveness is shown in the case in which $S l$ is a click. There is no significant difference as a discriminative stimulus between a light which remains on during reinforcement and a sound which is presented *before* reinforcement but never simultaneously with it. If we write the sound as $S\lambda \rightarrow l \rightarrow \lambda$, its relation to the present 'change' will appear. A different formula-

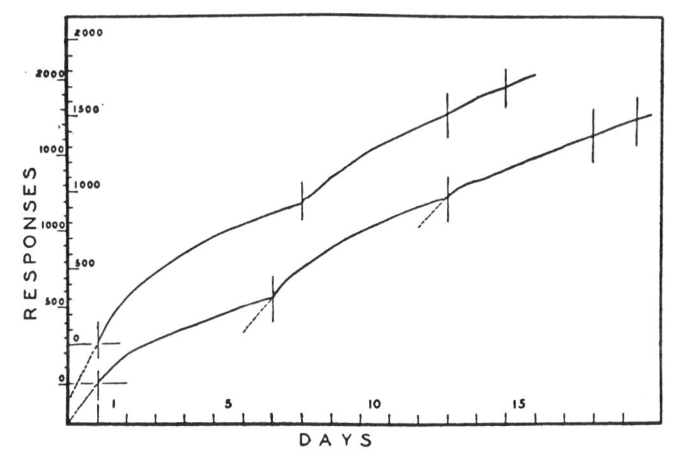

FIGURE 75 (15)

AVERAGED CURVES FOR TWO EXPERIMENTS ON THE REVERSAL OF A DISCRIMINATION

tion is required in the case of the change because of other possible bases for the discrimination.]

In another series of experiments an attempt was made to separate the factors in the curves for the day of reversal by interpolating a day of periodic reconditioning. The records for a group of eight rats (90 days old at the start) have been averaged to give the lower curve in Figure 75, in which the average curve for the eight rats in the preceding experiment is also given. To aid the comparison the interpolated days of periodic reconditioning have been omitted from the main curve and inserted beneath the last days of the preceding discriminations. The broken lines give average slopes only, as no attempt has been made to show the initial positive acceleration. The curve may be described in detail as follows. The first day shows the average slope for the last of four days of periodic reconditioning (of $sS\lambda . R$) at intervals of five minutes. The slope is somewhat less than for the other group. The following six days of the figure show the development of a discrimination between $sS^{D}l . R$ and $sS^{\Delta}\lambda . R$. On the seventh day Sl is present continuously, and responses to it are periodically reinforced; the recovered slope is indicated with a broken line on the sixth day of the figure. The separate records resemble Figure 35, the positive acceleration approaching a straight line.[3] The total slope is somewhat less than that observed before discrimination, but this is mainly due to the time required for acceleration rather than to the retarded acceleration noted in Figure 73. The reversed discrimination begins on the following day from the new periodic slope, and is well developed at the end of six days. Another day of periodic reconditioning (of $sS\lambda . R$) follows, and the resulting slope is shown with the broken line on the twelfth day. The retarded acceleration is now beginning to be felt.

This second interpolated day of periodic reconditioning forces the reconditioning of $sS\lambda . R$, which could not occur in the first experiment because of the lack of induction. Accordingly, when the discrimination is reversed again, a significant increase in slope is observed. The average number of responses for the three days prior to the day of periodic reconditioning, the average for the latter (in parentheses), and the average for three days of the new discrimination are as follows:

85, 72, 62 || (192) || *99, 85, 80*

The extra responses after reversal are chiefly due to the periodic reconditioning of $sS\lambda . R$ on the interpolated day. That they are not supplemented by any considerable concurrent induction from $sSl . R$ is clear from the quick adjustment to a constant slope, though this slope, as I have already noted, is increased by the reversal.

The average latencies for this series of experiments are given as the solid circles in Figure 70. There is a significant increase at

[3] All records of periodic reconditioning, particularly on the first day, are subject to chance variations depending upon whether or not the rat responds soon after the magazine has been turned on. If it does not (and this is not under control) an orderly periodic reinforcement is, of course, impossible. Only rarely is the general course of the record obscured.

the first reversal, as before, but none at the second. The increase at the first is much less than in the previous experiment, because of the day of periodic reconditioning.

As a control, the conditions were reversed a third time without returning to the periodic slope. The result is shown in Figure 75 at the fourth vertical line and fully confirms the previous finding. On the following day the conditions were reversed in the middle of the hour, as already described (Figure 74).

A gradual increase in the final slopes attained in successive discriminations is clearly indicated in Figure 75. It is of considerable importance. The best evidence for the breakdown of all induction during an original discrimination is that the rate of responding to S^{Δ} approaches zero and does so as rapidly as if $sS^{D} . R$ were not being periodically reinforced. If there were any persistent induction from this periodic reinforcement, the rate should approach a constant value greater than zero, which would yield an extinction ratio expressing the number of responses of $sS^{\Delta} . R$ induced by one reinforcement of $sS^{D} . R$. No evidence for such a ratio (and hence no evidence for a persistent induction) has been found for the value of S^{D} here used, until the present experiment, where it makes its appearance after the first reversal. The curves in Figure 75 show very little tendency to reach a horizontal asymptote. On the contrary they seem to be stabilizing themselves at slopes considerably above those at the ends of the original curves. The constant output of responses to S^{Δ} cannot be due to any prior reserve and must indicate concurrent induction. It is necessary to qualify the second rule of induction given above (page 220), to this extent: the reversal of a discrimination seems to establish a small permanent induction in both directions.

The Absence of a Stimulus as a Discriminative Property

In one or two other experiments described in this chapter it was noted that S^{D} was the *absence* of a stimulus. Cases of this sort were occasionally introduced as controls against a possible difference between the effect of the correlation of S^{1} with the presence of a discriminative stimulus and of that with its absence. The possibility arises from the fact that the rat is less active in the light, from which it follows that the rate will be lower during a discrimination in which S^{Δ} is Sl and is present most of the time.[4] In the preceding experiments a control was supplied by dividing the animals between the two types. Half the cases in the experiment reported above were actually of the opposite sort, and in these cases l should be read λ and *vice versa*. In Figures 68 and 74 all records were as described. In Figure 69 the two lower curves were as described, the two upper were the opposite; the apparent correlation, which is deceptive, is due to the periodic slope which

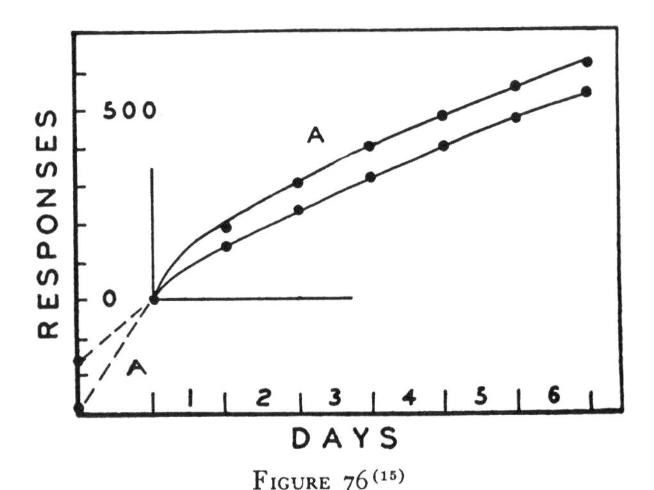

FIGURE 76 (15)

COMPARISON OF DISCRIMINATION CURVES WHERE THE S^D's ARE THE PRESENCE AND THE ABSENCE OF A LIGHT

The presence of the light was S^D in Curve A; hence, most of the curve was recorded in the dark.

happened to be greater in the latter case. In Figures 70 and 75 each point or curve represents an equal number of records of each type. The cases of Figures 71 and 72 can be found from Figure 69.

No significant difference is to be observed between the two cases which is great enough to disturb the present conclusions. The eight cases in the lower curve in Figure 75 are convenient for comparison because each case includes three discrimination curves of equal length (six days). Sorting these 24 curves into the two

[4] The effect of the light upon the rate does not mean that the reserve is being abnormally increased by a reduced output. The effect of the light may be to modify the drive, and as I shall show later the input and output of responses under a lowered drive is nicely balanced to maintain the state of the reserve.

kinds and averaging, we obtain Figure 76. In the curve at A the light was off during periodic reconditioning and during most of the discrimination. In the other curve the opposite was true. The difference in slope during periodic reconditioning (dotted lines) demonstrates the depressive effect of the light, although it may be to some extent due to sampling. The beginning of the discrimination shows a difference corresponding to these slopes, which is quickly lost as the rate falls. The effect of the light may be said to be approximately proportional to the rate of responding. The greater parts of the curves are nearly parallel. There is no significant difference in latency. Where the change prior to reinforcement was $l \rightarrow \lambda$, the average latency was 5.86; where it was the opposite the latency was 5.39.

Summary of the Relation of Induction to Discrimination

I shall now try to bring together the preceding data as they apply to the question of induction and its relation to the discrimination of the stimulus in an operant. A discrimination is defined as the process of creating a difference in strength between two related reflexes. The process would not arise except for induction; for the extinction of an independent reflex that has been independently conditioned is not a discrimination. The facts so far accumulated are as follows:

The repeated continuous reinforcement of an operant in the presence of S^D creates a reserve affecting the same response in the presence of S^Δ, although the extinction curve in the presence of S^Δ is not quite as great as in the presence of S^D.

For the gross value of S^D used in these experiments the extinction of $sS^\Delta . R$ may take place with little interference from the concurrent periodic reconditioning of $sS^D . R$, when the discrimination begins with a considerable reserve. For the sake of uniformity with the following facts it may be assumed that some induction takes place after the process of discrimination has begun but that the extensive extinction curve obscures it. When the initial reserve of $sS^\Delta . R$ is not so great, the effect of induction from concurrent conditioning of $sS^D . R$ is obvious. The induction is broken down apparently very early in the process of discrimination.

When neither reflex has previously been conditioned, there may be no interference whatsoever. In the optimal case the rat may make the required distinction from the start, with practically no induction between the processes and consequently without need of extinction. From this optimal mean position the result may diverge in two directions: (1) toward inductive conditioning, when the strength of $sS^{\Delta}.R$ is built up slightly and must be extinguished, and (2) toward inductive extinction, when both reflexes eventually disappear in spite of the reinforcement of $sS^{D}.R$.

The breakdown of the induction from S^D to S^{Δ} does not affect the reciprocal induction from S^{Δ} to S^D.

The separate breakdown of induction from S^{Δ} to S^D does not restore the induction from S^D to S^{Δ}, but a small permanent induction operating in both directions is created.

In no case has evidence been obtained that the rat comes to distinguish between two stimuli with respect to which its behavior did not already differ prior to the differential correlation with a reinforcing stimulus. This may be due to the fact that I have used a value of S^D great enough to permit detection of an original difference between direct and inductive conditioning. Whether the same laws hold for stimuli differing less grossly cannot be determined from the present evidence. But I may point out that failure to detect a difference in behavior prior to discrimination is no argument against the present interpretation of the process as essentially an accumulation of small differences, at least until techniques have become rigorous enough to make the failure a reasonable indication of the absence of such differences.

A formal expression of the process of discrimination may be given as follows:

THE LAW OF THE DISCRIMINATION OF THE STIMULUS IN TYPE R. *The strength acquired by an operant through reinforcement is not independent of the stimuli affecting the organism at the moment, and two operants having the same form of response may be given widely different strengths through differential reinforcement with respect to such stimuli.*

Discrimination and the Reflex Reserve

The preceding discussion has been given in terms of reflex strength only, and some recapitulation of its more important points may be required in terms of the reflex reserve.

In conditioning an operant we create a reserve which has the dimensions of a number of responses eventually to be emitted. The form of the response is independent of the discriminative stimulation that may happen to be active at reinforcement, and the total number of available responses may be assumed to be similarly independent. With respect to the stimulating field we observe that responses occur more readily in the presence of the exact stimuli that were present at reinforcement. The discriminative field at the moment of emission acts as a sort of patterned filter: if it matches the field existing at the time of reinforcement, the rate of responding is maximal; if it does not, the rate is depressed. Figure 53 suggests that even with a sub-optimal filter all responses in the reserve would be emitted if time allowed, even though a certain number of responses tend to be restrained. Restrained responses may be quickly evoked by changing to the optimal field, as at the arrows in Figure 53.

In a discrimination, regarded as a mere widening of the difference in strength in two fields, we build up the restrained part of the reserve which is under the control of S^D while permitting the responses available under S^Δ to be dissipated. Figure 60 is thus closely related to Figure 53. The breakdown of induction described above is not essential to this process, but it contributes to it by sharpening the filtering action of the stimulating field. The following statements may clarify this question:

Inductive conditioning. Responses accumulated under S^D are available under S^Δ.

Inductive extinction. Responses emitted under S^Δ are subtracted from the reserve available under S^D.

Discrimination. Reinforcement under S^D and extinction under S^Δ increase the number of responses available chiefly under S^D.

Breakdown of induction. Responses acquired under S^D may become less readily available under S^Δ.

The notion of a filter is, of course, merely a convenient device for representing the observed dependence of the rate upon external discriminative stimuli. In spite of differences in rate under different external fields a single reserve is involved so long as the form of the response remains the same. This is important enough to be given a formal statement.

THE LAW OF THE OPERANT RESERVE. *The reinforcement of an operant creates a single reserve, the size of which is independent of the stimulating field but which is differentially accessible under different fields.*

The lack of an eliciting stimulus in operant behavior together with the law of the operant reserve throws considerable weight upon the response alone, and this may seem to weaken any attempt to group operants under the general heading of reflexes. It is well to allow for a possible originating event (*cf.* the small *r* in our paradigms) and to provide for the other ways in which a response may be related to a stimulus. Nevertheless, it should be understood that the operant reserve is a reserve of *responses,* not of stimulus-response units. Whether the same can be said for respondents is not clear. If there is a similar law of the respondent reserve, following the form of the response, it would explain much of the heightened induction between homogeneous reflexes reported by Pavlov and account in another way for such a phenomenon as a 'transfer of inhibition' from one 'reflex' to another.

The formulation and the general experimental approach which I have described here are not traditional, but their idiosyncrasies arise from a difference in purpose rather than from any actual disagreement with traditional work. Most experiments involving discrimination are concerned with evaluating the least differences between stimuli discriminable by the organism. They are directed toward the measurement of a sort of capacity, and the process of discrimination enters into the experiment only as part of its method. No further inquiry is made into the process itself than is necessary in obtaining some convenient indication of the existence of differential behavior. Here I have not been concerned with the *capacity* of the organism to form a discrimination. The stimuli show relatively gross differences, and my main object has been rather to follow the course of the development of a discrimination and to determine as accurately as possible the properties of the process.

The present analysis does, I think, indicate a certain inadequacy in traditional methods so far as the dynamics of the process of discrimination are concerned. As a typical example consider the case in which a rat is to take the right turn at an intersection when a tone has a certain pitch and a left turn when it has another pitch. According to the preceding analysis there are four reflexes to be taken into account:

(1) sS^D : *Pitch A. R : right turn*
(2) sS^D : *Pitch A. R : left turn*
(3) sS^D : *Pitch B. R : right turn*
(4) sS^D : *Pitch B. R : left turn*

The experiment is designed so that (1) and (4) are reinforced while (2) and (3) are not. The strengths change in such a way that in the presence of Pitch *A* a right turn is made, in the presence of Pitch *B* a left. If the pitches are not widely separated (according to the discriminative capacity of the organism), the 'problem is solved' slowly. Induction from (4) upon (2) and (1) upon (3) (and reverse induction) keeps the strengths approximately at their original relative values.

The difficulty with the method is that changes in strength cannot be easily followed. The frequencies of reinforcement and extinction depend upon the behavior of the organism and are thus not under convenient control. As the experiment progresses the organism tends to make more and more correct 'choices,' as the strengths of the two reflexes draw apart, but no direct measure of strength is provided. The datum is merely the eventual 'choice,' and the course of the change in strength is not indicated by the increase in the number of correct choices. Some arbitrary measure is usually needed to obtain a 'learning' curve, as, for example, the number of correct choices in groups of twenty trials. The shape of the curve depends upon the value of twenty. For the purpose of establishing a strong discrimination so that one reflex will be prepotent over the other upon practically every occasion, a discrimination box will suffice; but for any study of the nature of the process it is a relatively crude instrument.

Chapter Six

SOME FUNCTIONS OF STIMULI

A Discriminative Stimulus as 'Inhibitory'

The discrimination of a stimulus provides another case to which the notion of 'inhibition' is frequently extended. If a discrimination is established between two composite stimuli differing with respect to their membership, such that the reflex $S^A . R$ is always reinforced while $S^A S^B . R$ is not, then S^B acquires an apparent power to suppress the action of $S^A . R$. The effect of S^B is called by Pavlov conditioned or differential inhibition. The case resembles true inhibition more closely than simple extinction because it involves a second stimulus; but according to the present interpretation a discrimination is only a modified form of extinction, and no concept of inhibition is needed to account for it. S^B does not act to inhibit the reflex of $S^A . R$ in a way comparable with, say, the inhibition of eating or of salivation by a loud sound. It is the differentiating property of a composite stimulus.

The proof offered by Pavlov that the action of S^B in reducing the magnitude of R in response to S^A is really inhibitory and 'not merely a passive disappearance of the positive conditioned reflex owing to the compound stimulus remaining habitually unreinforced' is based primarily upon the transfer of 'inhibitory' power from one stimulus complex to another. A tactile stimulus, for example, which has become 'inhibitory' when combined with the stimulus of a rotating object is found to 'inhibit' a response to a flashing lamp, in connection with which it had never gone unreinforced. But all evidence of this sort is weakened by the fact that different conditioned reflexes of Type S based upon the same reinforcing stimulus are not independent entities. The very important problem of their inductive interrelation has not been worked out. It may be connected with the necessity of some amount of discrimination in conditioning of this type and with the community of reserve of discriminative reflexes having the same form of response. For the moment the weakness of the evidence of transfer from one reflex to another within such a group may be indicated simply by citing Pavlov's demonstration that the simple extinction of the reflex *S: metronome . R: salivation* produced the complete extinction of *S: tactile stimulus . R: salivation* and a weakening of *S: buzzer . R: salivation*. This apparently establishes the inductive continuity of these rather diverse stimuli, which would account for the apparent transfer of inhibitory power in the cases cited by Pavlov. The necessary extinction of $S^A S^B . R$ may affect another reflex $S^C . R$ directly and account for the apparent effect of S^B when presented with S^C. Pavlov's examples from heterogeneous reflexes do not escape this criticism because the same effector is still involved.

Perhaps the simplest way to demonstrate the illegitimacy of assigning inhibitory powers to S^B is to reverse the conditions of the discrimination so that $S^A . R$ is extinguished and $S^A S^B . R$ reinforced. The activity of $S^A S^B . R$ is then reduced when S^B is *removed*. To be consistent we should say that the *absence* of S^B has acquired inhibitory power, but it is difficult to see how the absence of a stimulus may be introduced into a new experimental situation in order to show transfer. Other confusing or absurd cases may be generated by basing the discrimination upon the value of a single property. For example, we should be required to show that the response to a light could be inhibited by changing the pitch or odor of the light in order to demonstrate transfer from experiments in which such changes had acquired inhibitory power through discrimination. It is only when the differentiating component has the status of a stimulus that inhibitory power is assigned to it. When it is a single property or a change in a property, the analogy with true inhibition is less compelling.

It is possible that the member of a composite stimulus correlated with non-reinforcement may acquire a true conditioned inhibitory power quite aside from the process of discrimination. Failure to reinforce a response is one of the operations depressing reflex strength through an emotional change, and there is little or no distinction to be drawn between inhibition and one kind of emotion. In simple extinction the effect of failure to reinforce produces the cyclic fluctuation which characterizes the process. In a discrimination the presence of S^B upon each occasion when the reflex is unreinforced may give it a conditioned emotional power, which may be transferred to another situation if S^B is itself transferable. But S^B then depresses the strength of a new reflex, not because it has acquired inhibitory power from having been the property correlated with non-reinforcement, but because it has previously been correlated with an emotional operation necessarily bound up with non-reinforcement. A failure to reinforce has two effects: a change in reflex strength through conditioning and an emotional state. In 'conditioned inhibition' the transfer should be due to the former, but an indication of transfer may in reality be based upon the latter.

The use of the concept of inhibition in accounting for the effect of the discriminative stimulus not correlated with reinforcement is partly due to the narrowness of the traditional conception of a stimulus. The term has the unfortunate connotation of a goad or spur to action. In its traditional use it refers to a force which drives the organism, a meaning which has been congenial to writers who wish to prove that the occurrence of a bit of behavior under a given set of external circumstances is inexorable.[1] By extension the same *active* function of the stimulus has been made to apply to inactivity through the notion of inhibition as a suppressing force. While the present system presupposes the lawfulness of behavior and recognizes the rôle of the environment, it does not necessarily appeal to the environment as a driving force. The function of a stimulus in establishing a discrimination, for example, has nothing of the character of a goad. The distinction between a discriminative and an eliciting stimulus has already been referred to several times and may now be elaborated. It will be convenient at the same time to review other ways in which the action of the environment may enter into a description of behavior.

Various Kinds of Stimuli

A: THE ELICITING STIMULUS

The eliciting stimulus was defined in Chapter One as 'a part, or modification of a part, of the environment' correlated with the occurrence of a response. The notion of elicitation is here confined to that of correlation. The term describes the fact that presentation of a stimulus is followed by a response, and it is not neces-

[1] Cf. in this respect the practical identification of 'stimulus' and 'drive' by Holt (45).

sary to assign any transitive character to the rôle of the stimulus. The notion of elicitation is applicable only to respondent behavior and is, I believe, clearly enough understood not to require any elaboration here. What is meant by a part of the environment is a part of the energies or substances directly affecting the organism. It is elliptical, but often convenient, to speak of their sources—as, for example, when a bell is called an auditory stimulus or a book a visual one. The practice is dangerous, since the stimulation arising from such a source is highly variable, but it is frequently successful because of the generic nature of stimuli. In the case of a proprioceptive stimulus it is necessary to appeal to the activity of the organism responsible for the stimulation, and it is implied that a given movement (*e.g.*, flexing a limb) always produces the same stimulation within generic limits.

The slight ambiguity in the phrase 'a part, or a modification of a part, of the environment' needs to be removed. The phenomenon of adaptation in end-organ and nerve-fiber and the resulting critical rate of presentation of a stimulus have led to a conception of a stimulus as a 'change in the environment of an excitable tissue' (24). But where adaptation or fatigue takes place slowly, a prolonged period of responding may prevail during which it is difficult to see that anything is changing. The postural modification due to the position of the labyrinths in space or to extirpation of one labyrinth show a sustained response correlated with the sustained and unchanging stimulus supplied by the gravitational field. What is meant by a correlation of stimulus and response is that when the stimulus is present the response is also present (or will follow shortly). Usually it is required to present the stimulus in demonstrating the correlation, and presentation is, of course, a change in the stimulating field, but the response is correlated with the *presence,* not the presentation.

In this sense, then, a stimulus is a continuous agent, but a change may also act as an eliciting stimulus in certain reflexes, and the definition must allow for this possibility. As a continuous agent a stimulus must have a certain minimal number of properties—location, intensity, quality, duration, and perhaps others. As a change it may affect only one property and may also be simply the *withdrawal* of a stimulating force. A dog may prick its ears if a tone is presented as a continuous agent, but the response may also be elicited by a change in the pitch or intensity of a continuous tone or by its cessation. The number of unconditioned reflexes in response to changes in single properties is probably few, but such changes are freely available as conditioned stimuli of Type S.

B: THE DISCRIMINATIVE STIMULUS AND THE PSEUDO REFLEX

A pseudo reflex is a relation between a stimulus and a response which superficially resembles a reflex but depends upon or involves other terms than those expressed in the relation. Although it exhibits a similar topographical correlation of stimulus and response, it differs from a true reflex in many ways, and the distinction must be maintained if confusion is to be avoided in the study of the static and dynamic laws. The commonest example involves a discrimination of the stimulus in Type R. Let a discrimination be established between $S\lambda$ and Sl by reinforcing responses only in the presence of the latter. When this has been done, $[sSl \,.\, R]$ is greater than $[sS\lambda \,.\, R]$ and at any value of the underlying drive such that $[sSl \,.\, R]$ usually occurs but $[sS\lambda \,.\, R]$ does not, the following condition exists. Given an organism in the presence of $S\lambda$ ordinarily unresponsive, presentation of Sl will be followed by a response. For the sake of comparison an example may be selected and a paradigm written in imitation of Type S as follows:

The relation between the light and the response to the lever is a pseudo reflex. It has some of the distinguishing characteristics of a conditioned reflex of Type S: the original response to S^0 is irrelevant, the relation $S^0 \,.\, R^1$ may be absent prior to the 'conditioning,' the strength changes in a positive direction only, and measurements of latency, threshold, and the R/S ratio are possible. In all these respects it differs from a reflex of Type R, although the example is based upon operant behavior. In other respects it differs from both types. The response requires not only S^0 but S^1; the light is only one part of the discriminative stimulus in the presence of which the response is made. The stimulus S^1 is not withheld when the effectiveness of the 'conditioned stimulus' is tested; instead the response to S^1 alone is extinguished—a characteristic which has no parallel in either type.

In spite of these differences it would commonly be said that the light becomes the 'conditioned stimulus' for the response to the lever, just as it becomes the stimulus for, say, salivation. This confusion with Type S obviously arises from a neglect of the extinguished reflex. The relation of pressing the lever to the lever itself is ignored and only the relation to the light taken into account. The lever comes to be treated, not as a source of stimulation, but as part of the apparatus, relevant to the response only for mechanical reasons. When the discrimination is based upon a response not requiring other discriminative stimuli, the chance of this neglect increases enormously. If we substitute 'flexion of a limb' for 'pressing a lever' and continue for the moment with Type R, no external stimulus is necessary for the execution of the response or for its reinforcement. Having established *s* . *R* as a conditioned operant of some strength, we introduce discriminative stimulation as before, reinforce *sSl* . *R* and extinguish *sS*λ . *R*. We then have a condition in which an organism is ordinarily unresponsive but immediately responds with flexion upon presentation of *Sl*.

Superficially this pseudo reflex resembles the conditioned reflex usually regarded as of Type S that is established by allowing the presentation of *Sl* to be followed by a shock to the foot until it elicits flexion when presented alone. There is perhaps no way of determining the difference from the topography of the correlated events alone. But the static and dynamic properties of the pseudo reflex are quite different from those of the true reflex, as I shall note shortly.

In a pseudo conditioned reflex based upon Type S the distinction is much less clear. Here we are invariably able to neglect the extinguished member because the available responses do not require external discriminative stimuli. For example, given the conditioned reflex *S: tone* . *R: salivation*, we may establish a discrimination between *S* and *SSl* (where *Sl* is, say, a light) by reinforcing only the latter. Then the organism, ordinarily unresponsive in the presence of the tone, will respond upon presentation of the light. (This will be seen to be the reverse of the case in which *Sl* is correlated with the absence of reinforcement and in which it is called 'inhibitory.') The only difference between the present relation of the light to the response and a true reflex of Type S is the extinction of the response to the tone, which shows that a discrimination has taken place. The reinforcement of *SSl . R* should condition responses to both of these stimuli separately through induction, but we observe that the organism is unresponsive in the presence of the tone alone. This difference may be reduced at will by reducing the significance of *S* in the basic reflex of the pseudo type. If we lower the intensity of the tone or choose a less important stimulus, we may approach as closely as we please to a conditioned reflex of Type S. We cannot actually reach Type S in this way, but we may easily reach a point at which the pseudo reflex is identical with any experimental example of that type because some amount of discrimination is apparently always involved in cases of Type S, as I have already shown.

The position of the pseudo reflex may be summarized as follows. When a pseudo reflex is based upon a reflex of Type R and when other discriminative stimuli are necessary for the elicitation of the response, there are important practical and theoretical reasons why a separate formulation is demanded. When the response does not require external 'support,' there are fewer differences, but a separate formulation is needed in order to clarify the differences in the static and dynamic properties. When the pseudo reflex is based upon Type S, the distinction is weakened but should still be made except when *S* can be reduced to a very low value relative to *Sl*. The static and dynamic properties then approach those of a true reflex, although it is doubtful whether a conditioned reflex of Type S ever appears experimentally without being disturbed by the necessary extinction of the spread of its reinforcement. In the limiting case the distinction between the pseudo and the true reflex of this type is impossible, not because the cases are identical, but because Type S fails to appear experimentally in a pure form.

The reader may object that in holding to the level of a correlation of a stimulus and a response in defining a reflex I have no right to call the relation between a light and the response to the lever pseudo. It is true that at any moment a pseudo reflex may have the properties of the kind of correlation that is called a reflex, but with respect to the static or dynamic properties, which are called into play in defining a reflex as a unit of behavior, the two cases differ widely. It is for the sake of a simple classification of these phenomena that I am insisting upon the distinction. In a discriminated operant there is no relation between the discriminative stimulus and the response that satisfies the static laws of the true reflex. The introduction of S^D reinstates the meaningfulness of the terms latency, threshold, after-discharge, and the *R/S* ratio, but the laws that describe them differ.

Perhaps the most important example is the *R/S* ratio. In a discriminated operant the magnitude of the response is relatively if not wholly independent of the magnitude of S^D provided S^D is above the threshold. The intensity of the operant is highly stable, except when it is deliberately differentiated (see Chapter Eight). The strength of the representative operant described in this book varies one hundred-fold as measured by its rate of occurrence. The actual force with which the response is executed may not differ at all or at most be doubled. The lack of a relation between the intensity of the response and S^D has long been a perplexing matter in the study of operant behavior, especially when the attempt has been made to apply concepts derived from the traditional work on respondent behavior. A very simple example is the behavior of a child in reaching for a block. Since the stimulus emitted by the block is visual, its intensity can presumably be changed only by varying the illumination. But within a fairly wide range, the intensive properties of the behavior will be only feebly if at all related to this variable. Nor can we measure dynamic changes in terms of the relation of the intensity of the response to that of the stimulus. The force with which the child reaches is a very imperfect measure of the degree of conditioning, for example, or of the state of the drive. The behavior is operant and is to be studied with its appropriate measure—namely, the rate of occurrence of the response. The apparent *R/S* ratio is probably useless.

Similarly, the apparent 'thresholds' and 'latencies' of a pseudo reflex do not obey the static and dynamic laws established for respondents. The threshold must here be defined as the lowest value of a stimulus capable of being used in setting up a discrimination. The values are in general lower than for true reflexes, and they do not vary linearly with changes in strength. Latencies are in general much longer for pseudo reflexes. For example, in Chapter Five an average value of twenty seconds was described. Like the threshold, the latency of a pseudo reflex is not a simple function of the strength.

Some exploratory experiments have been performed in an attempt to establish intermediate values of the discriminative latency either by reducing the drive or by reducing the reserve through extinction. No grading of the latency comparable with that in a simple respondent has been discovered. The latency is not prolonged during either kind of reduction in strength until extreme values begin to emerge that are scarcely to be regarded as latencies at all. As the drive is reduced, a point is reached at which the rat does not respond after presentation of the discriminative stimulus, but so long as it continues to respond, the latencies are of the original order. The case for extinction is tested by presenting S^D at regular intervals without reinforcement. Eventually

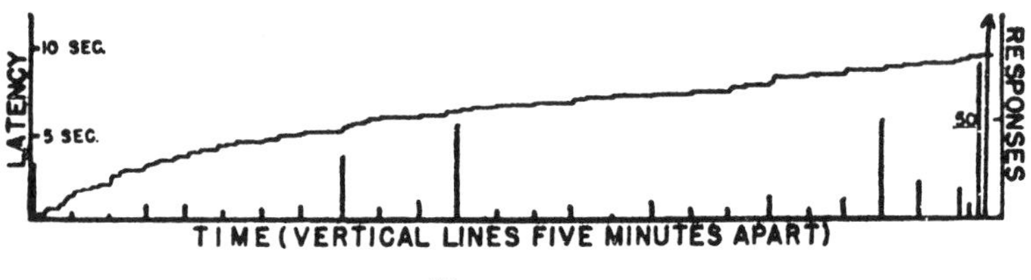

FIGURE 77

EXTINCTION OF THE REINFORCED REFLEX IN A DISCRIMINATION SHOWING NO PROGRESSIVE LENGTHENING OF THE 'LATENCY'

The heights of the vertical lines indicate the latency at each (unreinforced) presentation of S^D. The rate of responding declines smoothly but the latency shows no progressive change.

no responses are made that can be said to be correlated with the presentation of the stimulus. After extinction a single reinforced presentation of S^D serves to restore the latency to its original value. Neither extinction nor reconditioning curves are to be obtained from the latency.

Figure 77 shows the extinction of a discrimination that had been thoroughly established. On the day represented by the figure the discriminative stimulus (a buzz) was presented every five minutes but no responses were reinforced. The procedure was essentially like that of Figure 61 and the curves are obviously similar. In Figure 77 the latencies are also recorded with vertical lines, the height of each line representing the time elapsing between the presentation of S^D and the next response. The curve given by the frequency of responding falls off smoothly during the two-hour period, but there is no progressive lengthening of the latency. At the end of the period, in order to hasten the extinction, three presentations of S^D were made quickly. The last of these failed to evoke a response during more than a minute. It is doubtful whether this can be considered a latency, but in any event it shows the abrupt kind of change obtaining during extinction. A grading of the latency is clearly lacking.

The discriminative stimulus has a very different status from that of the eliciting stimulus. It is less likely to be regarded as a spur or goad and is perhaps best described as 'setting the occasion' for a response. Whether or not the response is to occur does not depend upon the discriminative stimulus, once it is present, but upon other factors. The case as here stated applies particularly to operant behavior. It is not so easily expressed in the case of respondent behavior because conditioned respondents seem always to involve discrimination and are therefore to some extent always pseudo.

Strictly speaking we should refer to a discriminated operant as 'occurring in the presence of' rather than 'elicited as a response *to*' S^D. The analogy with the true reflex is almost too strong to be resisted in casual speech, however, and little difficulty should arise from the extension of these terms, provided a general intermediate meaning is assigned to them with respect to the mere temporal and topographical correlation of stimulus and response. In the preceding chapters I have tried to be specific wherever there seemed to be any danger of confusion, but elsewhere 'elicitation' and 'responding *to*' are occasionally used in this broader sense.

In distinguishing between an eliciting and a discriminative stimulus I am simply contending that a stimulus may have more than one kind of relation to a response. The relation known as elicitation is the simplest to demonstrate and perhaps for that reason has been looked upon as unique and universal. But serving as the basis for a discrimination is also an important function, and it is actually the more common. The same temporal order of *S* and *R* obtains in both cases but the same quantitative properties are not to be expected.

C: THE REINFORCING STIMULUS

The effect of the reinforcing stimulus upon behavior has been described at some length and will not be elaborated here. A reinforcing stimulus is at the same time either an eliciting or a discriminative stimulus, but its action in reinforcing a reflex is a separate effect that must be listed among the various functions of stimuli.

A special case arises in the withdrawal of a negatively reinforcing stimulus, which yields another kind of pseudo reflex. For example, let a tetanizing shock to the tail of a dog be discontinued as soon as the dog lifts its left foreleg.[2] The discontinuance of the negative reinforcement acts as a positive reinforcement; and when conditioning has taken place, a shock to the tail will be consistently followed by a movement of the foreleg. Superficially the relation resembles a reflex, but the greatest confusion would arise from treating it as such and expecting it to have the usual properties.

D: THE EMOTIONAL STIMULUS

Another kind of stimulus, which will be discussed in detail in Chapter Eleven, has the function of setting up an emotional state. An emotional state is not a 'response' as here defined, and the stimulus is therefore not eliciting. The effect is upon reflex strength, as will be shown later, and in this respect the action of the stimulus resembles that in the case of reinforcement. But while the reinforcing stimulus affects the reserve, the emotional stimulus affects the proportionality of reserve and rate. Facilitating and inhibitory stimuli may be included in this class.

Curiously enough, the notion of a goad may apply to three of these cases, although it is usually confined to the special case of reinforcing stimuli. When the original response to a goad is the exact form desired, the goad may be (1) an eliciting stimulus. For example, a horse is made to start running through the use of a whip. When the response is already in progress, a goad may be used as (2) a facilitating (emotional) stimulus to increase the

[2] A case of this sort in which a rat was conditioned to press a lever whenever a continuous shock was administered through a grid on the floor, has been reported by Mowrer (63).

strength. For example, a running horse may be made to run faster by whipping. It is difficult to separate the eliciting and facilitative action in this example, but the distinction is clear when the response is of a form not elicited by the goad—for example, when a trained bear is made by whipping to grind a music box more rapidly. The most common use of a goad is as (3) a negatively reinforcing stimulus to be withdrawn as a positive reinforcement. An example is the maintenance of running in a horse by whipping it when it stops or drops below a certain rate. A light touch of a whip may, by preceding a stronger blow, become a conditioned negative reinforcement and have the same effect.

The various functions of stimuli may be summarized in this way: a stimulus may

(1) elicit a response ('elicitation')
(2) set the occasion for a response ('discrimination')
(3) modify the reserve ('reinforcement'), or
(4) modify the proportionality of reserve and strength ('emotion,' 'facilitation,' and 'inhibition').

These are the ways in which the environment enters into a description of behavior. They are all of very great importance, and it is clear that the traditional notion of the stimulus as a driving force is too simple. Distinctions of this sort must be insisted upon if an orderly quantitative science of behavior is to be achieved.

In spite of these differences all stimuli are alike in being isolable parts of the energies or substances affecting the organism. The procedure of analysis depends in each case upon the function. The generic nature of the stimulus was demonstrated only for eliciting stimuli in Chapter One, but the argument holds as well for the other three kinds. The correlation with a response, which (together with the dynamic changes in the correlation) defines the eliciting stimulus, is lacking in the case of a discriminative stimulus, but the essential defining property may be determined by comparing the strengths of an operant in the presence of various particular instances. Thus, if a rat has developed a strong operant $sS^D : tone . R$ the strength may be found to be independent of the pitch of the tone over a considerable range, although as in the case of the eliciting stimulus extreme values will affect the result, and the possible relevance of less extreme values is thereby indicated.

Similarly, it is possible to show that an emotional or reinforcing stimulus (either conditioned or unconditioned) is effective without regard to various minor properties.

I may add that emotional and discriminative stimuli, and possibly reinforcing stimuli, also exhibit temporal summation, as well as other temporal properties to be investigated in the following chapter.

The Various Functions of Conditioned Stimuli

Conditioning of Type S is based upon the approximately simultaneous presentation of two stimuli. It may take place even when S^1 is not an eliciting stimulus, and the additional possibilities arise of (1) conditioned discriminative stimuli, (2) conditioned reinforcements, and (3) conditioned emotional stimuli. In order to express cases of this sort we may set up a paradigm as follows:

S^0 . **R^0**
S^1 . (discriminative, reinforcing, or emotional effect)

in which the effect of S^1 is substituted for R^1 and becomes correlated with S^0.

A: CONDITIONED EMOTIONAL STIMULI

A conditioned emotional effect has already been appealed to (page 155). It is relatively simply expressed by saying that an incidental stimulus accompanying an emotional stimulus acquires through conditioning of Type S the power to set up an emotional effect.

B: CONDITIONED REINFORCING STIMULI

According to Pavlov a conditioned stimulus of Type S may be substituted for S^1 to establish a 'secondary' conditioned reflex. For example, when the stimulus S^A: *sound of metronome* has been strongly conditioned, it may be used to condition some 'more or less neutral' stimulus S^B if the latter accompanies it. $S^A . R$ must not, of course, be reinforced when S^B is presented, because the reinforcement would then act directly upon $S^B . R$. During secondary conditioning the strength of $S^A . R$ must be maintained by interpolated separate reinforcements. We have, then, a series of reinforcements of $S^A . R$ broken occasionally by the unreinforced combined presentation of S^A and S^B. But this is the exact procedure for establishing a discrimination based upon the membership of a component stimulus. In the case of secondary conditioning the result should be an increase in the strength of $S^B . R$, but in the case of discrimination the response to the simultaneous presentation of S^A and S^B should be extinguished. According to Pavlov either of these two directly opposed results may occur, depending upon the temporal relation of S^A and S^B, and in rare cases both may occur together. In order to establish secondary conditioning 'the new stimulus should be withdrawn some seconds before the primary stimulus is applied [(64), p. 33].' The critical point for stimuli of ordinary intensity is about 10 seconds. With strong stimuli the interval may be as long as 20 seconds and still give a discrimination rather than secondary conditioning.

When secondary conditioning and discrimination occur together there must be an increase in $[S^B . R]$, a decrease in $[S^A S^B . R]$, and presumably a maintenance of $[S^A . R]$. There is nothing contradictory about this, but the development of secondary conditioning under such circumstances is surprising. I am inclined to doubt the reality of secondary conditioning of a respondent in general. In any event the very arbitrary time limit makes the case of extremely limited application outside the laboratory, as I have already said. The actual data given by Pavlov are of small magnitudes, and there are several possible sources of artifact. For example, the lack of specificity of the salivary response to unconditioned stimuli raises (as in the case of 'disinhibition') the possibility of salivation due directly to the new stimulus. Moreover, the pairs of stimuli used by Pavlov (*e.g.*, a tone of 760 d.v. and the sound of bubbling water) are not certainly free of inductive interaction.

The use of a conditioned reinforcing stimulus in Type R raises no similar difficulty. Such a stimulus may be a conditioned stimulus of Type S or a discriminative stimulus of Type R. As an example of the former, let a tone be correlated with the presentation of food. Then any operant reinforced by the tone will increase in strength. The tone must not be reinforced upon such occasions, but its separate reinforcement may be effected without disturbing the conditioning of the operant and without establishing a discrimination. A similar case of negative conditioning might be established, as when a tone which has preceded a shock is produced by an operant and the operant declines in strength. As an example of the second case let a light be made a discriminative stimulus correlated with the reinforcement of the response to the lever. Then any response producing the light will increase in strength. It will be seen that both of these processes are intimately connected with the chaining of reflexes described in Chapter Two.

A minor experiment in which another member is added to the present representative chain by letting a discriminative stimulus act as a reinforcement may be described here. A discrimination was first established by reinforcing every response in a stimulus complex which included the lever and a differentiating stimulus *Sl* and by extinguishing all other responses to the lever in the absence of *Sl*. A situation was then set up in which some arbitrarily chosen response led to the presentation of *Sl*, after which the discriminative response in the presence of *Sl* was immediately reinforced. This could have been done by putting another lever in the experimental box or by using some other kind of response, but the result for another lever would be unclear because of induction and it would be hard to obtain any response that was different enough to be certainly free of this complication. In the present experiment the same operant was used. The experiment thus represents the case in which the interference between Reflexes III and IV is maximal. A first response to the lever produced *Sl*; a second response in the presence of *Sl* produced the sound of the magazine; and the response to the tray following the sound of the magazine led to food. The chain may be written

$$_sS^{IV} . R^{IV} \rightarrow {}_sS^{III}\, l . R^{III} \rightarrow {}_sS^{II} . R^{II} \rightarrow S^{I} . R^{I}$$

A group of twelve white rats approximately 140 days old at the beginning of the experiment were put on the usual schedule of periodic reconditioning at intervals of five minutes.[3] All the rats

[3] Through faulty technique only seven completed series were obtained with this group. Three series were eliminated when the differentiating stimulus was accidentally left on continuously for a few minutes during the first day of the discrimination. In one other case *Sl* remained absent during the last day of the discrimination because of a fault in the apparatus, and the rate returned to that characteristic of periodic reconditioning. The records in these four series are entirely compatible with the present conclusions but cannot well be included in an average. In the remaining case the rat developed a labyrinth disorder occasionally observed in this strain (see Chapter Twelve).

assumed an approximately constant rate of responding. After two days of periodic reconditioning a discrimination was begun by sounding a buzzer (as the discriminative stimulus) whenever the next response was to be reinforced. After four daily hours the rate of responding in the absence of the sound had fallen to less than half its original value. On the fifth day the order of events was as follows. When the rat was released, the buzzer was off. The first response turned it on and at the same time completed the circuit to the food magazine, making it ready for the next response. The second response, in the presence of the sound, was followed by a discharge of a pellet of food into a tray where it was accessible to the rat. The buzzer was then turned off and the magazine disconnected. Responses during the next five minutes were ineffective. At the end of five minutes the apparatus was again set (without supplying any stimulus to the rat) so that the next response turned on the buzzer and the next after that discharged the pellet of food. The light and magazine were then turned off again. This was repeated at five-minute intervals during the hour. All responses to the lever were recorded.

After the discrimination had been begun, all the responses which produced food were in the presence of the sound. Any inductive effect upon the operant in the presence of the sound was presumably lacking, according to the evidence presented in Chapter Five. The responses still observed during the intervals of silence were due to the original periodic reconditioning and should have continued to disappear as the discrimination curve was carried out, if the newly arranged correlation of one response every five minutes with the production of the sound were having no effect. The experimental result, however, was that immediately upon changing to the new condition the rate of responding in silence increased until the original rate was reached, or at least approximately.

The result is shown in Curve A, Figure 78, which gives the average for the seven completed series. In this figure no attempt is made to follow the rate during each hour. The rates have been plotted as horizontal bars as if they were constant. In order to indicate the course of the change from day to day, smoothed curves have been drawn to the center of these bars. It will be seen from this figure that the average rate of responding fell during discrimination from 202 responses per hour at the beginning of the first day to 89 per hour on the fourth day. On the fifth day, when responses in the presence of *Sl* then produced the differentiating stimulus periodically (this fact is indicated in the figure with the expression $R^{IV} \rightarrow L$), the rate increased to 121 on the first day and to 163 on the second. The rate dropped to 143 on the third day, indicating that 163 might have been a compensation for the low value of 121.

The average for the three days is 142, significantly below the previous rate of 201 for periodic reconditioning. We expect some

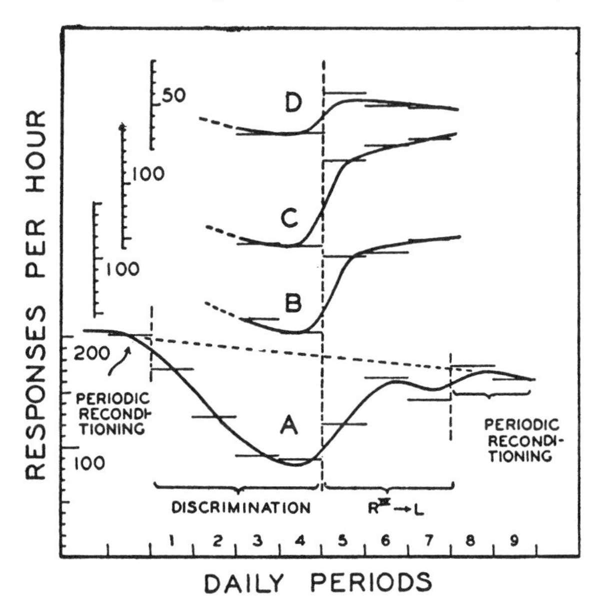

FIGURE 78 (17)

THE PRODUCTION OF S^D AS A REINFORCEMENT

decline in this rate with time (see Figure 29), but by returning to the procedure of periodic reconditioning it can be shown that the periodic rate was not quite fully attained. On the eighth day *Sl* was omitted altogether; periodically responses to *S*λ then produced, not *Sl*, but the sound of the magazine. The resulting rates on two successive days were 173 and 161 or an average of 167, which still shows a (more reasonable) decline in rate during the 16 or 18 chronological days of the experiment. The difference between 142 and 167 is, however, subject to a correction to be made later.

Another group of eight rats approximately 160 days old had been through the extensive series of reversed discriminations discussed in the preceding chapter in which the extra stimulus was a light but in which the actual differentiation had come to be based upon neither the presence nor the absence of the light but the change from one to the other in either direction. The average rates on the last two days of the series were 96 and 83 respectively. The rate was not falling as rapidly as this single difference would indicate. The procedure was then changed so that responses in the dark periodically produced the light, after which a response was reinforced. The increase in rate was in this case practically complete on the first day, as shown in Figure 78, Curve B. No recent periodic rates were available for comparison, nor was the procedure changed to periodic reconditioning with this group.

A third group of four rats approximately 170 days old had established a discrimination to the light without previous periodic reconditioning as discussed in the preceding chapter. On the third and fourth days the average rates were 44 and 42 respectively. On establishing the relation of responses in the dark to *Sl* the rate increased to 120, 133, and 139 responses per hour, as shown in Curve C, Figure 78.

A fourth group of four rats, approximately 150 days old, had established a discrimination without previous conditioning also as discussed above, in which the differentiating component was the absence of a light. On the fifth and sixth days of the discrimination, the average rates were 24 and 25 responses per hour respectively, the slight increase being insignificant. Upon changing the procedure so that periodically a response in the presence of the light resulted in its disappearance and the next response was reinforced the rate increased to 61 responses per hour, dropping to 50 and 49 on the following days (Curve D, Figure 78). The unusually low values for these animals and the relatively slight increase are partly due to the method of discrimination but also to the depressive effect of the light.

The important point in such a series is the first day of the relation of the response in the presence of *S*λ to the production of *Sl*. In the vernacular: 'the rat is in the course of learning to press the lever only when *Sl* is present; it must now learn that it is itself producing *Sl*.' The result appears in the present analysis as a change from the low value of *s*Sλ . *R* obtaining under the partially completed discrimination to that obtaining under periodic reinforcement by the differentiating stimulus. It is clear from the data for the first day that the change begins very soon, but we need to turn to the daily records themselves for its actual course. We are interested in any increase shown on the day of the change over the rate that would be expected from the continuation of the discrimination. The latter could be calculated by extrapolation from the preceding part of the curve, but for our present purposes

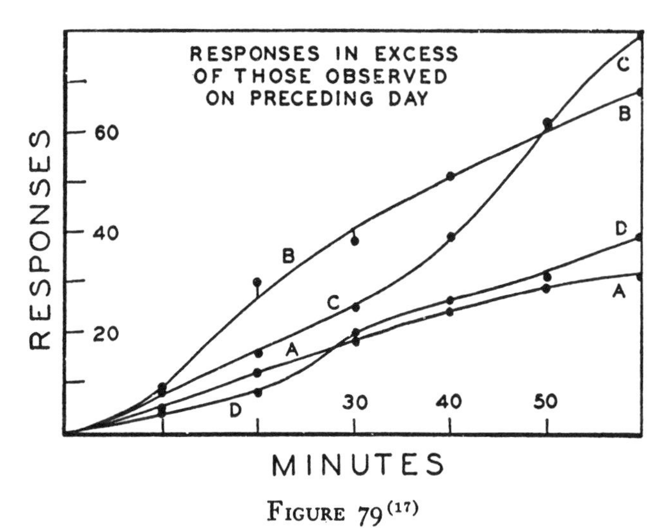

FIGURE 79 (17)

THE FIRST EFFECT OF S^D AS A REINFORCING STIMULUS

Data for the groups in Figure 78, showing the increase in the number of responses on the first day when the periodic appearance of S^D is correlated with a response. The curves were obtained by subtracting the values for Day 4 in Figure 78 from the values for Day 5.

we may suppose that the rate to be expected on the day of the change is identical with that of the preceding day. Any error introduced by this assumption works against the present argument. If we subtract this expected rate from the rate actually observed, any increase as the effect of *Sl* will be clearly shown. Measurements of the height of the curves were made at intervals of ten minutes during the hour. Averages of these points on the last days of the discrimination were subtracted from those for the first days of the change. The results are given in Figure 79. Here the base line is in each case the average summation curve for the responses of the group on the preceding day. The curves show the increases above this base line as the result of periodic reinforcement with *Sl*.

It is obvious that the effect is immediately felt. In all four groups a significant increase is evident at the end of ten minutes when only two reinforcements have occurred. In Group A this increase represents very nearly the full value: successive periods each add about the same number of responses. It may be noted from Figure 78, however, that the average value for this group for the first day of the change is low, and that a further increase in rate occurs on the following day. In the other three groups a significant increase is apparent, but it is followed sooner or later by a more marked acceleration. In Group B the greatest increase is observed between 10 and 20 minutes after the beginning of the hour; in Group C between 40 and 50 minutes; and in Group D between 20 and 30 minutes. These differences probably reflect the various histories of the four groups. Group A began with some periodic reconditioning and its only discrimination had not been carried very far. Group B had a very thoroughly established discrimination. Group C had had no previous periodic reconditioning, which may account for its delayed acceleration. A typical series of actual records is reproduced in Figure 80 (page 252).

These four experiments show, without exception, that the production of a differentiating stimulus had a marked reinforcing effect and that this was felt immediately. It is not possible, however, to estimate the value of the effect very closely. It is probably of the same order as the effect of the sound of the magazine, but a mere comparison of the two periodic rates will not yield a fair estimate for the following reasons:

1. Since *Sl* is produced by R^{IV} and since R^{III} follows immediately (within two seconds), S^{II}: *sound of magazine* follows R^{IV} closely enough to have a considerable reinforcing effect upon it. Under the conditions indicated by '$R^{IV} \rightarrow L$' it can be said that once every five minutes a response to S^{IV} is followed within two seconds by S^{II} (no attention being paid to the intervening events $S^{III} . R^{III}$). As shown in Chapter Four, an interval of two seconds between a response and the reinforcing stimulus reduces the effect of the reinforcement by about one-third. Consequently part of the periodic rate developed under '$R^{IV} \rightarrow L$' may be due not to *Sl* as a reinforcement but to S^{II}: *sound of magazine*.

2. It is also not clear whether the rate remaining as part of the discrimination must be added to the periodic rate under '$R^{IV} \rightarrow L$'. We do not know whether 'pressing the lever in order to turn on the light' is the same reflex as 'pressing the lever in order to make the magazine sound.' If we must regard them as discrete

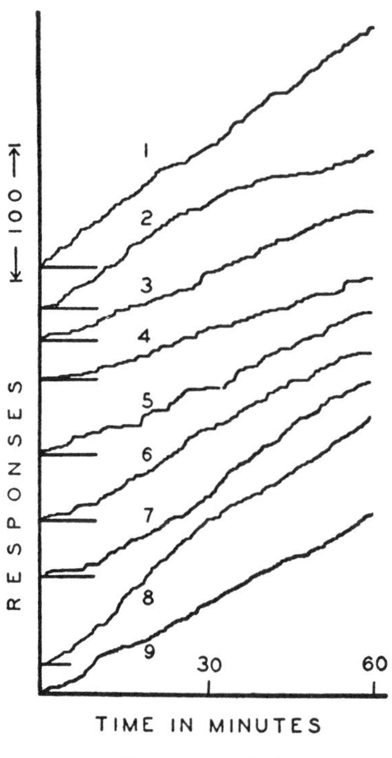

FIGURE 80 (17)

REINFORCING EFFECT OF A DISCRIMINATIVE STIMULUS

A set of records for one rat in Figures 78 and 79, Group A.

entities, their separate rates of elicitation must be supposed to summate. If the presentation of Sl has a reinforcing effect equal to that of S^{II}, then $sS^{IV} . R^{IV}$ should by itself achieve a value nearly equal to that observed at the beginning of the experiment. But $sS^{III} . R^{III}$ is by no means near zero and, on the assumption that separate reflexes are involved, the total observed rate should be greater than that originally observed for $sS^{III} . R^{III}$ alone. This is not in agreement with our result but the assumption is not thereby ruled out. It may be that the effect of *Sl* is even less than we had supposed.

It is simpler, however, to regard $sS^{III} . R^{III}$ and $sS^{IV} . R^{IV}$ as the same reflex. We thereby avoid the systematic question of how to distinguish them if they are not, and we may easily account for the present data. The return to a higher rate is comparable with that observed during the abolishment of a discrimination through reinforcement of the previously extinguished member (Figure 58) where we do not expect summation. Under this interpretation we may say that the summed effect of *Sl* upon $sS^{IV} . R^{IV}$, of S^{II} upon $sS^{IV} . R^{IV}$, and of S^{II} upon $sS^{III} . R^{III}$, through induction from $sSl . R^{III}$, is approximately equal to that of S^{II} upon $sS^{III} . R^{III}$ directly under periodic reconditioning. But this does not make possible a close estimate of the effect of *Sl* alone.

The experiment throws some light on a technical problem connected with discriminations in general. Where a differentiating stimulus is repeatedly presented to the organism, its occasional effect as a reinforcing stimulus must be taken into account. In the present method, if a light is presented every five minutes as a differentiating stimulus, it will have an added effect upon the rate whenever its presentation coincides with the elicitation of a response. In some of the preceding experiments the differentiating stimulus was set by hand, and the practice was followed of waiting until the rat was not responding very rapidly. In this way coincidences have generally been avoided. However, in any close analysis of the discrimination curve, a possible undesired result of this procedure must be considered. Only responses following a short period of no-responding are ever reinforced; consequently a discrimination may be developed that will have for its effect the elimination of closely grouped responses (see Chapter Seven). Although it is probable that no effect on the average rate will here be felt, the procedure cannot be regarded as without some special effect. In Group A in the preceding experiment the entire experiment was conducted automatically. The differentiating stimulus was introduced periodically by a clock and, of course, without respect to the momentary behavior of the rat. The discrimination curves do not differ significantly from those obtained with other methods, so far as the present degree of approximation is concerned, but in a closer examination the present result shows quite definitely that allowance must be made for an occasional coincidence.

A similar case of a conditioned negative reinforcement could be established in Type R by allowing an operant to produce the withdrawal of a discriminative stimulus. It would be more difficult, but not impossible, to use the same operant in this case also.

It will be convenient to report here some experiments on a separate point which were performed with the same animals and have some bearing upon the foregoing conclusions. A significant difference has previously been noted between a discrimination in which S^D is the presence of a light and one in which it is the absence. The light depresses the rate of elicitation. If the initial periodic reconditioning occurs in the absence of the light, the introduction of the differentiating stimulus will cause a sudden drop in rate where the absence of the light is the differentiating stimulus, because during the intervening periods of responding to S^D the light is on. In Group A in the preceding experiments a control was introduced against the possibility that a discrimination in which S^D was the sound of a buzzer differed from one in which S^D was the absence of the sound. In one-half of the cases the discrimination was really the reverse of that described. Of the successful series S^D was the presence of the buzzer in three cases. In the other four cases the buzzer was absent during the periodic reconditioning only and was therefore present most of the time. The curves of this group have been separated into two parts on this basis in Figure 81 A. The groups are small but give some indication of a difference. The horizontal solid lines are for the three rats with which S^D was the sound of the buzzer. The broken lines are for the others. There is no depressive effect on the rate during the discrimination but the process is retarded; the decline in rate is slower for the group in which the differentiation is the removal of the sound of the buzzer. When the relation between R^{IV} and S^D is established, a much more significant difference is to be seen. Where the rat is normally in silence but produces the sound of the buzzer periodically, the effect of the reinforcement is great. When the buzzer is sounding continuously, except when it is momentarily silenced by a response of the rat, the effect of the reinforcement is slight. Upon changing to simple periodic reconditioning the buzzers were silenced entirely, and the depressed group responds with a very significant increase in rate. Because of the low values for these four rats on Days 5, 6, or 7 it is impossible

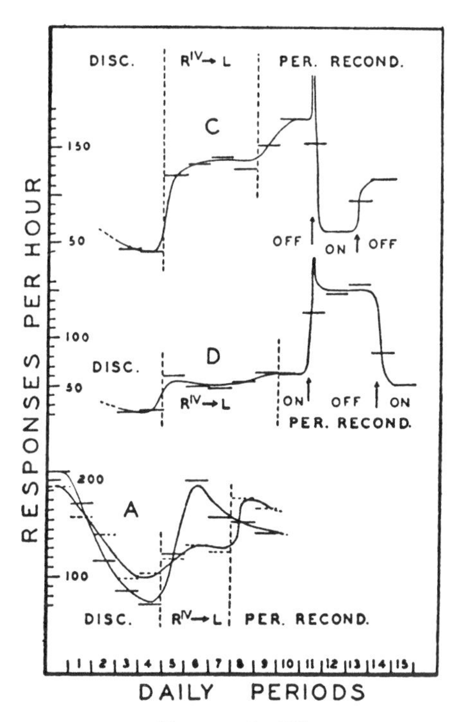

FIGURE 81 (17)

SHOWING THE DEPRESSIVE EFFECT OF STIMULI USED FOR DISCRIMINATION

The curves previously described have been separated according to whether S^D was a light or the absence of a light (or buzz). The presence of either stimulus depresses the rate.

to compare the average rate for the group with the rate on Days 8 and 9 under periodic reconditioning, as was noted above.

In Group B all eight cases were as described, so that the present problem does not arise. In Group C the differentiating stimulus was the presence of a light; in D, its absence. These experiments correspond very well with those in Group A, Group D showing only a slight increase when the relation of R^{IV} and Sl is established. Groups C and D were subsequently tested directly for the effect of the light, which was found to be relatively great. The completed series for both are given in Figure 81. In the case of Group C (top curve in Figure 81) the return to periodic reconditioning (ninth day) yielded a probably significant increase over the previous rate under reinforcement from Sl. This group is homogeneous and constitutes the only very clear evidence in the experiment that the chain

$$sS^{III} . R^{III} \ldots \text{etc.}$$

is elicited somewhat more rapidly than

$$sS^{IV} . R^{IV} \ldots \text{etc.}$$

During this simple periodic reinforcement the light was off. On the 11th day it was turned on continuously beginning at 25 minutes. No other condition was changed. Since the light is at this point still a differentiating stimulus, and since responses in the presence of the light are now no longer reinforced, except periodically, an extinction curve follows. A typical record is shown in Figure 82, Curve C. The convexity at the beginning of this curve is an incidental effect. The horizontal line through the curve marks the beginning of the extinction curve for $sSl . R$. The curve falls rapidly as the depressive effect of the light emerges over its excitatory character as a differentiating stimulus. The average rate for this day, including the extinction curve, is already lower than that of the previous day, and on the following day an unusually low value is maintained (record omitted in Figure 82). On the next day the light is turned off after twenty minutes. The rate again rises, although it does not quite return to its former high value. In Figure 82, Curve C′, the vertical bar marks the change to 'light off.' The average for the group, plotted as rate rather than as number *vs.* time, is given in Figure 81 C.

An important characteristic of the record C′ in Figure 82 is the complete absence of any compensatory increase in rate following removal of the depressive stimulus. The depression is clearly different from cases previously reported where any tendency to suppress the rate is followed by marked compensation.

In the case of Group D (Curve D in Figure 81) the light is on during the simple periodic reconditioning (tenth day on the graph), and the rate shows no increase as the result of changing from $R^{IV} \rightarrow Sl$. On the following day it is turned off after 25 minutes. An extinction curve follows for the same reason as in the case of Group C. Here, however, it leads to a greatly increased rate, a typical example of which is given in Figure 82, Curve D. After two days of this rate (omitted in Figure 82) the light is again turned on and the rate falls to its previous low level. In the particular case of Curve D′ (Figure 82) the effect is especially great immediately after the light is presented (at the vertical bar). Some slight recovery is evident toward the end of the hour.

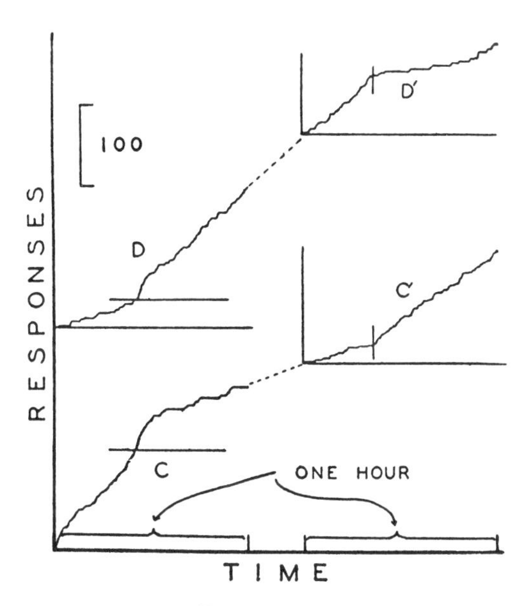

FIGURE 82 (17)

DAILY RECORDS SHOWING THE DEPRESSIVE EFFECT OF A LIGHT

C: Periodic reinforcement in the absence of a light. At the horizontal line the light was presented. Its first effect was as a discriminative stimulus (cf. Figure 60) yielding an extinction curve. Following this the rate was depressed by the light. C′: on the second day following C the rate is still depressed by the light and rises immediately when the light is turned off (at the vertical line). D: periodic reinforcement in the presence of a light. In the discrimination which had preceded this experiment the absence of the light had been S^D. When the light is now turned off, therefore, an extinctive curve is obtained as at C, except that an increased rate of responding now supervenes. D′: on the second day following D the increased rate still prevails but the effect of the light, presented at D′ is almost complete suppression.

C: CONDITIONED DISCRIMINATIVE STIMULI

The conditioning of discriminative stimuli in Type R raises the same difficulty as in secondary conditioning of Type S. Suppose, for example, that a light has become a discriminative stimulus for the response of pressing a lever. Then if a tone precedes the light sufficiently often, it should come to serve as a discriminative stimulus also. Here again the tone must not precede the light when the operant is reinforced or the effect of the reinforcement will be direct. But a series in which sS^D*: light* . R is occasionally reinforced but sS^D*: tone* S^D*: light* . R is not should establish a further discrimination in which sS^D*: tone* S^D*: light* . R declines. So far as I know, there is no available experimental evidence on this point.

The Pseudo Reflex and the Attempt to Extend Type S to Operant Behavior

The various kinds of pseudo reflexes described in the preceding pages throw some light on the difficulties that have been encountered in attempting to extend the Pavlovian type of conditioning (Type S) to conditioned behavior in general. The impossibility of a successful extension should have been obvious from the comparison of the two types given in Chapter Three. In many cases the skeletal 'conditioned reflexes' set up on the analogy of Type S are pseudo reflexes based upon Type R. The current literature contains descriptions of many devices for the study of conditioning which pretend to parallel the Pavlovian system but which utilize examples of this sort. Conditioning of Type R may enter into an experiment which is by intention only of Type S because most of the stimuli that elicit skeletal respondents (flexion, startle, winking, and so on) function as negative reinforcers in Type R. The cessation of the stimulus acts as a positive reinforcement, which may be correlated with some aspect of the response to produce conditioning of Type R. In describing the establishment of a conditioned skeletal respondent it is important to examine the temporal relation of the response to the cessation of the reinforcing stimulus. In the typical example of a conditioned flexion reflex two possibilities are presented, or, if we assume that the response to the shock is not only flexion but a mild clonus in a flexed position, there are three possibilities.

(1) In the first case the cessation of the shock is not correlated closely with any part of the movement of the limb. The shock persists long enough not to reinforce the original flexion very strongly when it ceases, and is so arranged that its cessation coincides at random with flexive and extensive stages of the clonus. Under these conditions the flexion reflex will persist if the stimulus is strong and if it is not elicited often enough to produce fatigue. (2) In the second case the cessation of the shock follows immediately upon flexion. This may be brought about mechanically if the foot of the animal is pressed against the electrode, so that the contact is broken by the response. The required circumstances arise also if the duration of the shock is just above the latency of the reflex. If the stimulus is strong, the conditioning of Type R may be obscured, but if it is weak and the respondent adapts out, the conditioned response will remain. It will take a form which is dictated by the conditions of correlation of the reinforcements. A common example is the conditioned response to a hot plate; the first response may be a strong respondent, which adapts out to leave a simple operant, the magnitude of which is great enough only to break the contact with the plate. (3) In the third case the cessation of the shock is contingent upon an *extensive* phase of the clonus. Flexion-followed-by-extension is reinforced as an operant. The strong extension that develops is not so easily obscured by the respondent flexion, and even though the stimulus is strong and the respondent intact, the strong conditioned operant composed of the extensive movement will be observed. If the respondent is weak, flexion occurs as a necessary part of the operant. This kind of contingency is not unusual outside the laboratory. In the well-known 'natural' example of the flexion reflex in which a dog steps upon a thorn, suppose the thorn to remain in the foot until it is dislodged by its own inertia as the result of a sudden downward movement of the foot. The case is the acquired response of *shaking off* a noxious source of stimulation.

These cases arise simply from stimulation with a shock and are distinguished by the different times of its cessation. An unconditioned stimulus (say, a tone) has not yet been introduced and conditioning of Type S is therefore not yet possible. When a tone is introduced the following possibilities arise. If the tone is correlated with a shock terminated as in (1), a true conditioned reflex of Type S may perhaps be established. The tone may eventually produce a brief flexion of the leg. When it is correlated with a shock terminated as in (2), the tone becomes a discriminative stimulus for the resulting operant. When the response is made before the shock, there is actually no reinforcement, positive or negative, and such a discriminated operant cannot maintain its conditioned status. The repeated presentation of the tone will extinguish the response until a shock is again received because a response has not been made. When a tone is correlated with a shock terminated as in (3), a discriminated operant arises in which the response is flexion and extension. As in Case 2 the repeated elicitation of the response will extinguish it because no reinforcement follows when the response is made. A fourth case is possible if the cessation of the tone is correlated with the response. Because of the contiguity of the tone and shock the tone becomes negatively reinforcing according to Type S and its cessation therefore positively reinforcing. The response is made because it has previously been followed by cessation of a conditioned negative reinforcement.

The second case is the commonest form of an intended extension of Type S which involves Type R. It appears in a majority of the techniques designed to study conditioned motor behavior. Four examples may be listed: (a) Watson's apparatus for conditioned finger withdrawal (75), where the movement of the finger breaks the shocking circuit, (b) Hunter's apparatus for use with rats (52), in which a shock follows a buzzer, provided no response has been made, (c) Hilgard and Marquis's technique for conditioning reflex closure of the eyelid in the dog [described in (48)], where the closure of the lid cuts off the unconditioned stimulus of a puff of air to the cornea, and (d) Brogden and Culler's device for the motor conditioning of small animals (34), in which a 'cat, upon turning the cage an inch or more when [a] sound begins, escapes the shock by breaking the . . . circuit.'

The actual behavior obtained in any of these procedures does not show the substitution of a stimulus in an unconditioned reflex such as is required by the Pavlovian formula. The behavior is a discriminated operant, and the sequence of response and reinforcement are obviously of Type R. One consequence of this operant nature is that the form of the conditioned response need not be identical with that of the unconditioned reflex in response to the reinforcing stimulus. The form of the response that cuts off a shocking current need not be the same as the unconditioned response to the shock. Even when the procedure does not provide for a correlation between the response and cessation of the negatively reinforcing stimulus, the same result may follow if the stimulus is one which is frequently cut off by the response in the life of the organism outside the laboratory (see the section above on Conditioned Discriminative Stimuli), as is the case in the examples just cited.

Overlap of Operant and Respondent

When an operant and a respondent overlap topographically, an experimental case may be set up that seems to have unique properties. It has been described by Konorski and Miller (56, 57). A shock to the foot of a hungry dog elicits flexion and food is then given; eventually flexion occurs in the absence of a shock. The case may be interpreted in the following way. The operant $s \,.\, R$: *flexion* is weak and appears only occasionally. The strong respondent $S \,.\, R$ has more or less the same form of response, and we must assume that its elicitation brings out at the same time the operant, which sums with it. We have in reality two sequences: S^0 : *shock* . R^0 : *flexion*, and $s \,.\, R^0$: *flexion* $\rightarrow S^1$: *food*. The respondent $S^0 \,.\, R^0$ need not increase in strength because it is followed by S^1 but may on the contrary decrease (see page 63). But the operant $s \,.\, R^0$ increases in strength to a point at which it is capable of appearing without the aid of the respondent. The existence of the two independent components is demonstrated by the fact that the operant eventually appears without the respondent when it has become strong enough through conditioning and that it may even be conditioned without the aid of the respondent although less conveniently because at least one unconditioned occurrence must be obtained.

There is thus an important difference between the sequence 'shock, flexion, food' and the sequence 'lever, pressing, food.' The first contains a respondent. Since there is no eliciting stimulus in the second sequence, the food is correlated with the response but not with the lever as a stimulus. In the first sequence the food is correlated as fully with the shock as with the flexion because of the necessary (*i.e.*, the eliciting rather than the discriminative) connection between the stimulus and the response. The Konorski and Miller case does not fit either type of conditioning so long as the double correlation is maintained. Conditioning of Type S will occur (the shock-salivation discussed in Chapter III), but so far as the economy of the organism is concerned, there is no reason why conditioning of Type R should occur so long as there is always a correlation between the reinforcement and an *eliciting* stimulus. Nothing is to be gained in such a case; the original sequence operates as efficiently as possible. In nature the correlation of a reinforcing stimulus with a respondent is very rare, if it can be said to exist at all. It may appear to exist when the form of a respondent overlaps that of an operant correlated with a reinforcement.

A special case of the overlap of respondent and operant arises when the latter is reinforced by renewing a negative reinforcement. It is traditionally described as the voluntary control of an involuntary response. For example, let some negative reinforcement be correlated with closing the eyes (so that opening the eyes, or keeping them open, is positively reinforced), and let a stimulus for a wink be presented. The result may be treated in terms of algebraic summation. If the operant of keeping the eyes open is strong, a response may be lacking. That is to say, the 'involuntary' response will be 'controlled.' In this example, however, it is probably always possible to find a stimulus for the respondent of an intensity that will invariably elicit the response in spite of the conflicting operant.

Chapter Seven

TEMPORAL DISCRIMINATION OF THE STIMULUS

Time as a Discriminable Continuum

The scientific study of sensory discrimination was encouraged by the philosophical movement known as British empiricism, which emphasized the importance of 'sense data' in an understanding of the human mind. It has been carried on principally as part of a science of mind based on the doctrine of 'mental elements.' Throughout, it has been concerned very largely with the mechanisms employed in the reception of stimuli and especially with determining their liminal capacities.

Now, in a science of behavior the vigorous investigation of the limen as a single aspect of the process of discrimination would be suggested only at a relatively advanced stage. It is initially of greater moment to know what is happening when a discrimination is being made—a subject that may be avoided by a mental science because of its metaphysical premises concerning the activity of the experiencing organism. There is one question of capacity, however, that cannot be ignored in the early study of discrimination as an aspect of behavior. The formulation of this important process must depend to some extent upon the nature of what is discriminated, and it is necessary to know the various properties of stimuli to which an organism is capable of responding differentially. The list given in Chapter Five of the sensory continua upon which pairs of stimuli could differ was not intended to be exhaustive, and I shall now make certain additions concerning temporal properties. Stated more generally the problem is how time as a dimension of nature enters into discriminative behavior and hence into human knowledge.

There are certain temporal 'discriminations' which are not properly referred to as such. They arise because behavior necessarily takes place in time. By altering the temporal conditions of any one of the dynamic processes defined in Chapter One, it is possible to change the resulting state of the behavior and hence to demonstrate what might be called a 'differential response to time.' For example, in Chapter Four it was demonstrated that the rate of responding during periodic reconditioning was a function of the period. The state of the reflex quickly adjusts to a change from, say, five to six minutes between reinforcements, and the organism might therefore be said (inaccurately, I am contending) to distinguish between five- and six-minute intervals. Again, it was shown in Chapters Three and Four that the effect of a reinforcement in conditioning of Type R was a function of the time elapsing between it and the correlated response. A difference in the resulting state was demonstrated when periodic reconditioning was delayed four and six seconds, and it might be said that the organism 'distinguishes' between these intervals. Experiments designed to test the 'temporal limen' of the rat often make use of dynamic processes of this sort and do not actually involve a temporal discrimination as it will be defined here. For example, a rat may be retained for different lengths of time while running along several different paths to food. The rat comes to take the path along which it is retained the shortest time. The effect of a given retention is to delay the reinforcement of the response of moving into that path. A comparable case would be an arrangement of two levers, responses to one of which were reinforced after two seconds and to the other after eight. The rat would come to respond to the two-second lever, not because it had made a discrimination between two and eight seconds, but because the response to the two-second lever was more strongly reinforced.

Another use of the term 'temporal discrimination' which goes beyond the definition to be given here arises in the treatment of *eliciting* stimuli. The temporal properties of an eliciting stimulus are, of course, important. According to the law of Temporal Summation the prolongation of a stimulus has the same effect as an increase in its intensity, and a sort of temporal limen might be defined as the least change in the duration of a stimulus necessary to produce a detectable difference in the magnitude of a response. Such a limen does not require conditioning for its demonstration, and it differs fundamentally from the discriminative limen to be considered here. A spinal frog, for example, may 'distinguish' between a stimulus lasting three seconds and one lasting four by giving responses of different magnitudes, but this is not what is ordinarily meant by a temporal discrimination.

The phenomena that are referred to here as temporal discriminations may be formulated in the following way. In the establishment of a discrimination (as contrasted with the elicitation of a response) the temporal properties of a stimulus acquire a new significance. In both conditioning of Type S (which in practice always involves a discrimination) and the kind of discrimination of Type R described in Chapter Five, a stimulus is temporally correlated with another event—the presentation of a reinforcing stimulus. The temporal correlation makes it possible to single out a given point on the continuum established by the sustained presentation of a stimulus. For example, let a tone be presented and maintained for some time. So far as the elicitation of a response is concerned, the only importance attaching to the prolongation is the resulting summative effect. But when we establish a coincidental relation between a second event and some *point* in the course of the prolonged stimulus, the organism may begin to distinguish between the stimulus momentarily at that point and the same stimulus momentarily at some other point by reacting differently to the two in some other way than cumulatively. This is a temporal discrimination, as the term will be used here.

The successive parts of a continuous stimulus will be indicated as follows. The continuum is divided into parts of arbitrary length through the use of a clock. We may follow the accepted division into seconds, minutes, and so on, or use some other convenient unit. For example, it may be convenient to divide a stimulus lasting two minutes into eight parts of fifteen seconds each. When first presented, the stimulus may be written St_0. At the end of one arbitrary division it may be written St_1, at the end of a second St_2, and so on. If a stimulus is presented and immediately withdrawn, it may subsequently function in much the same way as a prolonged stimulus. (Here there is no parallel in the case of elicitation because there is no possibility of a summative effect.) Such a stimulus may be written at the time of its presentation as $S + t_0$. After one unit of time has elapsed (the stimulus having been withdrawn) it may be written $S + t_1$, after a second $S + t_2$, and so on.

Temporal Discrimination of Type S

The two outstanding cases of a temporal discrimination of Type S are the so-called 'delayed' and 'trace' conditioned reflexes of Pavlov [(64), pp. 40–41, 88–105]. The 'reflexes' are pseudo, as I shall note later. The basic observation in the case of the 'delayed reflex' is as follows. Let a conditioned reflex first be established to St_0, and let the reinforcement then be delayed so that $St_0 . R$ is extinguished but $St_1 . R$ reinforced. When the discrimination has been established, presentation of the stimulus is not followed by a response until one unit of time has elapsed. By gradual steps the interval may be increased until the response occurs only when a maximal stage of St_n is reached.

With such an organism as a dog the property of duration is not highly significant in the establishment of a discrimination. There is considerable induction between neighboring points on the continuum, and reflexes differentially reinforced draw apart in strength only very slowly. Because of the strong induction, a gradient is usually obtained, so that the strengths of the reflexes at successive points on the continuum may vary somewhat as follows:

$$[St_0 . R] = 0$$
$$[St_1 . R] = 0$$
$$[St_2 . R] = 1$$
$$[St_3 . R] = 3$$
$$[St_n . R] = 7 \text{ (maximal value)}$$

A temporal discrimination of this sort may presumably be established without beginning with the reinforcement of $St_0 . R$. The reinforcing stimulus could originally be presented only after some arbitrary interval, but according to Pavlov the discrimination is difficult to execute in this way. Some response to St_0 develops through induction, although it is never reinforced according to the conditions of the experiment. Extinction takes place, and eventually the delayed response is obtained. If $St_0 . R$ is originally conditioned and a long interval then introduced at once without a progressive approach through smaller steps, the reflex may disappear altogether. According to Pavlov, the response appears again at the point of reinforcement and advances to an 'intermediate position between the commencement of the conditioned stimulus and its reinforcement (p. 89).'

The possibility of a discrimination of this sort is of great importance in the study of conditioning of Type S. Unless S^0 and S^1 are simultaneously presented, a temporal discrimination will sooner or later develop, and it is then meaningless to use the 'latency' of the conditioned reflex as a measure of its strength. The criticism is important because an interval of time is frequently introduced between S^0 and S^1 in order to observe the strength of $S^0 . R$ without withholding S^1. Aside from the invalidation of latency as a measure, the procedure cannot be successful for the purpose for which it is designed, in view of the possibility of a temporal discrimination. Although some 'anticipatory' responding may occur through induction from St_n upon St_{n-1}, it is not the whole response. Part of the conditioned response must be obscured by the unconditioned response to S^1 if S^1 is not omitted.

In a 'trace' reflex the stimulus is presented and withdrawn, and the reinforcement follows at some later time. The same kind of discrimination is here involved. The induction between $S + t_0$ and $S + t_n$ is apparently much greater than that between St_0 and St_n, and the case is established with great difficulty when the reinforcement is originally correlated with a relatively late $S + t_n$. The most convenient procedure is to begin with $S + t_0$ and to introduce progressively longer delays.

A simple variant of this kind of discrimination is obtained when a reinforcing stimulus is presented to an organism at some set rate. Pavlov reports an experiment in which food was presented to a dog every thirty minutes. Eventually the dog began to secrete saliva at about the time at which food was to be presented. Here S is the presentation of the food itself, and the dog comes to make a discriminative response to $S + t_{30}$ alone. With this long interval Pavlov reports no 'anticipatory' induction to, say, $S + t_{29}$ but on the contrary an occasional delay to $S + t_{31}$ or $S + t_{32}$.

It has already been observed (page 191) that when the difference in the strengths of a reflex in the presence of S^D and S^Δ is taken as the measure of a discrimination, the state of the drive must be considered. In the very early stages of a discrimination a response may be obtained to S^D and not to S^Δ if the drive is low, although the actual difference in strength is small. This is particularly significant if the discriminative stimulation is slight and the discrimination therefore difficult. The essentially discriminative nature of what Pavlov calls a trace reflex is demonstrated by his observation that it is most easily developed at a low drive. For example, the averages of three series of successive responses to a whistle plus the lapse of time differed when the reflex was weak and strong as shown in the following table (unit of time = 30 seconds):

$S+$:	t_1	t_2	t_3	t_4	t_5	t_6
Weak	0	0	0	⅔	2⅓	5⅔
Strong	1	4⅓	3	3⅓	4⅓	5⅔

A given stimulus St_n or $S+t_n$ may stand not only as the sole factor in a discrimination but as a component. An example of a component not involving time is as follows. Let reinforcement be correlated with *SaSb* but not with *SaSbSc* or *SaSc*. Then in the presence of *SaSb* the organism will respond, but if *Sc* is added, it will not. Now, St_n or $S+t_n$ may be substituted for *Sc* in such a case. Pavlov describes an experiment of this sort in which *Sa* was the general stimulation from the experimental situation, *Sb* the sound of a metronome, and *Sc* the sound of a horn. *SaSb* was reinforced but $SaSbSc+t_n$ was not. As a result a response could be obtained upon presentation of the metronome provided the horn had not previously been sounded at a certain interval. In a similar experiment involving the combination of the sound of a metronome and another stimulus $S+t_n$, where *S* was periodic feeding as described in the preceding paragraph, a dog came to respond to a metronome when it occurred at the thirtieth minute but not at the fifth or eighth. Unlike the case described above, this experiment showed the expected induction from $S+t_{30}$ to *preceding* points on the continuum; after more than eight minutes the metronome produced an effect which increased as the length of time increased.

In 'trace' and 'delayed reflexes' we are dealing with discriminations, and we should not be surprised to find the concept of 'inhibition' again being offered in explanation. In this case the suppression of activity during the interval between $S^{\Delta}t_0$ and $S^{D}t_n$ is spoken of as the 'inhibition of delay.' The effect may be formulated without this notion, as the preceding paragraphs should indicate, and there are again logical and practical reasons for omitting it. The data for 'disinhibition' during delay are in general more convincing than during extinction or non-temporal discriminations, but other alternative explanations are here suggested. In a typical experiment Pavlov reports a 'disinhibition' of the delay to a tactile stimulus brought about by a metronome as follows (units of thirty seconds):

S:	t_0	t_1	t_2	t_3	t_4	t_5	*Total*
Tactile	0	0	2	6	13	16	37
Tactile + metronome	4	7	7	3	5	9	35

The figures for tactile stimulation alone are the averages of four series. It will be seen that in the presence of the metronome there is no significant delay. One possible explanation is the very slight discriminative efficiency of a difference in time which, in view of the gradualness with which S^D is presented, should facilitate the operation of the factors already advanced as producing the effect of 'disinhibition.' A special possibility is that of inductive overlap between the presentation of the 'disinhibiting' stimulus and the presentation of food, which is in accord with the fact that the total secretion is approximately the same in the two cases and hence indicative of a single reserve.

In no part of the preceding formulation does 'time' or 'an interval of time' enter with the status of a stimulus. Time appears as the single property of duration, comparable with intensity, wave-length, and so on. As I have already noted, the appearance of a single property in the position of a stimulus is a certain sign that the reflex is pseudo—that is, that the stimulus is discriminative rather than eliciting. In the present case there is a discrimination between the point St_n and the adjacent points St_{n-1} or St_{n+1} on a temporal continuum, just as in another case there is a discrimination between a given wave-length and adjacent wave-lengths. The discrimination arises because the reinforcement is correlated with a stimulus possessing this single value of the property, but the response is not correlated with the single property in isolation. We cannot write *S: 10 seconds . R* as a true reflex. Like *S: wave-length of 550 mμ . R,* the expression is incomplete. It is for this reason not a rigorous use of the term to call the preceding cases 'delayed' or 'trace' reflexes. In the extension of this formulation of time to the more general question of human knowledge, it is important to make this distinction. Time is frequently spoken of as a stimulus. For example, Pavlov says that 'the duration of time has acquired the properties of a conditioned stimulus (p. 40)' and speaks of 'time intervals in their rôle as conditioned stimuli (p. 41).' But time has not the proper dimensions of a stimulus. To regard it in this simple and inaccurate way is to raise a strong barrier to an understanding of the part that time plays in nature and in our knowledge of nature.

The insistence upon the distinction between real and pseudo reflexes is, I believe, not a quibble. It is directed toward obtaining some degree of order and regularity in a science of behavior. For example, the laws of latency, threshold, after-discharge, and so on, are intended to apply to reflexes generally, but if we permit ourselves to write *S: interval of time . R* as a reflex without qualification, they are meaningless when applied to such an entity. We have not only overlooked much of the process of establishing such a relation, but we emerge with an entity which has unusual properties and appears to behave anomalously.

Temporal Discrimination of Type R

The preceding examples of temporal discrimination concern conditioned reflexes of Type S. The same principle of dividing a temporal continuum into distinguishable parts through the correlation of an external event applies to Type R, although a special technical difficulty arises. In Type S any point on a temporal continuum (any moment during the continued presentation of a stimulus) may easily be singled out for correlation with a reinforcing stimulus because the latter is arbitrarily controlled. In Type R the reinforcement is contingent upon the occurrence of the response, and there is no certain way of obtaining a response at a given time during the presentation of a discriminative stimulus. The correlation between some point on the continuum and the reinforcement may simply be established and the organism allowed to respond freely. Thus, reinforcement of the response to the lever might be conditional upon its occurring between 15 and 20 seconds after presentation of a light. If a simultaneous discrimination has previously been established, all responses to $S^{\Delta}t_0 \ldots t_{15}$ will be unreinforced and should disappear, while those to $S^{D}t_{15} \ldots t_{20}$ will be reinforced and maintained. If no discrimination has previously been established, one will presumably arise through induction of $st_{15} \ldots t_{20}$ upon $St_0 \ldots t_{15}$ which will disregard the temporal factor, but the behavior should eventually be narrowed down as the temporal discrimination progresses.

I have no specially designed experiments to report on this subject, but the behavior of the rat during periodic reconditioning (Chapter Four) involves two temporal discriminations of Type R. In the first case, which may be called the *Temporal Discrimination from the Preceding Reinforcement,* the discriminative stimulus is the complex stimulation arising from the presentation of a pellet of food and its ingestion plus the lapse of time and is therefore of the sort occurring in Pavlov's 'trace reflex.' If we divide the usual interval of five minutes into ten parts of thirty seconds each, then it may be said that all occurrences of $sS + t_0 \,..\, t_9 \,.\, R$ go unreinforced, while certain occurrences of $sS + t_{10} \,.\, R$ are reinforced. That a temporal discrimination develops is clear from the results given in Chapter Four. The first evidence is the flattening of the separate extinction curves following periodic reinforcements. Because of the periodic procedure $sS + t_0 \,.\, R$ is weakened and $sS + t_n \,.\, R$ strengthened, while intervening reflexes are affected according to their proximity to these extremes. When the interval between reinforcements is not too great (say, three minutes), the strength of the reflex immediately after reinforcement may reach zero, in which case the discrimination produces the third order deviation described in Chapter Four.

The action of the receipt and ingestion of food as a discriminative stimulus may be demonstrated, even when the discrimination is not far advanced, by reinforcing the reflex at intervals of two and eight, rather than five and five, minutes. The slope of the curve is not seriously affected, but the local discriminative effect of the adjacent reinforcements is clearly indicated by the fact that the resulting curve is wave-like rather than linear in character. Figure 83 (page 272) is of this sort. Adjacent reinforcements are accompanied by a depression in rate, which is due to the summation of their discriminative functions. It will be recalled from Chapter Four that when the reinforcements are evenly spaced, the rate is approximately constant throughout the hour.

The maintenance of a constant extinction ratio during periodic reconditioning may seem to weigh against this interpretation. If a discrimination is established between $S + t_n$ and $S + t_{n-x}$, why does the rate of responding not decline as it does during the typical discrimination described in Chapter Five? In other words why does the rat not simply learn to wait until the time of reinforcement? There are several answers. In the first place the discriminative difference between 'light-on' and 'light-off' is greater than that between even remote points on the temporal continuum. It is quite possible that although a temporal discrimination is established in the sense that a separation in strength takes place between $sS + t_n . R$ and $sS + t_{n-x} . R$, there may be little or no inductive breakdown. The Law of the Operant Reserve should then hold, and the periodic reinforcement of $sS + t_n . R$ should yield a con-

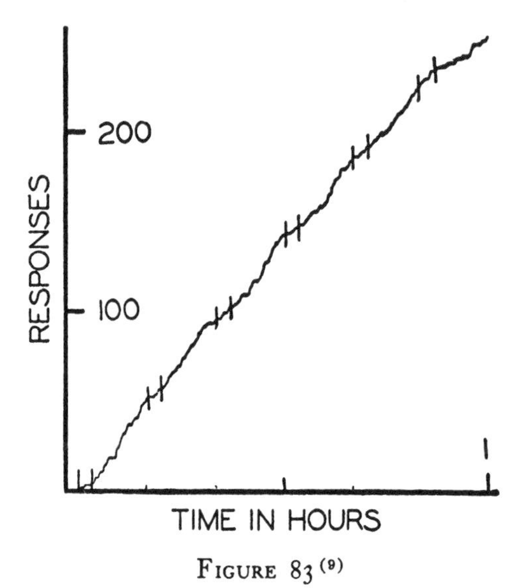

FIGURE 83 (9)

WAVE-LIKE RECORD OBTAINED BY GROUPING SUCCESSIVE REINFORCEMENTS

Responses were reinforced alternately at two- and eight-minute intervals. The discriminative effect of the ingestion of the pellet is additive.

stant extinction ratio even though most of the occurrences of the response actually take place in the presence of $sS + t_{n-x}$. The slight decline in rate observed during the twenty-four days of Figure 29 may show a breakdown of induction commensurate with the inductive properties of the continuum. But there is no need to appeal to the slightness of the discriminative difference. Although it is clear that the rat discriminates between $S + t_{10}$ and $S + t_1$, so that it responds in the presence of the former but not in the presence of the latter, it does not discriminate between $S + t_{10}$ and

$S + t_9$. But responses in the presence of $S + t_9$ are not reinforced. There is an essential difference between a discrimination in which the change from S^{Δ} to S^D is gradual and one in which it is abrupt. A comparable experiment in which some such stimulus as a light takes the place of (or, rather, supplements) the lapse of time would require that the light slowly increase in intensity from zero immediately after reinforcement to a maximal value at the end of the interval. A somewhat cruder parallel might be obtained in which a light is alternately off and on for periods of two and three minutes respectively, and the response is reinforced once during each of the latter periods. In such a case a discrimination should be established between Sl and $S\lambda$, but the extinction ratio should be maintained during the period of Sl. This is essentially what takes place during a temporal discrimination of this type. The rat discriminates between the ranges $S + t_0 \ldots t_x$ and $S + t_x \ldots t_n$, but its responses in the presence of $S + t_x \ldots t_n$ are only *periodically* reinforced. This characteristic seems inevitable in a temporal discrimination because S^D cannot, of course, be presented in any other way than gradually.

This interpretation of the rate under periodic reconditioning has a bearing upon the subsequent extinction curve which may conveniently be noted here. It may be argued that the curve for extinction obtained after periodic reconditioning has peculiar properties (particularly the large area that it encloses) not because of the creation of a larger reserve, as I have argued in Chapter Four, but because of a temporal discrimination. In contrast with its behavior in original extinction the rat may continue to press the lever because a period of no reinforcement has previously been followed by reinforcement. This objection is valid within certain limits. The extinction curve obtained after periodic reconditioning is for $sS + t_{n \ldots n+x} . R$. There are several consequences. First, the curve for extinction should not be regarded as beginning until the first reinforcement is *omitted*, in order to dispense with the few responses to $S + t_0 \ldots t_n$ that would be included if the curve were measured from the last reinforcement. This condition was observed in Chapter Four when the curves were begun on a fresh day and also (as in Figure 36 B) when the reinforcement was suddenly discontinued. A second consequence is that the extinction curve may begin at a higher rate than that prevailing under periodic reconditioning, as in Figure 38, Chapter Four. This is in accord with the fact that the curve is for $sS + t_n \ldots . R$. The curve for discrimination should not show a comparable increase in rate because the periodic stimulation from the ingestion of food is still received, and that this is the case may be seen by comparing Figure 38 with Figure 54. A third consequence is that the discriminative component continues to change as time elapses, but according to the Law of the Operant Reserve the same number of responses should be obtained. The height of the curve is therefore not affected, and its shape is very probably not significantly modified.

The procedure of periodic reconditioning also provides the basis for another kind of temporal discrimination. There are two kinds of events taking place during such an experiment: (1) periodic reinforcements and (2) the occurrence of unreinforced responses. The temporal discrimination just considered was based upon the first of these procedures, but the second is also available. Because of the extinction ratio and the occurrence of an approximately constant number of elicitations between reinforcements, there is a relation between reinforcement and the preceding behavior of the rat. This may be expressed in terms of the preceding rate, or better, of the interval of inactivity immediately preceding the reinforced response. This second discrimination may be called the *Temporal Discrimination from the Preceding Response.*

If the rat is responding at a precisely constant rate, the reinforced response will always be preceded by a constant interval of no responding. For example, if the extinction ratio is 20:1 and the interval five minutes, responses occur at the rate of four per minute and each reinforced response is preceded by an inactive period of fifteen seconds. Now, if there is any local variation in the rate, it is more likely that a reinforced response will be preceded by a *longer* period of inactivity. This follows directly from the fact that the establishment of the connection between the response and the reinforcement is wholly independent of the behavior. A schematic case is shown in Figure 84. If the rate is constant, there is only one possibility, which is maintained throughout the experiment, as shown by the shaded horizontal bar in Line A. Establishing the connection between the reinforcement and the response at any time during the period means simply that the next response will be reinforced and will have followed the preceding response at a constant interval. If, on the other hand, the rate shows local variations, there must necessarily be alternate crowded and vacant spaces as shown schematically in Line B, and the tendency toward grouping means that a reinforced response will more frequently follow a relatively long interval. In the particular case represented in Line B the establishment of the connection between the response and a reinforcement will result in a reinforced response following a *short* interval during only one-third of the time (indicated by the black bars). During two-thirds of the time (open bars) the connection leads to a reinforced response following an interval four times as long. Through a temporal discrimination the result of this

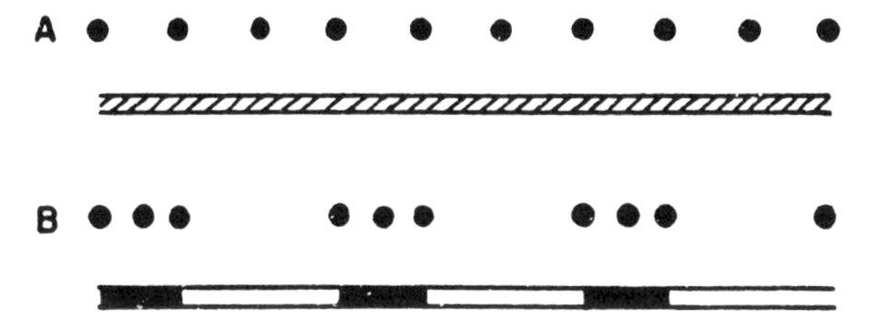

FIGURE 84
The dots represent responses occurring in time from left to right.

change is to strengthen the response following long intervals of inactivity and to weaken it following short intervals. The effect is a stabilization of the rate at an approximately constant value, which is determined by the extinction ratio.

This second kind of temporal discrimination probably contributes to the flattening of the successive small extinction curves when periodic reconditioning is first begun, especially at lower frequencies of reinforcement. The reinforcements tend to occur when the rat is responding relatively slowly toward the end of each curve, and consequently the rapid responding at the beginning is weakened. The discrimination works here in collaboration with that based upon the preceding reinforcement. That is to say, the rat stops responding rapidly just after each reinforcement because (1) responses following reinforcement are never reinforced and (2) responses following close upon other responses are seldom reinforced. The second discrimination alone is probably responsible for the prolonged maintenance of a constant rate during periodic reconditioning. The linearity of the curves can hardly be due to the discrimination based upon the preceding reinforcement, for it would in such a case be the accidental result of a given stage of development of the discrimination. Its continued maintenance as such is highly unlikely. The discrimination based upon the preceding response, however, produces a constant rate as its ultimate and stable effect.

It is an oversimplification to regard this second kind of discrimination as based upon only one preceding unreinforced response. Other antecedent responses must be supposed to contribute also, although less importantly. If the effect of an unreinforced response may be said to be to 'postpone' a later response, then when a re-

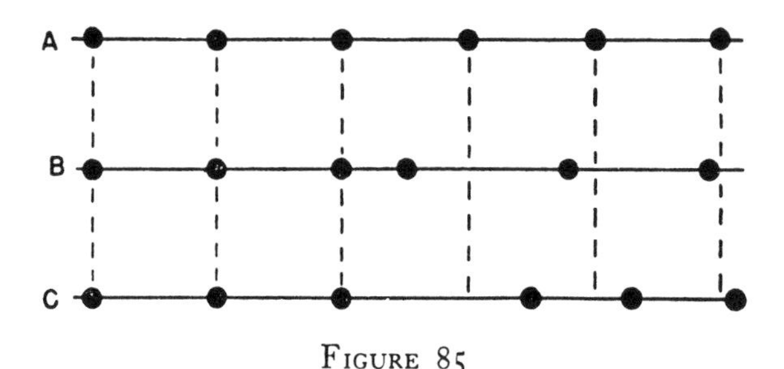

FIGURE 85

The dots represent responses occurring in time from left to right.

sponse occurs for some reason too soon, there is a summated effect toward postponement. Thus in Figure 85 the longer interval in Line B is due to an (incidental) grouping of the two preceding responses. If for any reason a response occurs too late, there is a reduction in the normal postponing action and the following response should occur sooner, as in Line C. The discrimination is based upon all the preceding responses (less and less significantly as the interval increases), but for the present argument its effect in producing and maintaining a constant rate is qualitatively the same as if only the preceding response were effective. Figure 85 shows a kind of compensation, which is similar in many respects to that of the mechanism responsible for the rate of responding in the absence of a discrimination, which was discussed in Chapter Four. The distinction is, I think, sufficiently clear and the evidence for a discriminative relation as the controlling factor during periodic reconditioning convincing enough.

The discrimination that is responsible for the more or less constant rate observed during periodic reconditioning is very common outside the laboratory. In man a much more rapid discrimination from the preceding reinforcement would normally develop. With an interval of five minutes, a period of inactivity of at least two or three minutes would usually soon appear. But during the latter part of the period such a discrimination would be ineffective (that is to say, there would be no 'cue as to whether it was time for a reinforcement to occur or not'), and a similar repetitive responding should take place. In the vernacular we should say that a person was making an occasional test of the lever. The rate at which the testing is made will depend, as in the case of the rat, upon the drive (see Chapter Ten), upon the frequency with which reinforcement has been made in the past, and so on. We might also say roughly that the person realizes that the reinforcement is a function of time, that he knows he has only to wait long enough and a response will be effective, and that having tested the lever once unsuccessfully he soon feels that it is time to try again. The last part of this statement applies to the discrimination here being discussed, although the terminology goes far beyond the observed facts.

Reinforcement at a Fixed Ratio

Although the time at which a given response occurs under the procedure of periodic reconditioning is eventually determined by a discrimination, the total number of responses was shown in Chapter Four to be a function of the reinforcement. This function, or the notion of an extinction ratio, does not involve the distribution of the separate responses; but the existence of a definite ratio, when considered a little more broadly, raises an interesting problem, part of the solution of which lies in the nature of the temporal discrimination that I have just described. Suppose, for example, that the extinction ratio of a given reflex is 20:1. Does this mean that the organism is incapable of surviving in an experimental environment in which a pellet of food may be obtained whenever a lever is pressed, say, twenty-five times? According to the notion of an extinction ratio this should be the case, for the receipt of a pellet of food every twenty-five responses should not maintain the reserve of the reflex, and its strength should therefore decline to zero. But this seems an improbable state of affairs. A solution suggests itself in terms of the variation of the extinction ratio with the drive. But although the ratio increases as the organism becomes hungrier, as will be shown in Chapter Ten, a value is not reached that would avail the rat if the reinforcement were made to depend upon the completion of as many as, say, one hundred responses. A satisfactory solution requires an analysis of the behavior of the rat when the ratio of reinforced to unreinforced responses is not the result of the rate of responding at a given periodicity of reinforcement but is externally fixed.

When the extinction ratio at a given drive has once been ascertained, three cases may be set up by changing the program of reinforcement from that of periodic reconditioning to reconditioning at a fixed ratio. The fixed ratio may be set at a value (1) less than, (2) greater than, or (3) equal to the extinction ratio. The first effect upon the organism in each case is predictable from the relation previously demonstrated between the rate of responding and the frequency of reinforcement. These changes do not involve the temporal discrimination that we are about to consider, but I shall deal with them first. Various experimental examples of the three cases have been obtained from a group of eight rats. The extinction ratios were first determined during periodic reconditioning at intervals of five minutes, and reconditioning was then carried out at fixed ratios less than, greater than, or equal to the values so obtained. At the degree of drive used in these experiments the group gave relatively low extinction ratios, which were better suited to exhibiting the principal kind of change taking place.

Case 1. Fixed ratio < extinction ratio. If a rat has been responding at a rate of, say, ten responses per interval between reinforcements, and if the ratio is then fixed at, say, eight, it is obvious that the frequency of reinforcement will increase. If the response has been reinforced every five minutes, it will now be reinforced every four. But since the extinction ratio remains constant, an increase in the frequency of reinforcement must bring about an increase in the rate of responding. This means that the frequency of reinforcement will still further increase in turn, and so on. A limiting rate of responding will eventually be reached.

If we assume that the full effect of the premature reinforcement in the first interval (when it occurs at four instead of five minutes) is felt as an immediate increase in the rate of responding, the records obtained should have the character indicated in Figure 86. In these theoretical curves an extinction ratio of 10:1 has been assumed, which with a period of five minutes should yield the straight line in the figure. The results to be expected from changes to fixed ratios of 4:1, 6:1, 8:1, and 9:1 have been indicated with curves as marked. Thus, at a fixed ratio of 6:1 the second reinforcement is obtained in three rather than five minutes. The rate of responding then increases, and a second pellet is obtained

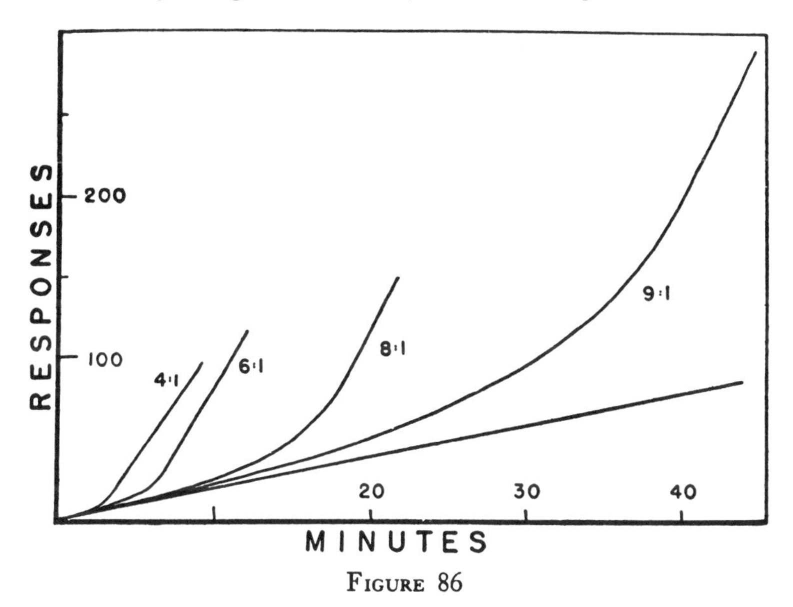

FIGURE 86

CALCULATED ACCELERATIONS FROM VARIOUS FIXED RATIOS LESS THAN THE EXTINCTION RATIO (ASSUMED TO BE 10:1)

in 1⅘ minutes. The rate increases again, and the third pellet is obtained in 1 3/25 minutes, and so on. The constant slope at which the acceleration stops has been set arbitrarily at thirty responses per minute, ten seconds being allowed for the ingestion of each pellet. The final slope is lower at the lower fixed ratios because a greater share of the time is taken up with eating.

The assumption that a premature reinforcement has an immediate effect cannot be wholly allowed. It was shown in Chapter Four that a change from one rate to another does not follow immediately upon a change in the conditions of the reinforcement. When the frequency of reinforcement (or any other factor controlling the total amount of reinforcement) is reduced, the rate of responding does not drop to a new level at once. In the limiting case of a change to a zero frequency of reconditioning the typical curve for extinction is obtained. When an intermediate frequency is adopted, the rate of responding changes along an extinction curve. A comparable effect observed during any *increase* in the total amount of reinforcement is shown in Figure 34. Experimental curves exactly comparable with the curves in Figure 86 are not, therefore, to be expected, but they may be approximated

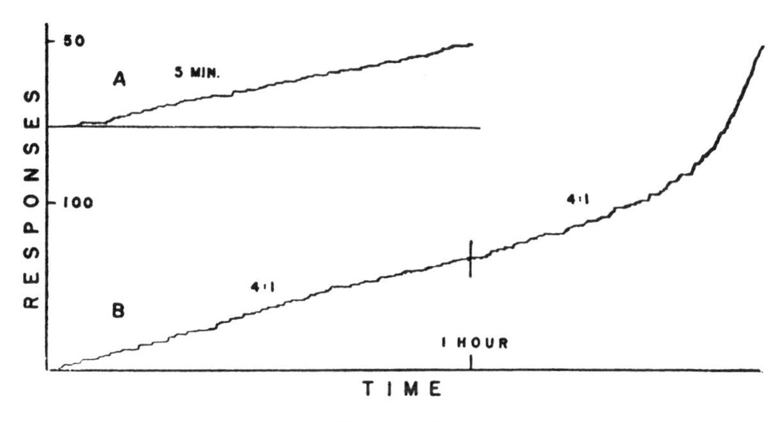

FIGURE 87

ACCELERATION FROM A FIXED RATIO SLIGHTLY LESS THAN THE EXTINCTION RATIO

The extinction ratio is calculated from Curve A which was recorded with five-minute periodic reinforcement.

so closely that the lag in the effect of the automatically increasing frequency of reinforcement must be supposed to be very slight. Figure 87 gives a representative experimental curve. The record at A is for periodic reconditioning at intervals of five minutes. The extinction ratio for this hour was $4\frac{1}{6}$:1. Previous determinations of the ratio for this rat gave an average slightly above 5:1. The records at B are for the following two days, when the response was reinforced at the fixed ratio of 4:1. On the first of these days the rate increased slightly but not excessively above the average for the rat. During the hour eighteen reinforcements were received. On the following day an acceleration began to appear, and a constant maximal rate was reached before the end of the hour.

The curve is very similar to the calculated curve in Figure 86 for a fixed ratio equal to 80 per cent of the extinction ratio. That the acceleration was not obtained on the first day may have been due to an incidental change in the extinction ratio, such, for example, as accompanies a change in drive (see Chapter Ten). The ratio on that day may have been only 4:1. Because of slight irregularities in the

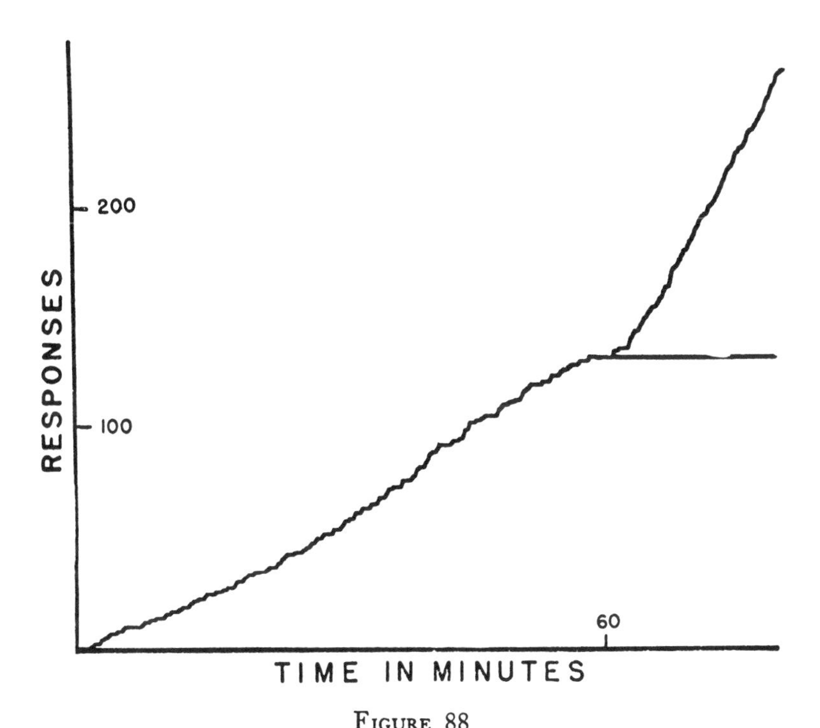

FIGURE 88

RAPID ACCELERATION FROM A FIXED RATIO APPROXIMATELY ONE-HALF THE EXTINCTION RATIO

rate of responding and the lag in the effect of an increase in reinforcement, it has not appeared to be profitable to test the present interpretation by describing curves of this sort with an equation. Curves of comparable smoothness are practically invariably obtained when the base-line is comparably uniform.

In the case represented in Figure 87 the fixed ratio was not less than 80 per cent of the extinction ratio. When it is relatively lower, a more rapid acceleration may be observed. In Figure 88 the first curve is for periodic reconditioning at five-minute intervals. The second curve taken on the following day shows the effect of a change to a fixed ratio equal to approximately one-half the extinction ratio. Here the acceleration to a maximal slope is very rapid, corresponding to the case at 4:1 in Figure 86.

When the two ratios are near together, a temporal discrimination to be described below enters to modify the result, so that curves showing a slow acceleration similar to that for a ratio of 9:1 in Figure 86 are not obtained.

Case 2. Fixed ratio > extinction ratio. When the fixed ratio exceeds the extinction ratio, so that in comparing the input and output for each reinforcement the rat may be said to 'operate at a loss,' the result depends upon the degree of excess. If the fixed ratio is very high, the response may be wholly extinguished before a second reinforcement is reached. Even when a reinforcement is occasionally received, the reconditioning effect may not be enough to sustain a given rate of responding and the response will therefore disappear through extinction. It is thus possible for a rat to starve even though a supply of food would be available if the strength of the reflex could be maintained. If the fixed ratio only slightly exceeds the extinction ratio, the rate of reinforcement is at first reduced and the rate of responding may decline, but if the extinction is not too rapid, the discriminative effect to be described shortly may enter.

When a fixed ratio less than the extinction ratio has led to a maximal rate of responding as in Figure 87, a sudden change to a larger ratio may extinguish the reflex, more or less rapidly according to the magnitude of the new ratio. Figure 89 shows the record for the day following the series in Figure 87, when the fixed ratio was suddenly changed from 4:1 to 12:1. The considerable irregularity in the record is typical. The final effect in extinguishing the reflex is clearly shown. This result may be obtained (as will be noted later) only when the maximal rate has not been long sustained, or in other words before the discriminative effect to be described in a moment has entered.

Case 3. Fixed ratio = extinction ratio. If the rate of responding is strictly uniform, reinforcing one response every *m* minutes is equivalent to reinforcing one every *n* responses, if *n* is the extinction ratio. There would seem to be no difference between the two programs, yet they have widely different consequences. In the first case no change in the behavior of the rat will improve its condition, but in the second the rat has only to respond more rapidly to receive a more frequent reinforcement. The question is whether or not the rat can 'tell the difference' and adjust its behavior accordingly, and, if so, through what processes. Since at a constant rate of responding there is no actual difference in the procedures, no change in behavior is to be expected, and with optimally uniform rates of responding over short periods of time none is observed, as will be shown in a moment. But it is difficult to prolong an experiment of this sort for two reasons. First, there

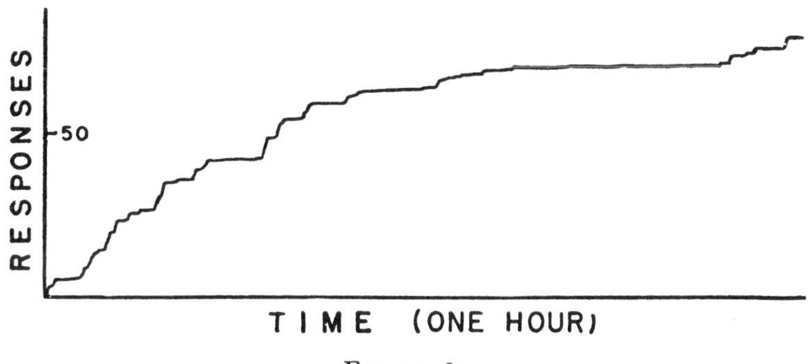

FIGURE 89

EXTINCTION RESULTING FROM A FIXED RATIO MUCH GREATER THAN THE EXTINCTION RATIO

is some slight variation in the extinction ratio from day to day, so that an exactly equivalent fixed ratio cannot be arranged. If on a given day the rat is slightly hungrier than usual, Case 1 may arise; if it is less hungry, Case 2. Second, the rate of responding is not in fact wholly uniform, and hence the discrimination already mentioned may enter. The discrimination enables the rat to take advantage of the fixed ratio and increase the rate of delivery of food.

This new discrimination is very important, especially with respect to the concept of a reflex reserve. It arises because a program of reinforcement based upon the completion of a number of responses has properties which distinguish it from the program responsible for the second temporal discrimination described above. The essential difference is that a reinforcement based upon the completion of a given number of unreinforced responses (*i.e.*, according to a fixed ratio) favors the reinforcement of responses following relatively *short* intervals, rather than long as in the case of simple periodic reconditioning. Thus in the particular case represented in Figure 84, Line B, when reinforcement is provided according to a temporal schedule, there is one chance that a reinforced response will follow an interval of one-half unit as against two chances that it will follow an interval of two units. Under reinforcement according to a fixed ratio in the same case there are two chances that a reinforced response will follow an interval of one-half unit and only one chance that it will follow an interval of two units. Here again the relation of the magnitude of the discriminative effect of a preceding response to the subsequent elapsed time is not known, but the effect upon behavior is clear. It is to produce a discrimination in the direction of an *increased* rate of responding. By virtue of the irregularity of its responding the rat can make this discrimination and hence adjust itself efficiently to the fixed ratio. It is only because its rate of responding varies that the rat can feel the correlation of the reinforcing stimulus with the completion of a number of responses and hence distinguish between the two programs of reinforcement.

In Figure 90 four sets of records are reproduced to show the result of an attempt to match fixed and extinction ratios. The first day in each set shows the more or less constant rate obtained under periodic reconditioning at five-minute intervals, from which an extinction ratio could be calculated. On the following days a fixed ratio was adopted in each case as nearly as possible equal to the extinction ratio so obtained. The matching was successful in three cases and resulted in the maintenance of approximately the same rate of responding for at least three experimental hours. In the fourth case an acceleration began near the end of the first hour at a fixed ratio, and although on the following day the ratio was doubled in an attempt to hold the rate down, a maximal value was maintained as shown in the figure in broken lines. The other records show an irregularity during the three experimental hours that is greater than that occurring under normal periodic reconditioning, and all three series eventually accelerate on the third day. In Series B there are three apparent starts toward a maximal rate. Series D showed signs of extinction on the fourth day (third day at a fixed ratio), and the ratio was therefore dropped slightly on the following day, when the final acceleration occurred.

In view of the possibility of slight changes in the extinction ratio from day to day, it is not possible to say certainly that this eventual acceleration is not due to such a difference between ratios as exists in Figure 87, but it is probable that it represents the development of the discrimination for which the procedure of reinforcement according to a fixed ratio provides the necessary con-

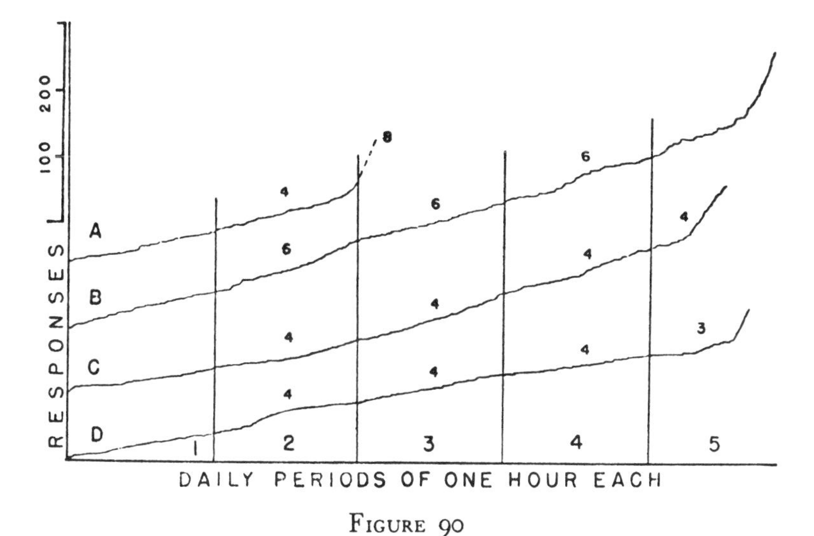

FIGURE 90

EVENTUAL ACCELERATION WHEN THE FIXED RATIO IS SET AS NEARLY EQUAL TO THE EXTINCTION RATIO AS POSSIBLE

On the first day the response was periodically reinforced.

ditions. In any event the present result is that, whether because of technical difficulties in maintaining a constant drive or because of the establishment of a discrimination, the prolonged constant rate obtained under periodic reconditioning cannot be maintained through reinforcement at a fixed ratio.

In spite of this new discrimination and the resulting increase in rate the extinction ratio has, so far as we have gone, not been exceeded. When the fixed ratio equals the extinction ratio, there is no change; when it is less than equal, the contribution to the reserve from the periodic reinforcement is greater than is needed to maintain the rate. But a further problem now arises. When the fixed ratio is greater than the extinction ratio, the rate may also rise to a maximum as in the preceding cases, even though there should be a loss to the reserve with each reinforcement. This will come about if opportunity for the development of the new temporal discrimination is provided before extinction takes place. The initial difference between the extinction and fixed ratios must not be too great, but fixed ratios of higher values may be reached through progressive steps. In the experiments now to be reported a ratio of 192:1 is reached by rats that showed an extinction ratio of the usual value of about 20:1 at the same drive. This is more than a discrimination. The rat makes a vital adjustment by expending more responses than the periodic reconditioning is supposed to supply, and its behavior in this respect would seem at first glance to invalidate the whole conception of an extinction ratio and of a reserve. But it is at this point that the special nature of reinforcement at a fixed ratio becomes important.

A special apparatus was designed which automatically reinforced responses at fixed ratios. The lever was similar to that in the usual apparatus, but the recording and reinforcing were mechanical, rather than electrical, and entirely automatic. The movement of the lever turned a ratchet and produced the discharge of a pellet at ratios determined by the setting of certain gears and toothed discs. Ratios of 16, 24, 32, 48, 64, 96, and 192 responses to one were available. Four such pieces of apparatus were used simultaneously in a sound-proof room. There was no soundproofing between them, since the degree of control was to some extent sacrificed to convenience. To accommodate considerably higher rates a smaller excursion of the writing point was chosen.

The major part of the experiment consisted of one-hour tests made upon two groups of four rats each for fifty-four consecutive days. The total number of responses recorded in that time exceeded 387,000. The rats were conditioned in the usual way and a ratio of 16:1 was established for three days. Progressively higher ratios were chosen until ratios of 48:1 and later 64:1 were maintained. By the twelfth or thirteenth day six rats were ready to be advanced to ratios of 96 and 192:1. A program was then followed according to which each rat was reinforced for periods of three days at the same ratio, which was either 48, 96, or 192 to 1 in random order. Two rats failed to adjust to these higher ratios. One suffered extinction at a ratio of 64:1; the other barely maintained that ratio and was extinguished at any higher value. These exceptions will be omitted in the following discussion.

Since the apparatus did not permit going above a ratio of 192:1 the experiment did not reach the limit at which a fixed ratio could be maintained. It was clearly demonstrated, however, that a ratio of nearly 200:1 was possible. During each three-day series at that value there was no evidence that the response was undergoing extinction, even though the extinction ratio under periodic reconditioning was of the order of 20:1. This is the principal fact with which I am here concerned, and one which requires careful analysis if it is to be reconciled with the notion of a constant extinction ratio at a given drive.

It will be convenient to begin with a further examination of the discrimination by virtue of which the rat can be said to 'know that it has only to press a certain number of times to receive a reinforcement.' The behavior of the rat during a single hour at a high ratio shows clearly enough that a discrimination of the sort to be expected from a fixed ratio is actually developed. Typical records for one rat at the three principal ratios are given in Figure 91 (page 288). The significant aspect of each record is the relation between the rate at any given point and the immediately preceding rate, a relation that obtains during the interval between any two reinforcements. The effect is a series of short curves, convex downward, where the reinforcements coincide with the points of abrupt change. This curvature is in accord with the notion of a special temporal discrimination.

The difference between the discriminations under periodic reinforcement and under reinforcement at a fixed ratio may be stated more rigorously as follows. When responses following long intervals (that is, following low rates of responding) tend to be reinforced preferentially as in the case of periodic reinforcement, the response contributes proprioceptive and exteroceptive stimulation functioning as S^{Δ} but the discriminative stimulation beginning as $S^{\Delta} + t_0$ and progressing toward $S^{\Delta} + t_n$ becomes increasingly less powerful. The effect of the response is to weaken the operant; the effect of a lapse of time to strengthen it. On the other hand, when responses following short intervals (that is, rapid rates of responding) are preferentially reinforced, the stimulation from a single response functions as S^D rather than as S^Δ, but the stimulation beginning as $S^D + t_0$ and approaching $S^D + t_n$ becomes increasingly less powerful. Since both of these temporal discriminative stimuli are based upon previous responding, they may be indicated with the letters PR and the stimulation arising from a response written S^Δ_{PR} in the case of periodic reinforcement and S^D_{PR} in that of reinforcement at a fixed ratio.

In both types of experiment the discrimination from the preceding reinforcement is active, since one reinforcement never occurs

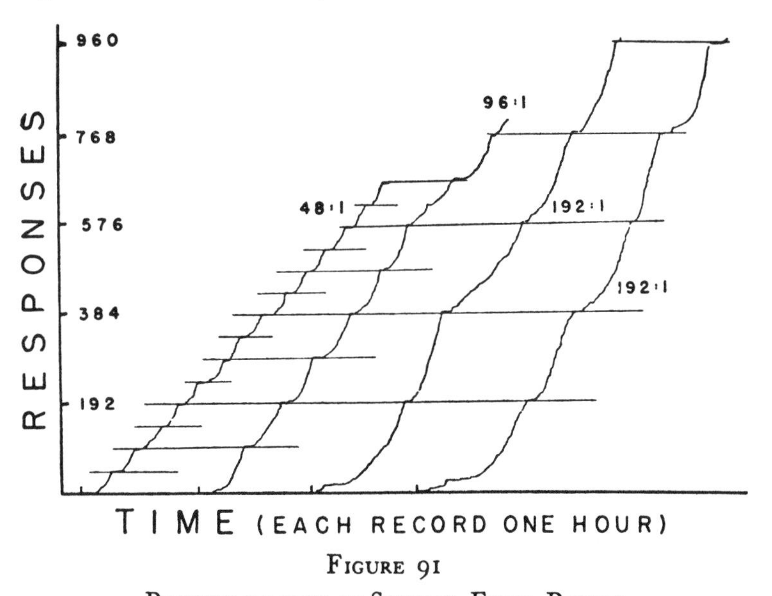

FIGURE 91

REINFORCEMENT AT SEVERAL FIXED RATIOS

The ratios are marked. Note the smooth accelerations between reinforcements (at horizontal lines).

immediately after another. A reinforcement therefore acts as S^Δ in both cases. Its effect may be written S^Δ_{reinf}. As the result of this discrimination the rat stops responding for a short period just after receiving and ingesting a pellet of food. But under periodic reconditioning S^Δ_{reinf} wanes as S^Δ_{PR} wanes and a response soon occurs; while under reinforcement at a fixed ratio S^Δ_{reinf} operates while S^D_{PR} is greatest and fails to operate as S^D_{PR} fails. In the latter case the operant may fail to gain as the pause after ingestion increases, and if these were the only factors to be taken into account no further responding might be expected. But aside from the weakening of S^{Δ}_{reinf} another factor tends to strengthen the operant during the pause, namely, the recovery of the reserve from the strain imposed upon it by the preceding run, a fact to be discussed shortly. Eventually, a response occurs. This strengthens the operant at once by contributing S^{D}_{PR} and another response soon occurs, again increasing the strength. The rate continues to accelerate until a reinforcement is received, and the section of the record assumes the curvature shown in Figure 91.

These more or less smooth accelerations are probably examples of what Hull (47) has called a 'goal gradient.' Although the notion of a goal is not part of the present conceptual scheme, its reinforcing effect may be replaced with the simple notion of reinforcement. A gradient, likewise, is descriptive of a certain aspect of behavior, but it is also not a fundamental concept since it may arise from the operation of many different kinds of factors. It has no parallel in the present system but may easily be introduced if convenient. In one case described by Hull a gradient appears because of progressive differences in the effect of a reinforcement due to the decreasing times elapsing between a reinforcement and the progressive steps of a complex act, such as the successive steps in running a maze. Such a series of responses (say, M_1, M_2, M_3 . . .) must be fairly similar if strengths are to be compared conveniently, but they must also be made in the presence of distinguishable discriminative stimuli in order to register the differential effect of the reinforcement. In the present case we may set the operants $sS + t_0 . R$, $sS + t_1 . R$, $sS + t_2 . R$ against the series M_1, M_2, and M_3 and so on. Both constitute series of responses by virtue of which the organism comes nearer to a 'goal' or reinforcement. In the present case the responses are identical, the only differentiating material arising from the accumulation of S^{D}_{PR}. The 'gradient' is due to the change in the discriminative material. (It may be added for the sake of those who are worried by the suggestion of a 'final cause' that it is not increasing nearness to the 'goal' that produces the increase in rate except in so far as nearness is, under the terms of reinforcement at a fixed ratio, a function of the distance already covered from the preceding reinforcement.)

There are considerable individual differences in the character of the 'gradients.' At one extreme the pause after ingestion may be relatively great and the subsequent acceleration to a maximal or near maximal rate very rapid. The record for one hour is step-like and sharply angular. At the other extreme the pause is brief, but the rate immediately following it is low and accelerated slowly. In this case there are almost no horizontal breaks in the record and the scallops are quite uniformly curved. Examples of these extremes and of intermediate cases are given in Figure 92. Given equal over-all rates (and hence presumably equivalent reserves), the difference may be expressed in terms of the 'friction' of the

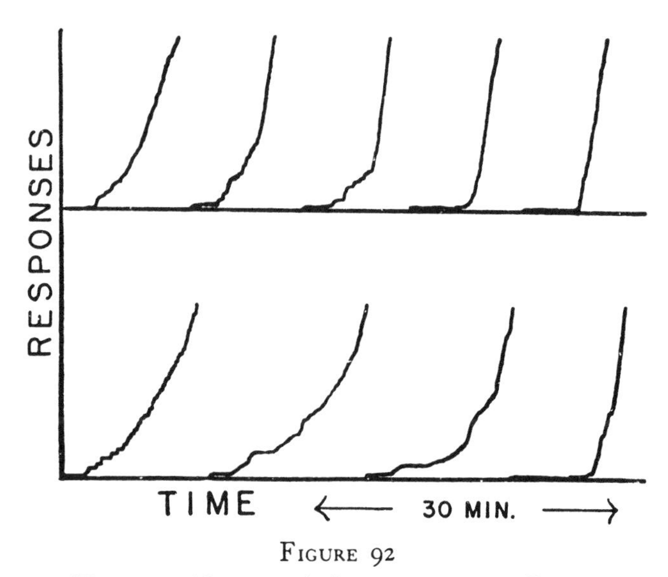

FIGURE 92

VARIOUS TYPES OF 'GRADIENT' OBTAINED UNDER REINFORCEMENT AT A FIXED RATIO OF 192:1

first response. If the response comes out easily, the discrimination based upon S^{Δ}_{reinf} is still strong and the rate is depressed. If it comes out late, the discrimination has grown weak and the S^{D}_{PR} from the response is effective in producing a rapid acceleration. The difference may also be due to some extent to the state of the reserve.

The part played by the accumulation of S^{D}_{PR} is evident when for any reason the acceleration is momentarily interrupted, as, for example, by incidental stimulation (especially, in this experiment, from the other apparatuses). If the interruption is sufficiently long to reduce the accumulation of S^{D}_{PR} considerably, a rapid rate is reached only through a second acceleration. Several examples from a single record are indicated in Figure 93.

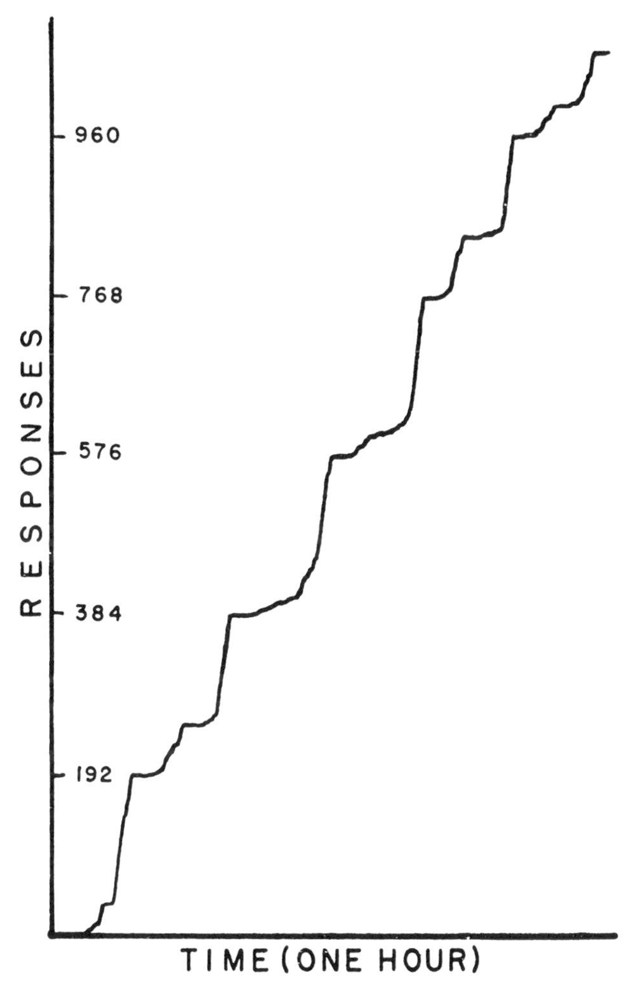

FIGURE 93

BROKEN GRADIENTS UNDER A FIXED RATIO OF 192:1

There is a relation between the amount of S^D added by a single unreinforced response and the ratio of reinforcement. If the ratio is low, one response adds a relatively large amount of S^D; if high, a relatively small amount. If the total effect of a reinforcement in establishing S^D is constant, the discriminative property acquired by one response would be that total effect divided by the number of responses leading to the reinforcement, although a uniform distribution to each response is not necessarily required. Hence, it may be said that, not only does the rat develop a discrimination of such a sort that a response produces as one of its effects an amount of S^D, but the amount of S^D is fixed. It is because of the dependence of the amount upon the ratio that the rat may be said to distinguish between ratios of, say, 48:1 and 96:1.

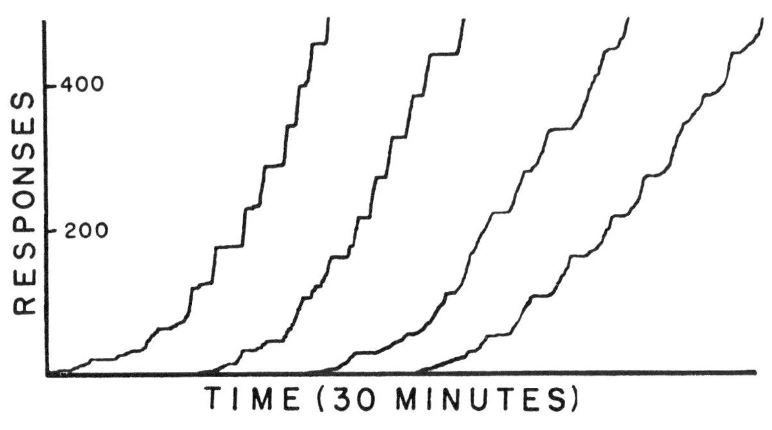

FIGURE 94

READJUSTMENT OF THE TEMPORAL DISCRIMINATION IN DROPPING TO A SMALLER FIXED RATIO

The response had in each case previously been reinforced at a ratio of 192:1 for three days. On the days in the figure the ratio was set at 48:1.

The differing values of S^D contributed by a response at different ratios were evident in the present experiment when the ratios were occasionally changed. In Figure 94 four typical records are given to show the transition from a ratio of 192:1 to 48:1. In each case the response had been reinforced at the higher ratio for the three preceding days. The amount of S^D contributed by one response should therefore have been 1/192 of the total effect of one reinforcement. Consequently, when the ratio is changed to 48:1, the curve begins at a low slope after the initial reinforcement and accelerates only gradually because of the relatively weak effect of each response. The second reinforcement occurs prematurely (according to the previous schedule) when the rate has not yet reached its maximum. The effect of this reinforcement is now distributed in 48 (rather than 192) parts, and there is therefore a tendency for the over-all curve to accelerate. After three or four reinforcements at the new ratio the proper value of S^D per response has been established, and the over-all curve is henceforth linear.

An example of a change from 192:1 to 16:1 is given below in Figure 99 (page 301), and a similar tendency is evident, though not so striking, in other changes in ratio. The change from a low to a high ratio produces the converse case in which the amount of S^D begins at too high a value for the case in hand. The ultimate value at the higher ratio is approached through a curve of negative acceleration.

It was noted above that one of the factors tending to cancel the effect of S^{Δ}_{reinf} was the recovery of the reserve from the strain imposed by the preceding run of responses. The relation of the reserve to the discrimination based upon S^D_{PR} must now be examined. The behavior of the rat under a fixed ratio is not independent of a reserve. The failure of two rats to exceed ratios of 64:1 is a sufficient indication that a limiting value may be reached at which the response undergoes extinction in spite of the new discrimination. Even when a high ratio is maintained without extinction, it is clear that a reserve (with some kind of relation to the frequency of reinforcement) is controlling the response. Several experimental facts point to this conclusion, the first of which is the shape of the extinction curve obtained after reinforcement at a fixed ratio.

Four typical extinction curves for the rats in the preceding experiments are given in Figure 95. They were obtained by simply withholding reinforcement during one hour. The smallest curve is for one of the rats that were incapable of sustaining high ratios and, because of the relatively small reserve, shows most clearly the effect upon the extinction curve of the discrimination based upon S^D_{PR}. When the rat begins to respond on the day of extinction it begins to produce discriminative stimulation which in the light of the preceding program of reinforcement increases the probability that a reinforcement will be received. Each response contributes to the situation a certain amount of S^D_{PR} and thereby strengthens the operant. The result is that the rate of responding persists at a maximum and the available reserve is drained within ten or fifteen minutes instead of being emitted at a gradually decreasing rate for the full hour as in normal extinction. The records

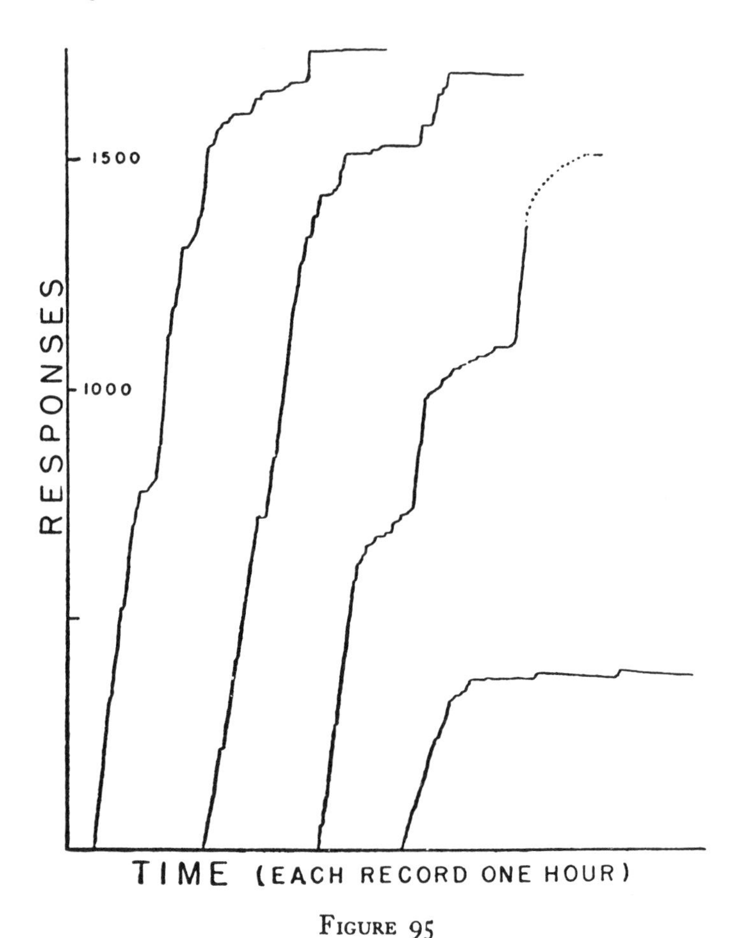

FIGURE 95

EXTINCTION AFTER PROLONGED REINFORCEMENT AT FIXED RATIOS

for the rats with more capacious reserves are similar with two exceptions: (1) a greater share of the hour is needed to drain the reserve and (2) a transitional stage occurs between the periods of rapid and of very slow responding. The actual dynamics of the reserve are not well enough known at the present time to make a plausible explanation of this transitional stage possible. The third curve in the figure shows two incipient transitions during which the total amount of S^D_{PR} is reduced but again built up as a remaining part of the reserve is tapped. Because of these interruptions no horizontal section is reached before the end of the hour as in the other cases. (The end of the third curve was poorly recorded, but the dotted lines indicate its course approximately.)

These curves are not extraordinarily uniform, but the principal effect of the reinforcement at a fixed ratio is clear. It is sufficiently accurate to say that if a response is going to be emitted at all, it will come out as soon as possible, because of the fact that under reinforcement at a fixed ratio an unreinforced response contributes discriminative stimulation which strengthens the operant. In normal extinction the operant is weakened by an unreinforced response. Here it is strengthened until the limit of the reserve is reached. The statement makes no allowance for responses not accessible in any case on the first day but which could appear in 'recovery' at a later time, but such allowance should be made.

This peculiarity of the extinction curve does not develop until reinforcement at a fixed ratio has had time to establish its appropriate discrimination. Extinction after reinforcement at low ratios, which do not require the discrimination in order to be maintained, has the usual properties. In Figure 96 A (page 296) an increase in rate is first shown as the result of changing from periodic reconditioning to reinforcement at a low ratio. At the arrow the reinforcements were omitted, and a normal extinction curve follows. In Figure 96 B a record of rapid responding under the low ratio of 12:1 is first shown. The second curve is the extinction obtained on the following day. It is clear that in these cases enough time has not been allowed for the development of the temporal discrimination peculiar to reinforcement at a fixed ratio and that the extinction curve is thus not affected.

It may be noted in passing that exceedingly high rates of responding are reached in the experiment represented in Figure 95. The first two curves show at least 1400 responses made in about twenty minutes. In both curves there are runs of several hundred responses each at the rate of at least 100 responses per minute. The records may be compared with those reproduced elsewhere (for example, Curve A in Figure 87) in which rates of less than one response per minute are quite uniformly maintained. The strength of the operant with which we are dealing may therefore vary by a factor of at least one hundred under the procedures described in this book. The significance of the fact in any considera-

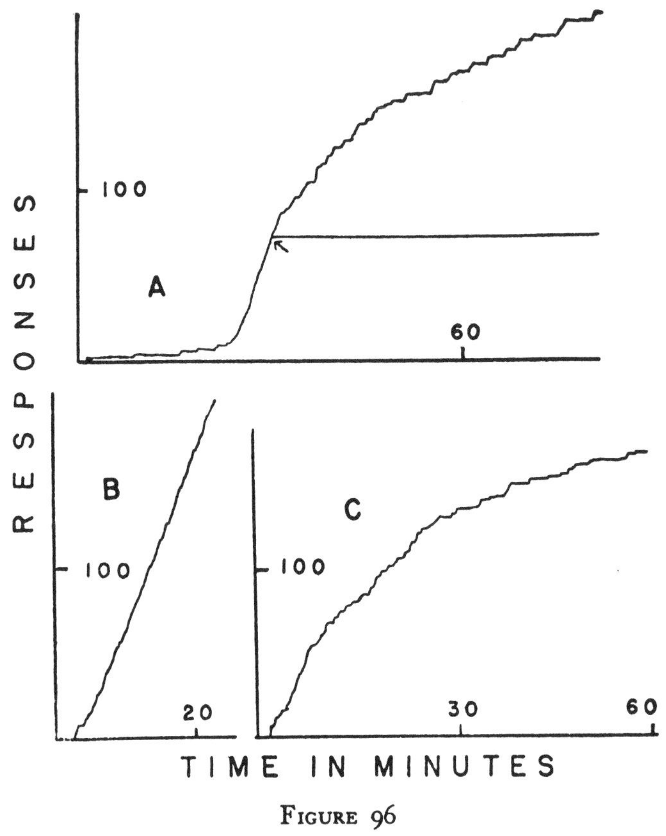

FIGURE 96

EXTINCTION AFTER BRIEF REINFORCEMENT AT FIXED RATIOS

The character of the curves in Figure 95 is lacking because the temporal discrimination has not yet developed.

tion of the dimensions of the strength of operant behavior has already been pointed out. It will be mentioned again in the following chapter.

From the shape of the typical extinction curve after reinforcement at a fixed ratio we may assume that the effect of the accumulation of S^D_{PR} is to strain the reserve by bringing out responses which under normal discriminative stimulation would have remained within the reserve for some time. The same effect is to be expected in the case of the higher ratios of reinforcement, and hence it was possible above to appeal to the state of exhaus-

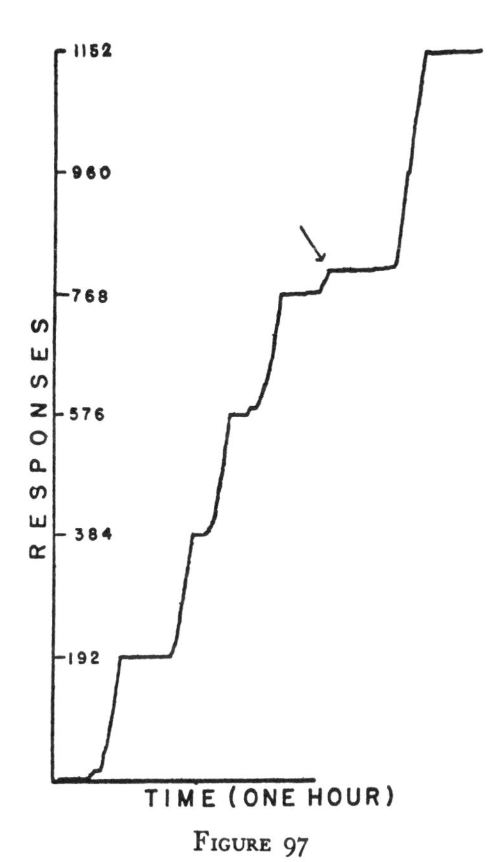

FIGURE 97

THE PERSISTENCE OF TEMPORAL DISCRIMINATIVE STIMULATION FROM ONE GRADIENT TO ANOTHER

tion of the reserve as one factor in the determination of the rate of responding just after a reinforcement. When the reserve is not strained, the effect of S^{Δ}_{reinf} is negligible; the rat returns to the lever very shortly after ingesting a pellet. But the preceding run which occurs under reinforcement at a fixed ratio places the reserve in a state of strain which acts with S^{Δ}_{reinf} to produce a pause of some length.

Two experimentally observed results follow. One is the nature of the subsequent behavior of the rat when a given run is for any reason interrupted before reinforcement. The interruption permits the reserve to recover to some extent, and when the reinforcement is finally received, it is followed by a shorter pause than usual. A typical example is given in Figure 97 (page 297). At the point indicated by an arrow the usual acceleration leading up to a reinforcement was interrupted for some time. The delay which follows the subsequent reinforcement is negligible, because the effect of S^{Δ}_{reinf} is less strongly supported by the drain on the reserve. This compensatory effect, similar in many respects to those already described in preceding chapters is fairly strong evidence of an underlying reserve wherever it occurs.

The second fact supporting the effect of the reserve in determining the rate after ingestion is the relation between that rate and the ratio at which the response is being reinforced. It is obvious that the state of strain imposed by the accumulation of S^{D}_{PR} will be greater the higher the ratio, and there should be a proportionately greater average pause after ingestion. This prediction is borne out by the experimental result. It is true, of course, that not only the first pause but the subsequent acceleration is affected, but measurements of the pauses will suffice for the present point. An estimate of the pause following ingestion was made by measuring the first five pauses in three records at each ratio for each rat, care being taken not to use records that involved transition from one ratio to another. Every rat showed the trend of the average, which was toward an increasing delay as the ratio was increased. The averaged delays were 73 seconds at a ratio of 48:1, 96 seconds at 96:1, and 120 seconds at 192:1, as shown in Figure 98. It will be seen from inspection of the records in Figure 91 that the differences would be still greater if the effect upon the acceleration were taken into account after the pause.

(The effect of the reserve is responsible for the fact that there is no simple relation between the ratio and the number of pellets received during an hour. The actual relation has been included in Figure 98 in dotted lines. Thus, at a ratio of 192:1 the rats received, on the average, 7.2 pellets during the hour; at 96:1, 13:1; and at 48:1, 20.0.)

So far I have shown simply that under reinforcement at a fixed ratio a temporal discrimination occurs which is different from that occurring under periodic reinforcement and which accounts for the shape of the resulting curve. The problem of how the extinction ratio is exceeded in such a case has not, however, been wholly solved. The fact to be explained is not the various changes in rate that occur during reinforcement at a fixed ratio but the

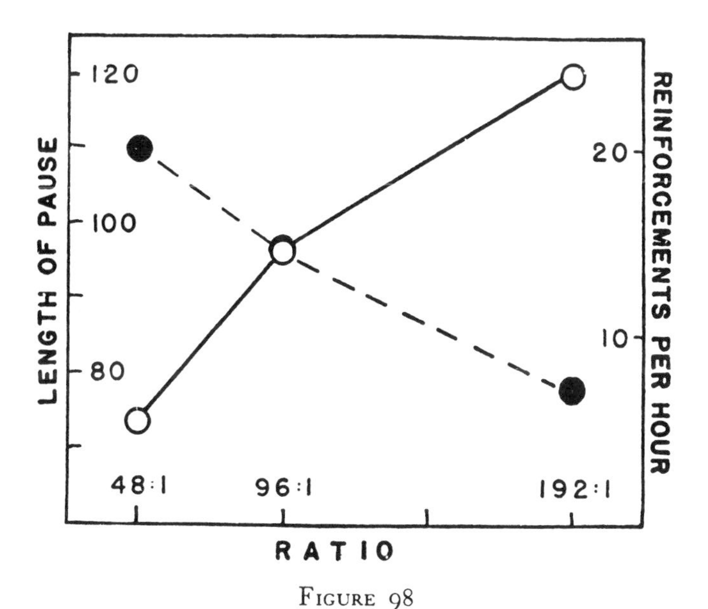

FIGURE 98

Open circles: length of pause in seconds following ingestion of pellet as a function of the fixed ratio.

Solid circles: number of pellets received per hour at different fixed ratios.

observed maintenance of a high average rate. With a demonstration that a special discrimination is in effect during reinforcement at a fixed ratio, an explanation is forthcoming from an appeal to the reinforcing effect of a discriminative stimulus described in Chapter Six (page 246). Under a fixed ratio a response produces discriminative stimulation correlated positively with reinforcement. The case is, therefore, comparable with that in which a light has acquired discriminative value and in which presentation of the light is then arranged to reinforce a response. When a rat presses a lever 192 times to obtain food, the food reinforces the last response (and perhaps a few preceding responses) directly. But there is another kind of reinforcement acting upon other responses: namely, the production of S^D_{PR}. The early responses in each run are, in common language, not made because they produce food but because they bring the production of food nearer. They produce the accumulated S^D_{PR} in the presence of which a response will be reinforced, just as (in the experiment described in Chapter Six) they produced the light.

This state of affairs may be clarified by considering the case in which the particular act reinforced by food is the pressing of the lever twice in close succession. If the response is the double pressing of the lever, we should expect a single periodic reinforcement to set up a reserve of, say, twenty *pairs* of responses—or forty pressings. Now, we may either count each pair as one response or appeal to the fact that the first pressing produces and is therefore reinforced by the S^D in the presence of which the second response produces reinforcement. This was almost the case described in Chapter Six, except that the stimulation supplied by the first act of pressing the lever was augmented by the stimulation from the light in the presence of which the second pressing was reinforced.

This is fundamentally a problem in the definition of a unit of behavior. As a rather general statement it may be said that when a reinforcement depends upon the completion of a number of similar acts, the whole group tends to acquire the status of a single response, and the contribution to the reserve tends to be in terms of groups. But the process through which this is brought about appears upon analysis to be the development of the kind of discrimination we have just been considering. With such an organism as the rat, a contribution to the reserve commensurate with a large group is probably never achieved. The unity of the group as 'a response' is never fully realized. And, unfortunately, it is difficult to test the case with small groups because the extinction ratio sets a lower limit. In this experiment the rat does not, of course, reach the point of pressing the lever 192 times as a 'single' response closely defined. But the tendency toward the establishment of such a 'response' is responsible for the apparent high extinction ratios here observed. The discrimination that is responsible for the intermediate reinforcement of responses is never fully developed, but it reaches the extent at which a ratio of 192 pressings to one reinforcement can be maintained.

Even at a ratio of 16:1, which is below the normal extinction ratio at this level of drive, it is clear that the rat does not press the lever 16 times 'as one response.' In Figure 99 four records are given for one of the rats in the preceding experiment which were placed at a ratio of 16:1 after having maintained 192:1. The first record shows the transition described above in which the

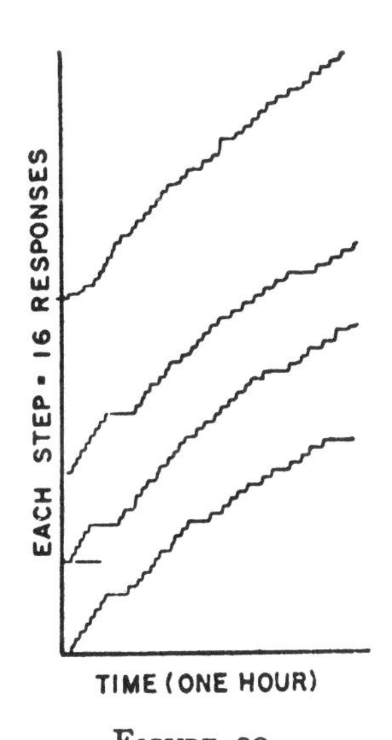

FIGURE 99

REINFORCEMENT AT A RATIO OF 16:1

The top record shows a readjustment of the temporal discrimination as in Figure 94. The negative acceleration in all records is due to a change in drive (see Chapter Nine).

new value of S^D becomes attached to each response. At this ratio pellets are received so frequently that a change in hunger produces a characteristic negative acceleration (see Chapter Nine). But the pauses between each group of responses are considerably greater than when every response produces food. The temporal discrimination is not complete enough to give sixteen pressings the status of a single response and there is a limiting effect imposed by the reserve. The delay during the first part of these records is typical and is unexplained.

When we speak of sixteen pressings functioning as 'one' response, we raise a problem in the definition of a response as a fundamental unit which will not be wholly solved here. There is an important parallel in the case of running, which consists of the repetition of a group of separate responses. When no external discriminative stimulation is provided (as when running takes place in an activity wheel), the reinforcement of a given amount of running acts directly upon the last group of steps but not necessarily upon the rest of the series. In what sense does the principle of a reserve apply here? Presumably, the answer depends upon the additional reinforcement provided through a discrimination similar to that in the case of repeated pressing. As such a discrimination develops, a secondary reinforcement spreads to the early members of the sequence of responses. Some experiments on the reinforcement of running are described in Chapter Nine.

The behavior of the rat under reinforcement at a fixed ratio explains some of the properties of records taken in earlier experiments on periodic reconditioning, particularly the development of the second-order deviations (page 123) and certain anomalies of the extinction curve after prolonged reconditioning. The explanation rests upon the fact that when the third-order deviation (which we now interpret as a temporal discrimination based upon the reception and ingestion of food) has established itself, there is a resulting indirect correlation between the reinforcement and the rate of responding similar to that prevailing under reinforcement at a fixed ratio. As soon as this kind of deviation develops, a long pause follows each reinforcement, but unless it is as long as the period of reinforcement itself it is never followed by reinforcement. This is contrary to the usual preferential reinforcement of long pauses obtaining under periodic reinforcement. Furthermore, the pause is followed by a period of rapid responding in which compensation is effected for the delay, and this usually continues until the reinforcement is received. Hence there is an indirect correlation between rapid responding and reinforcement. At higher frequencies of periodic reinforcement the combined result of these indirect correlations is enough to produce the converse sort of discrimination which characterizes reinforcement at a fixed ratio, and to yield a slightly scalloped curve. The discussion of second-order deviations in Chapter Six should be consulted and Figure

32 compared with Figure 91. In the former figure the scallops do not fully respect the times of reinforcement; the breaks occur at reinforcements but not at every one. The records indicate that the rat begins to be affected by the pseudo correlation and hence to accelerate its responding very much as in Figure 87. The correlation immediately breaks down and there is little or no net gain to the reserve. The reserve is, however, strained by the rapid

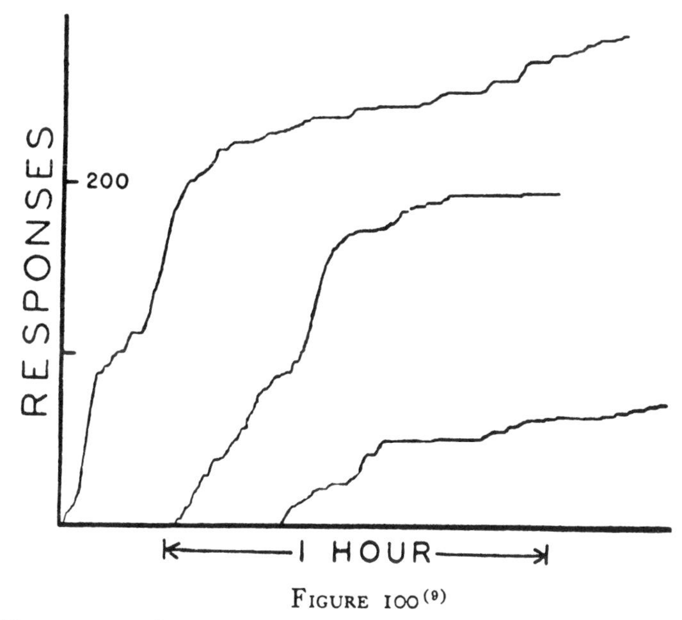

FIGURE 100[9]

EFFECT UPON EXTINCTION OF A PSEUDO TEMPORAL DISCRIMINATION DEVELOPED UNDER ORDINARY PERIODIC REINFORCEMENT

responding and a sudden drop appears when S^{Δ}_{reinf} is received. The process is subsequently repeated.

This qualitative speculation as to the cause of the second-order deviation is not in itself very convincing, but the contention that pseudo correlations between reinforcement and rapid responding and between lack of reinforcement and a pause actually set up a discrimination is supported by the extinction curve obtained after prolonged periodic reconditioning at the higher rates at which the effect is observed. In Figure 100 the first three days of the extinction of the reflex in Rat P7 are represented. The process was begun after the prolonged periodic reconditioning shown in Figure 29. It will be seen that during the first twenty minutes indicated in the first record the rat pressed the lever more than two hundred times. That this was, however, a temporary 'strained' state is shown in the period of compensation that followed, during which the rat responded only desultorily for a period of more than an hour, and at the end of which the curve was probably close to its theoretical position. (It is obviously difficult to infer the 'true' course of the record.) The transition from the high rate to the depression that follows is remarkably smooth. A similar overshooting occurs on the second day, and there is a trace of it on the third. These curves obviously resemble those for extinction after reinforcement at a fixed ratio (Figure 95). The responses are crowded at the beginning of the curve, a transitional stage intervenes, and a very low rate of responding is eventually reached. The incipient transitional stage closely resembles that of the third curve in Figure 95. In the present case further extinction on subsequent days was explored. Periods of rapid responding again occur, although the effect is delayed and less extensive. Similar curves could presumably have been obtained by permitting further extinction on later days for the eight rats discussed above, but this was not done because the rats were transferred to a final test of the present interpretation to be described shortly.

The pseudo correlation of reinforcement and rapid responding that may prevail under periodic reinforcement accentuates the third-order deviation begun by the temporal discrimination based upon the reinforcement. S^{Δ}_{reinf} is responsible for the initial low strength after ingestion, but this is eventually responsible for a weakening of the S^D from the preceding responses and hence tends to prolong the period of no responding following ingestion. It is for this reason that the third-order deviation appears first at the higher rates of responding where the indirect correlation is first felt.

It might be expected that the rat would be less inclined to respond just after eating if it is to wait, say, twelve minutes before eating again rather than, say, three. But this is not the case, as may be seen in Figures 30 and 31. The contrary effect is due to the fact that with periodic reinforcement every three minutes a high rate is maintained, at which the discrimination characteristic of reinforcement at a fixed ratio appears. The step-like character of the records is thus due first to the discrimination based upon the ingestion of food and second to the weakening of S^D brought about by the delay due to this discrimination. That the appearance of the steps at the high rates only is not due to the frequency of reinforcement rather than to the rate of responding can be shown by increasing the rate of responding while holding the frequency of reinforcement constant (as, for example, by changing the drive). In experiments upon such a change in drive in Chapter Ten, a marked step-like character is demonstrated soon after the rate reaches a sufficiently high value.

Periodic Reinforcement after Reinforcement at a Fixed Ratio

The pseudo correlation between rapid responding and reinforcement that may prevail under simple periodic reconditioning is especially likely to occur when a procedure of periodic reconditioning is returned to after reinforcement at a fixed ratio. Even though reinforcements are now temporally arranged, the rat will almost invariably continue to respond until a reinforcement is received, and the resulting relation between reinforcements and the long runs of responses is none the less real for being the indirect result of the previous training of the rat. The persistence of a discrimination which is not justified by the conditions of the experiment will depend largely upon accident. An occasional reinforcement after a pause may be expected if the behavior of the rat will admit of pauses at all. In most cases these do occur, and the pseudo discrimination is eventually broken down. But little consistency in the rate of breakdown is to be found in the following experiment on this subject.

After the single day of extinction described above, seven of the rats in the group were given a rest of thirteen days and were then returned to the procedure of reinforcement at a fixed ratio. The usual rates at 96:1 (64:1 in three cases) were obtained and used in calculating periods of reinforcement that would produce the same number of reinforcements per hour. For example, if a rat made 800 responses per hour at a ratio of 96:1, it received eight reinforcements in addition to that given to the first response (which was always reinforced). Periodic reinforcement at intervals of seven minutes would produce the same number, and hence the period was set at seven in such a case. This was simply a reversal of the change described earlier in the chapter where a ratio was calculated from the rate under periodic reinforcement. The significance of the change for the rat was also reversed. The high rate of 800 responses per hour was then no longer necessary. If the new conditions were clear enough to destroy the discrimination upon which the behavior of the rat rested, its rate should have fallen to that determined by the extinction ratio—or, say, 150 responses per hour.

The following results were obtained. The two rats that had previously been unable to sustain ratios above 64:1 and were then operating at that ratio showed no significant change during eighteen days under the procedure of periodic reinforcement. These rats had the least change to make, but no start in the right direction was observed. Two other rats showed an *increase* in rate following the change, which continued more or less irregularly for ten days in one case and four in the other. The reason for the increase was apparent in the actual behavior of the rats. Whenever a run was begun after less than the average delay, the former correct ratio was reached before the time for reinforcement. The rat continued to respond, and the effect was to introduce reinforcements at greater ratios than those previously in force. As in the case of the first two rats no breakdown of the discrimination occurred; instead, the procedure produced an increased average ratio of responses. The remaining three rats showed the progressive decline in rate that one would expect in the absence of disturbing influences.

A successful attempt was made to demonstrate the part played by the pseudo correlation between reinforcement and rapid responding by deliberately setting the apparatus for reinforcement only when the rat had not been responding for at least fifteen seconds. Under this condition no response was ever reinforced if it had been preceded within fifteen seconds by another response. The effect was quite clear in every case. The first day under the new procedure showed a slight rise in some cases because the rat was responding so rapidly that intervals of fifteen seconds could not always be found at approximately the time of reinforcement and hence some very long runs were emitted. The required number of reinforcements was achieved, however, and only after fifteen-second pauses. On the second day the rate fell and continued to fall on the following days until a relatively stable value was reached. On the last day of the experiment the original ratio was set up as a control, and the rate promptly rose to approximately the value reached before the fifteen-second interval was inserted.

Chapter Eight

THE DIFFERENTIATION OF A RESPONSE

The Problem

It is necessary to distinguish between the discrimination of stimuli and a process of differentiating between forms of response. The tendency to cast all behavior in the respondent mould, with the implication of a strict and ubiquitous stimulus-response relationship, is perhaps responsible for the neglect of this distinction in current work on discrimination. If for every response there were a rigorously corresponding stimulus, a discrimination between two forms of either term would necessarily involve the corresponding forms of the other, and there would be no need to consider more than one process. But in operant behavior the strength of a response may be independently varied and two closely related forms of a response may become distinguished by developing different strengths irrespective of discriminated stimuli. The process is quite different from that described in Chapter Five.

In conditioning an operant a reinforcement is made contingent upon the occurrence of a response having certain properties. For example, the presentation of food is made contingent upon any movement of the organism that will depress a lever. When a suitable response occurs, it is strengthened by the reinforcement. The problem of sensory discrimination arises because the reinforcing effect is to a considerable extent independent of the stimulation presenting itself at the moment of emission of the response. Through the process of induction the strengthening of the response may carry over to a different stimulating situation. But this 'sensory' induction has a parallel which concerns the response only. The subsequent elicitations of a response that are due to a given reinforcement, whether or not they are made in the same stimulating field, may differ in some of their properties. The lever may be pressed from a different position, or with a different hand, and so on. This is another kind of induction, and it gives rise to the problem to be considered in this chapter. To avoid confusion I shall speak of sensory discrimination using the terms already presented. The discrimination of the form of a response, however, will from this point on be referred to as Differentiation.

Let us suppose for the moment that a response may be completely described by enumerating three properties. A reinforcement is made contingent upon the occurrence of the response *Rabc*. The response occurs and is strengthened. Subsequent responses occurring as the result of this reinforcement are not exact replicas of *Rabc* but must be written *Rabd, Rafc*, and so on. Now, if the reinforcement is contingent upon only the property *a*, all of these responses will also be reinforced. This will produce an even wider inductive spread, with greater deviations from *Rabc*. A response may eventually be elicited through induction which does not possess the property required for reinforcement. Assuming a fairly closely circumscribed reinforcement, let *Rabc* be reinforced and *Rabd* not. When *Rabd* occurs because of induction from *Rabc*, it will be partially extinguished. There will be a reverse inductive action upon *Rabc*, but since the direct effect of both conditioning and extinction is greater than the inductive, the strengths of *Rabc* and *Rabd* will draw apart. Unless, as in the sensory case, there is an actual breakdown of induction, further reinforcement of *Rabc* will continue to produce an occasional occurrence of *Rabd*.

Where a certain latitude in form is allowed by the reinforcement, a response tends to narrow itself spontaneously and to persist with a fairly closely circumscribed set of properties. The two principal mechanisms responsible for this narrowing are as follows.

1. *Frequency*. Since direct strengthening is greater than indirect, the most frequently occurring form automatically strengthens itself preferentially. The first form reinforced has an initial advantage and may persist as a 'fixation.'

2. *Concurrent negative reinforcement*. The execution of a response may supply negative reinforcing stimulation tending to reduce the net reinforcing effect. 'Difficulty' and 'awkwardness' may be expressed in terms of the negatively reinforcing stimulation automatically produced by a response. If the various members of an inductive group of responses differ in awkwardness or difficulty, there will be a resulting differential effect of the reinforcement. The simplest and easiest form prevails, because it re-

ceives a positive reinforcement without emotional depressant effect. (See the section on Negative Reinforcement, Chapter Three.) This factor may act to correct the persistence of an awkward form of response due to the factor of frequency.

The preceding statements are based largely upon incidental observation, and I have no special experiments to report. The variation in the form of a response during successive elicitations is easily observed. The 'fixation' of a closely circumscribed form occasionally occurs. For example, a rat may continue to press the lever with its nose or teeth if it first responds in that way. The commoner case is the development of an efficient and easy response made with one hand or with both. If the required properties are changed, the rat adjusts quickly. The process involves the descriptive properties and may be designated as the 'topographical differentiation' of the response. The problem is similar to that of original conditioning. There is, however, another side to the problem of differentiation, which concerns the quantitative (intensive or durational) properties. The reinforcement may be made contingent upon pressing the lever *with a certain force* or upon holding the lever down *for a certain length of time*. Here the topographical properties do not change greatly, but there is still a definite process of differentiation. In coming to press the lever with a certain force, for example, the rat is 'learning something new,' but the process cannot be assimilated to the cases already described. It is to this quantitative differentiation that the present chapter will be devoted.

Additional pieces of apparatus are required to record the values of the properties being examined and to secure the preferential reinforcement of responses having a given value of the selected property. The differential reinforcement of responses according to intensity is secured by connecting the lever with a ballistic pendulum. The excursion of the pendulum is a function of the force with which the lever is pressed. An adjustable electrical contact is closed when the pendulum makes the required excursion. To measure duration a Telechron clock is started when the lever is pressed and continues to run so long as the lever is not released. At the end of the required period a reinforcement is effected through an adjustable contact. Release of the lever resets the contact arm of the clock at zero. In recording both intensity and duration a special kymograph is used, the drum of which turns a uniform distance *at each response.* The intensity or duration is represented by the vertical movement of a writing point, the values for successive responses summating. The value for a single response may be read directly from the record; the values for a group of responses form a step-like line, the slope of which gives the mean value. To convert the intensity into a linear movement a small fly-

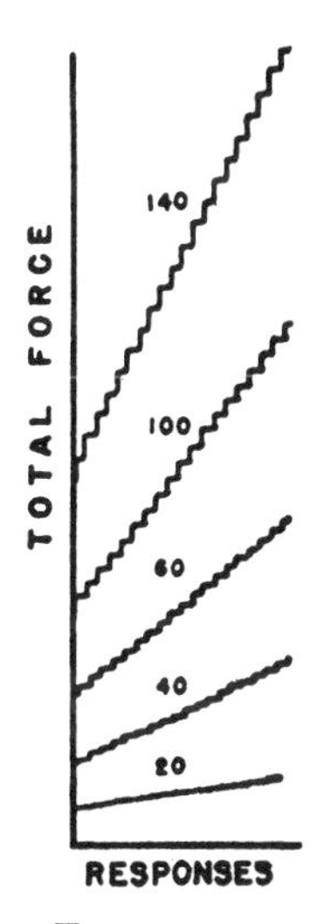

FIGURE 101

CALIBRATION CURVES FOR THE INTENSITY OF THE RESPONSE TO THE LEVER

The curves were made by repeatedly dropping the lever with weights attached as marked (in grams).

wheel is connected with the lever. A response spins the wheel according to its force. The fly-wheel is geared down to a shaft which winds up a thread attached to the writing point. To convert time in the same way it is only necessary to let the lever close a circuit to a clock, the shaft of which winds up a thread connected to the writing point.

All of these devices may be calibrated easily. In the case of intensity a weight is attached to the lever and the lever repeatedly dropped from its uppermost position. Records made by the rat are compared with calibration curves made with different weights. Calibration curves for durations are obtained by holding the lever down repeatedly for selected intervals of time measured with a stop-watch. In Figure 101 a few calibration curves for intensity are given. They were made by repeatedly dropping the lever with different weights attached as indicated (20, 40, 60, 100, and 140 grams). The increase in slope with weight is not linear because of the friction of the fly-wheel and other factors; but an adequate measure of any given force is provided. There is some slight irregularity, especially at the lower slopes, which restricts the use of the record in measuring a single response. For a group of responses the record is sufficiently accurate for all present purposes. The curves for duration are similar, except that the slope varies linearly with the duration and the records are valid for single responses. With the coordinate values used, the mean duration of a series of responses may be obtained by dividing the tangent of the angle of the record with the horizontal by 0.09.

The data to be reported were obtained with groups of from four to sixteen rats. No datum is reported for which any exception has been observed unless so stated. The experiments are not, however, described in detail nor is a résumé given for all the data obtained.

Differentiation of the Intensity: The Normal Force of the Response

The normal force with which the lever is pressed is shown in the parts of the curves in Figure 102 up to the points marked E. The rats were tested for short periods on several successive days. From fifteen to twenty responses were reinforced each day. In the figure the daily records are separated by vertical lines. The first record in each case shows the force on the day of conditioning. In Series A the exceptionally low force is barely sufficient to depress the lever. The slope of Series B is near the mean for all rats examined, and the calibration indicates the effect of a weight of 35–40 grams. Series C shows an exceptionally high force, although it may be seen that the first six or seven responses were quite weak. The significance of this will be pointed out later. Except for minor local deviations each rat maintained a constant mean force under these conditions for as long as the experiment was carried on (two to five days).

Figure 102 also shows the effect of extinction upon the force.

On the third day in Series A and C and the fifth in Series B ten responses were first reinforced in the usual fashion and the reflex was then extinguished, beginning at E. The extinction curves given by the change in rate had the usual form. In Figure 102 the curves represent the change in force. In practically every case some responses occurring during extinction show an increased force. Stronger responses generally occur near the beginning of the ex-

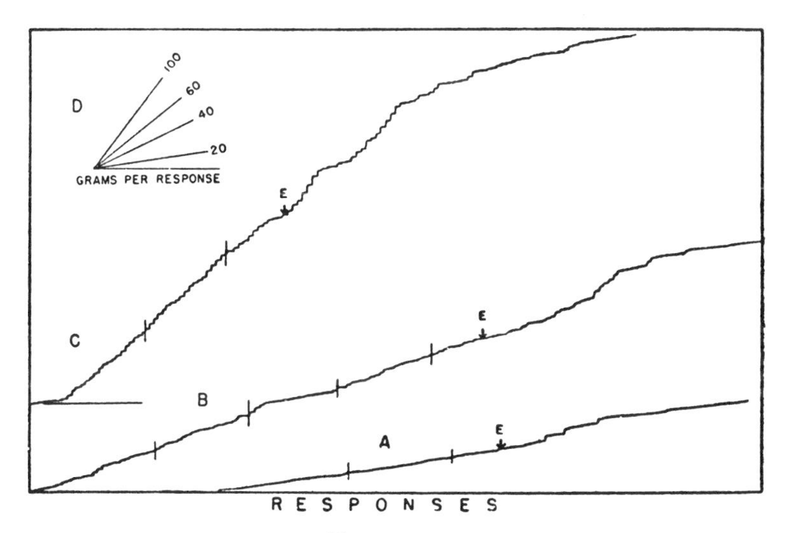

FIGURE 102

INTENSITY OF UNDIFFERENTIATED RESPONSES

The mean intensity is given by the slope. Calibration curves for comparison are given at D. Vertical lines separate daily records. All responses were reinforced until E, when extinction was begun. All three rats show some intensification during the first part of extinction and a minimal intensity at the end of extinction.

tinction and give way to an unusually low force which is then steadily maintained. This is clearly shown in Figure 102 in the relatively flat sections at the ends of the records. Even in Series A the final mean force is less than that under normal reinforced responding, and this is much more obviously the case in the other series. The low final slope might be regarded as a compensation for the earlier strong responses, but another interpretation will be advanced later.

CHANGE IN FORCE UNDER DIFFERENTIAL REINFORCEMENT

Although a relatively constant mean slope is maintained during reinforced responding, there is some variation in the force of single responses, and it is therefore possible to reinforce differentially with respect to this property. Even where there is little variation, differential reinforcement necessarily involves extinction, and some strong responses are made available.

When responses are differentially reinforced with respect to their intensity, the relative frequency of strong responses immediately increases. It is difficult to follow the process closely because of the occurrence of strong responses due merely to the occasional failure to reinforce. If a value is chosen for differential reinforcement that is likely to occur, say, once in ten times without differentiation, responses possessing that value begin to occur more frequently (say, once in two times) with what is apparently an instantaneous change. The stronger responses may exceed the required force, and hence it is possible to advance the critical value still further. In this way a progressively higher mean value may be obtained. The first critical value decided upon must, of course, be within the normal range of variation, but through progressive differentiations any value within the capacity of the rat may apparently be reached.

The rate of advance toward an extreme value of the intensity is slower than one might expect from the speed of the first differentiation. Two series of records showing the gradual attainment of a high intensity are given in Figure 103. Each section marked off by vertical lines represents a daily experiment, during which twenty responses were reinforced and unreinforced responses of inadequate force also occurred. The records are reproduced in the order in which they were taken, and the required value for reinforcement is also indicated for each. It will be seen that progress is slow, and that in no case does the rat reach a point at which all responses are strong enough to be reinforced. The principal result is the increase in the mean slope. These experiments were not carried on, but it would be very difficult to obtain further progress. In Figure 104 (page 316) the highest mean value for any rat is shown. Responses were reinforced only when they were equivalent to a weight of 100 grams. (The rat weighed less than 200 grams.) The character of the record will be seen to be important later in comparing the differentiation of the duration of the response.

The apparently final state reached in a differentiation of this

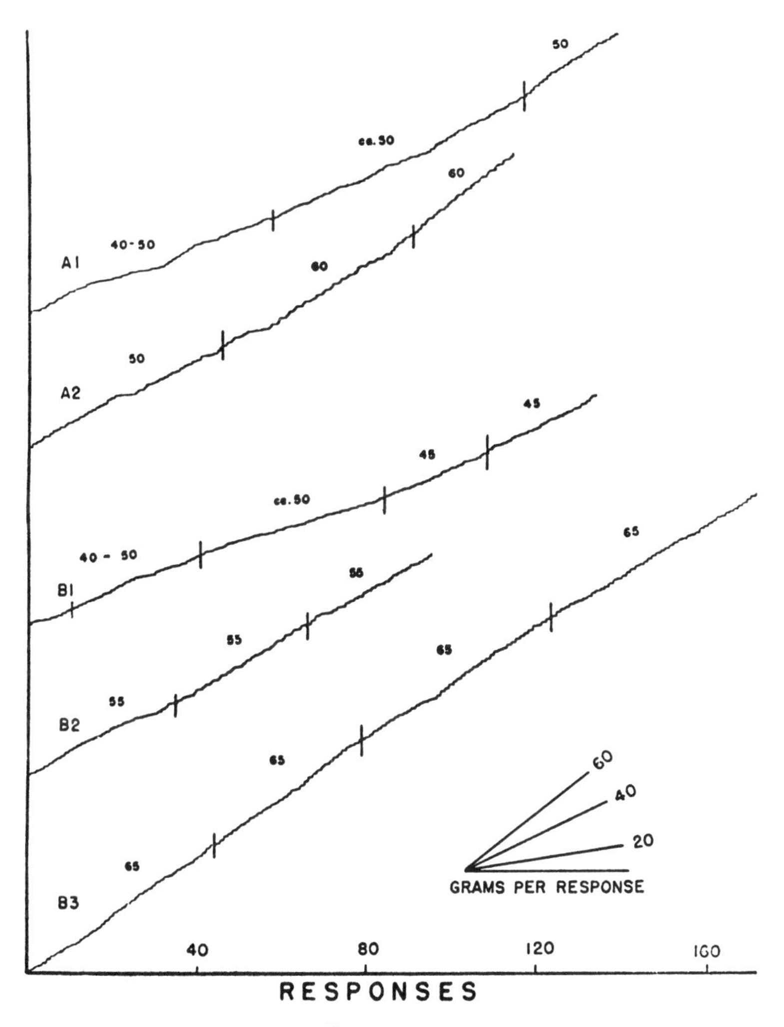

FIGURE 103

INCREASE IN MEAN INTENSITY WHEN RESPONSES ARE REINFORCED ONLY AT INTENSITIES ABOVE CRITICAL VALUES

The critical values are marked. Vertical lines separate daily records. Calibration curves for comparison are given.

sort at any value beyond the normal range is one in which alternate responses are generally strong enough to receive reinforcement. As will be shown in a moment, there is a constant tendency to reduce the force. When the differentiation is being made with diffi-

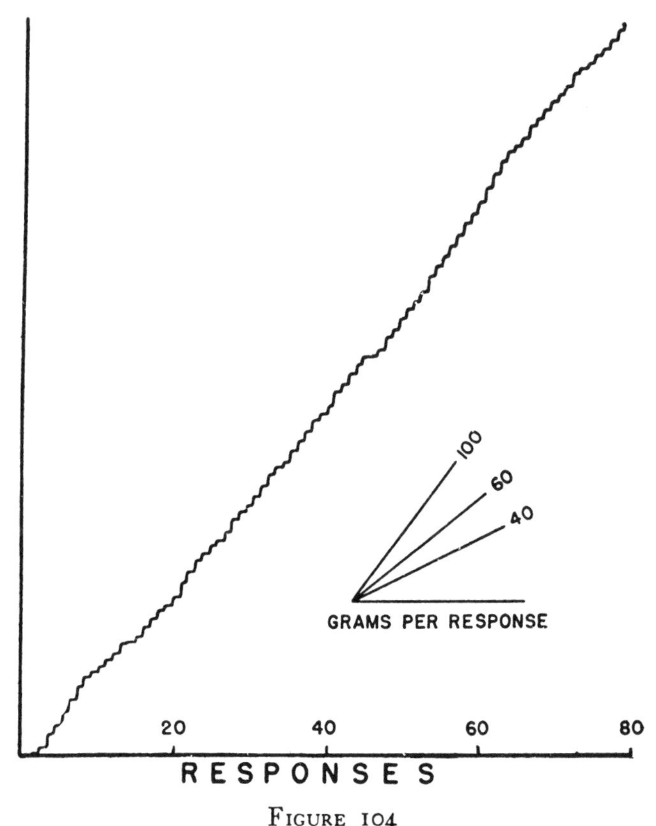

FIGURE 104

EXCEPTIONALLY HIGH MEAN INTENSITY

Only responses above 100 grams were reinforced.

culty, this tendency brings the force below the critical value immediately after a reinforcement. In extinction, on the other hand, as will also be shown in a moment, there is a tendency to increase the force. Hence the following order of events prevails: (a) the rat makes a successful response, (b) the force relaxes, and the next response is too weak, (c) because of the lack of reinforcement the force increases, and the next response receives reinforcement, and so on. Rats tend to adjust to a force which secures only slightly above the reinforcement of every other response. This is not true, of course, for very low critical values, when the proportion may be much higher. At the other extreme, if the proportion drops too far, the differentiation is lost entirely.

Figure 105 gives a typical example showing how the differentiation may be lost if the critical value is advanced too rapidly.

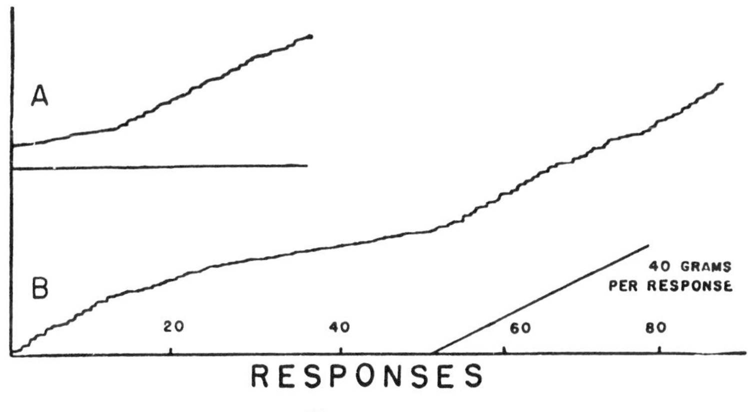

FIGURE 105

LOSS OF DIFFERENTIATION WHEN CRITICAL VALUE IS SET TOO HIGH

A: original differentiation developing a mean intensity of about forty grams. B: the differentiation was lost when the critical value was set at sixty grams but was subsequently regained when the value was set at forty. At sixty grams too few responses were reinforced to prevent extinction.

Record A is for the first day of the differentiation, when the apparatus was set so that approximately one-half the responses were reinforced. As the mean force increased, the critical value was advanced. The discontinuity shown in this record is occasionally obtained. The final mean force on this day was approximately 40 grams. On the following day an attempt was made to secure responses of 60 grams. The result is shown in Record B. A few responses at the beginning of the record were strong enough to be reinforced, and the initial mean force was slightly above the final value of the preceding day. But because of the predominance of unreinforced responses, extinction set in and with it the usual decline in force. (It will be shown shortly that the decline described above is also typical after differentiation.) From this point on no responses possessed the intensity required for reinforcement, and the operant would eventually have been extinguished. The critical value was, however, later reduced to 40 grams, at which one or two responses were reinforced and the differentiation reinstated. The critical value was then moved up gradually until a mean of over 40 grams was maintained. In general it may be said that the tendency of the force to decline in extinction sets the limiting rate at which the critical value may be advanced.

The decline in force during extinction is preceded by an initial increase. Because of this increase it is possible to maintain a relatively high intensity simply by reinforcing every other response, regardless of force. The unreinforced response in each pair increases the intensity of the next response, which is reinforced. There is thus an effective correlation of the reinforcement with stronger responses, not set by the apparatus but arising indirectly from the reinforcement of alternate responses. The result is an average intensity considerably above the normal base. For example, a rat that had established a slight differentiation, as described above, maintained an average slope of 31.7° for five days on which 20 reinforced responses were alternated with unreinforced. When every response was then reinforced on the sixth day the slope fell to 21° and averaged 18° for five days. This was equivalent to approximately half the intensity sustained by alternate reinforcement. The effect may be obtained without a preliminary differentiation (but see below for the possibility of an unintentional differentiation in all experiments of this sort). For example, a rat which had had no deliberate differential reinforcement maintained a mean slope of 31.9° for ten daily periods of alternate reinforcement and extinction. When every response was then reinforced, the slope fell to 28.5° on the first day and to 21° on the fourth day, with a mean for the four days of 23.9°. In one case in which several attempts to advance the force by differential reinforcement had failed, the use of alternate reinforcement was successful. The greatest slope obtained by differential reinforcement was 15°. A four-day series of alternate reinforcements gave slopes, respectively, of 15°, 21°, 20°, and 29°. The explanation seems to be that alternate reinforcement succeeds in obtaining a more delicate and more rapid differential reinforcement than the setting up of an arbitrary value.

LOSS OF THE DIFFERENTIATION

When reinforcement is again supplied to every response without respect to intensity, the rat may continue to respond with extra

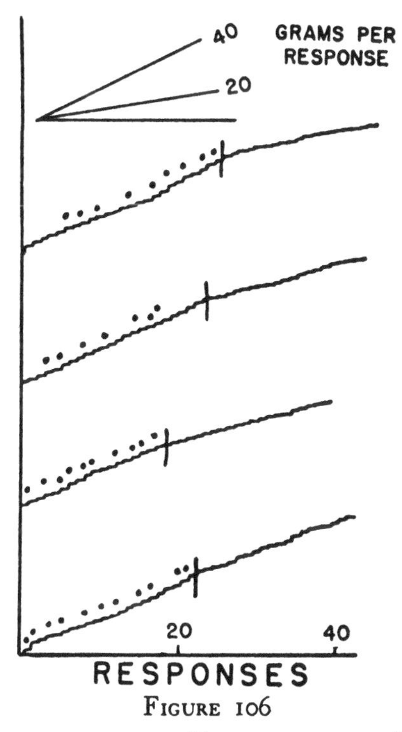

FIGURE 106

DECLINE IN INTENSITY AFTER DIFFERENTIATION WHEN ALL RESPONSES ARE AGAIN REINFORCED

Four successive daily records for one rat. In each case ten responses were first reinforced differentially at or above fifty grams. Beginning at the vertical lines, all responses were reinforced. The decline in mean intensity may or may not be rapid, probably depending upon accidental correlations of intensity and reinforcement.

force for some time, or the force may drop quite rapidly to a nearly normal value. Four records taken with the same rat on successive days are shown in Figure 106. In each case ten responses above 50 grams were first reinforced. Twenty consecutive responses were then reinforced at any value. The bottom record shows little change in force, while at the other extreme the top record drops to a nearly normal level after two or three responses. The other

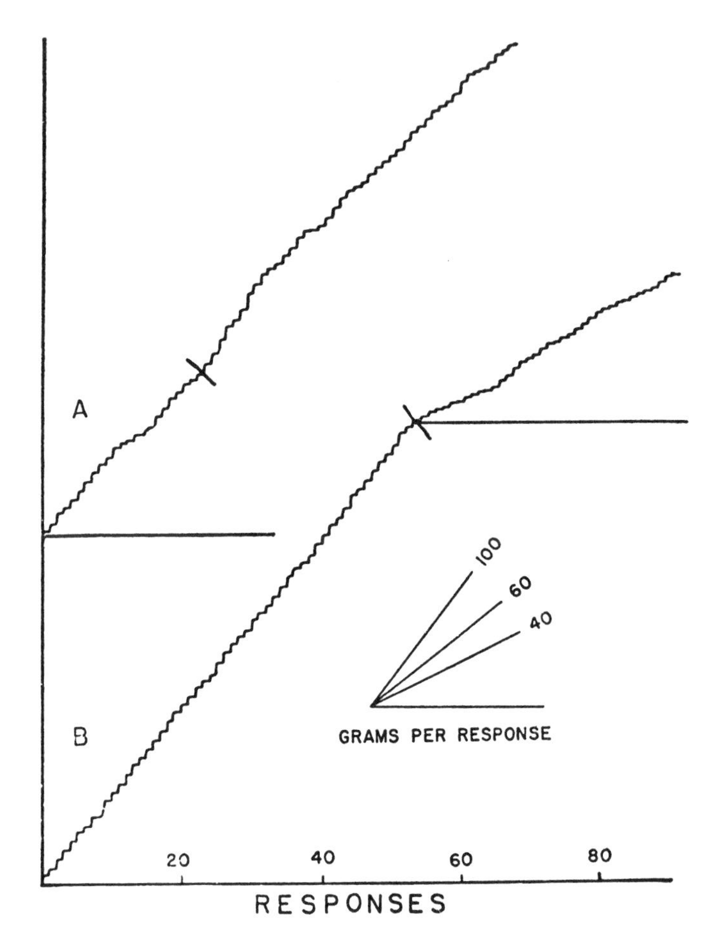

FIGURE 107

SPONTANEOUS LOSS OF THE DIFFERENTIATION OVERNIGHT

A: very slight drop to mean intensity when all responses were reinforced (cf. Figure 106). B: similar to A except that the short line marks an interval of twenty-four hours. The drop to a low intensity is immediate on the second day.

records show typical intermediate rates of change. The failure to obtain a consistent result is perhaps due to the accidental correlation of force and reinforcement which still obtains if the responses are strong, even though it is not demanded by the apparatus.

A drop to a nearly normal force usually occurs spontaneously overnight. In Figure 107 the record at A shows the very slight drop produced characteristically by this rat when the reinforcement is first contingent upon responses of 65 grams or more and

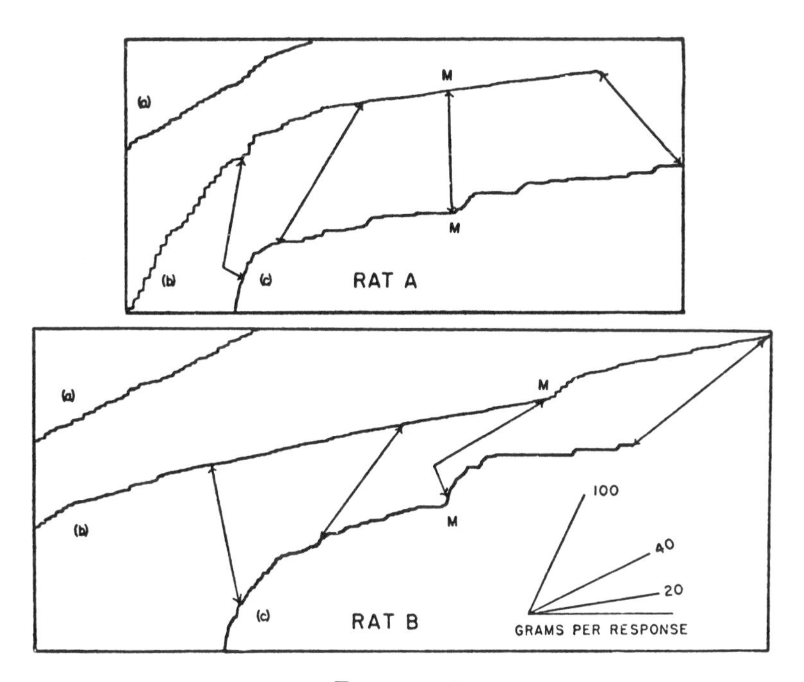

FIGURE 108

CHANGE IN THE INTENSITY OF THE RESPONSE DURING EXTINCTION AFTER DIFFERENTIATION

(a) Mean intensity under differentiated reinforcement at 50 grams. (b) Change in intensity during extinction. (c) Usual extinction curve given by the rate of responding. Same coordinates as heretofore. The curve for Rat A is 65 minutes long; for Rat B 70 minutes. Arrows show corresponding points on the two curves.

At the beginning of extinction the magazines were not sounding. They began to sound (but were empty) at M. The differentiation of the intensity is lost by Rat B before any very significant drop in rate has occurred.

is then provided without restriction. The record at B was taken on the following day when all reinforced responses were above 65 grams. The record ends with a third day when every response was reinforced regardless of force. The effect of the twenty-four hours in reducing the mean force is obvious.

EXTINCTION AFTER DIFFERENTIATION

I have already mentioned two characteristics of the change in force during extinction: an initial increase in force, followed by a decline to a very low value. Two examples are given in Figure 108. The short records at (a) show the mean force obtaining under differential reinforcement at 50 grams. The differentiation had been carried out for only a few days. The records at (b) show the course of the change in force on the following day when no responses were reinforced. At the beginning of the experiment the magazines were off; at M they were turned on but were empty. The usual extinction curve given by the change in rate (see Chapter Three) is also shown in each case at (c). Corresponding points in the two curves have been connected with arrows. It is characteristic of the force curve that a high mean force is quickly developed. In the case of Rat A a high force persists for about twenty-five responses, in the case of Rat B for only six. The low force ultimately reached is the lowest observed for each rat. When the magazines were turned on to produce the usual reinforcing sound, an increase in rate is obtained as described in Chapter Three. That the second curve is not of the same order as the first may be due in part to the fact that the sound of the force-recording kymograph could be heard by the rat at each response and to some extent resembled the sound of the magazine. The increase in rate due to the sound of the magazine proper is accompanied by an increase in force in Record Ab and by none in Record Bb.

During extinction both the force and the rate begin at high values and end at low. But there is otherwise no simple relation between them. In Figure 108 A the drop in force coincides with the drop in rate fairly closely, but no increase in force accompanies the acceleration in rate when the magazine is turned on. In Figure 108 B the major decline in force takes place long before the rate has fallen off appreciably, but the increase in rate when the magazine is turned off is accompanied by an increase in force. In general the evidence indicates that the change in force during extinction is relatively independent of the rate.

The ultimate low force displayed in this process obviously resembles that observed in extinction prior to differentiation (see Figure 102). The similarity suggests that some force-differentiation normally occurs even when 'every response produces reinforcement.' The normal reinforcement is made contingent upon the depression of the lever to approximately the mid-point of its excursion. But the rat may make responses short of this point, as may be easily observed when the response is first being conditioned. These incomplete weak responses are one of the causes of the recorded irregularities in conditioning curves. There is thus some differential reinforcement in the ordinary experiment. Only responses strong enough to depress the lever to the required point are reinforced. In Figure 102 this incidental differentiation is probably responsible for the excessive force in Series C, which develops at the fourth or fifth response, and to a lesser extent for the normal base-line in the other cases. The evidence for the differentiation is principally the facts that the ultimate force in extinction is lower than the normal force and that some strong responses usually occur in extinction that does not follow explicit differentiation.

If it is true that some differentiation is always set up and that it disappears in extinction, it is possible that the end of the normal extinction curve given by the rate involves an artifact and that incomplete weak responses are being made that are not recorded. This could be easily checked by watching the rat, but it is controverted by the nature of the contacts made toward the end of extinction. I have never observed any indication that the contact on the lever was not being firmly closed, except on the first day of conditioning. The duration of the contact does not change significantly during extinction. As a possible explanation it may be noted that the proprioceptive stimulation from pressing the lever acquires reinforcing value which persists even when the magazine is disconnected. It may be enough to prevent the reduction of the intensity of the response to a point at which the recording apparatus fails.

The curves in Figure 108 are representative of extinction after only small amounts of differentiation (2–4 days). When the differentiation is maintained for periods of the order of 10–15 days, the decline in force is much delayed. The force curves after prolonged differentiation show a flattening toward the end and incipient flattenings elsewhere with a tendency to fluctuate between periods of high and low intensity. The curves given by the change in rate are also considerably altered by the differentiation, but this is probably to be accounted for by the order of reinforcement during differentiation. Since many weak responses continue to be elicited and are not reinforced, there is an irregular periodic reconditioning with some resemblance to reinforcement at a fixed ratio. It is therefore difficult to say whether the force differentiation has an effect upon the reserve, although the evidence is probably against that possibility. The extinction curves obtained after slight differentiation show no effect, and those after prolonged differentiation are still of the magnitude to be expected from the irregular reinforcement.

Just as the normal extinction curve may be used to detect the presence of reserved responses after long periods of 'disuse' (see Chapter Three), the force curve during extinction may be used to detect the survival of the differentiation when reinforcement has for some time not been conditional upon intensity. For example, after a differentiation has been developed, let responses be reinforced for several days without restriction. Since the original change in intensity takes place quickly, it would be difficult to measure the amount of the differentiation surviving simply by reinstating the required conditions. But by extinguishing the response the differentiation may be made to manifest itself in the force-curve. Figure 109 gives the extinction-curve for a rat that had developed a 65-gram differentiation and for six days had then received 40 reinforcements per day without regard to force. On the day represented by the figure ten responses were first reinforced at any force. The magazine was then disconnected. The rate curve (B) shows considerable cyclic variation but is otherwise normal. The force curve shows very clearly the alternate variation in force characteristic of extinction after prolonged differentiation. There is a gradual reduction in mean force during the hour. From such a record it may be inferred that the effect of the differentiation has survived during the six days on which it has not been in use. The same procedure could be used when no responding takes place whatsoever and when the intervening period is great enough to be expected to produce 'forgetting.'

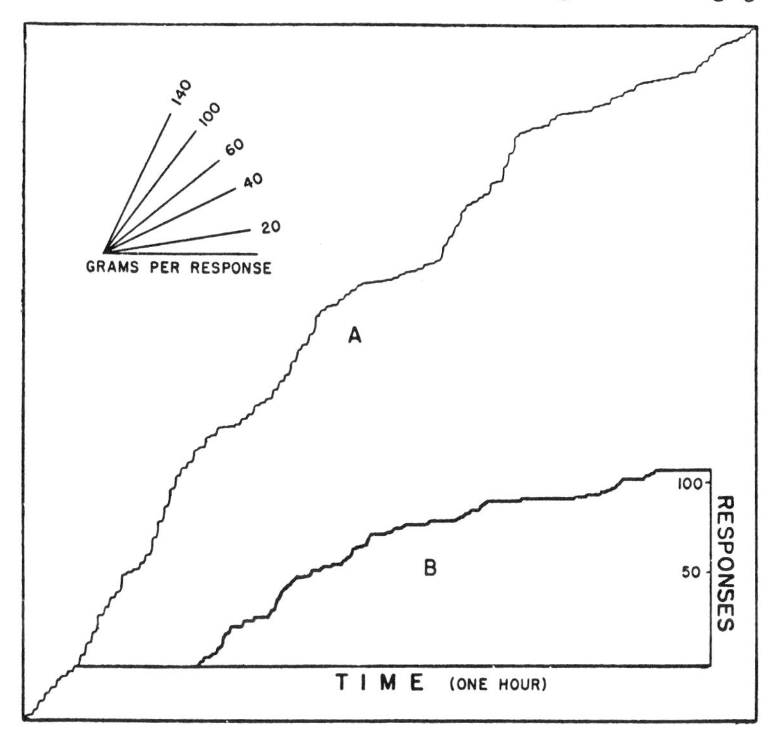

FIGURE 109

EXTINCTION AFTER PROLONGED DIFFERENTIATION FOLLOWED BY A PERIOD OF NO DIFFERENTIATION

The curve at A begins with ten responses reinforced irrespective of intensity. Extinction follows. Curve B is given by the rate of responding with the same coordinates as heretofore.

PERIODIC DIFFERENTIAL REINFORCEMENT

When responses above a given value are *periodically* reinforced, many responses above that value also go unreinforced. Nevertheless, the mean intensity of the response quickly rises and is maintained at about the same value as when all responses above the value are reinforced. A set of typical records is given in Figure 110 (page 326). Daily periods of one hour each are marked off by straight lines. Because of the large number of responses, the intensity was recorded on reduced coordinates. Units on the two axes were not reduced proportionately, and the resulting slope is somewhat greater at each intensity. The relative change in intensity will suffice here. The first day of the figure shows the normal rate under periodic reconditioning at five-minute intervals and the normal mean intensity. On Day 2 the first two reinforcements were given irrespective of force, but from that point on only responses

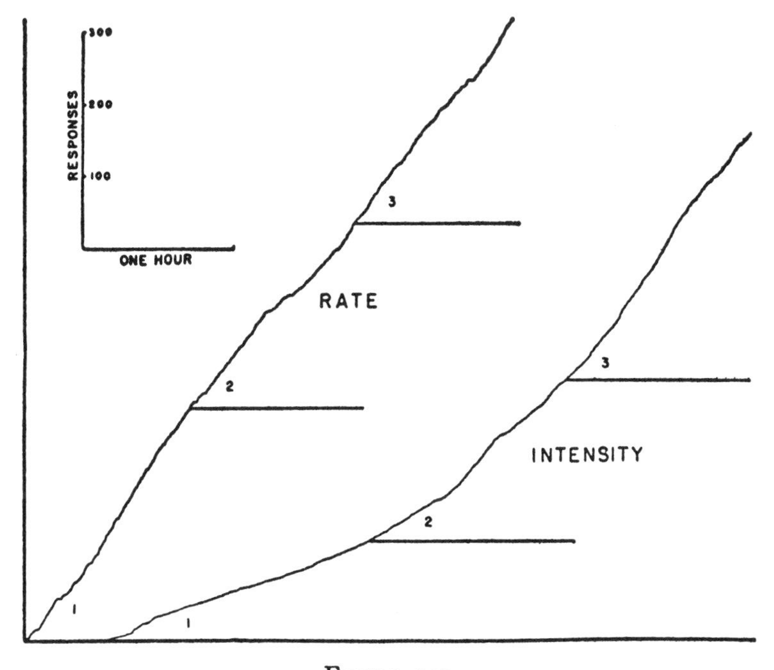

FIGURE 110

INCREASE IN MEAN INTENSITY WHEN RESPONSES ABOVE A GIVEN VALUE ARE ONLY PERIODICALLY REINFORCED

The slopes of the intensity-curve are somewhat higher than in the previous calibrations.

above 25 grams were reinforced. One response was reinforced every five minutes on the average, although it was necessary that the interval should vary somewhat in order to obtain responses of the required force. On the third day the required intensity was increased to 30 grams. No effect upon the rate follows from the differentiation but the mean intensity increases very much as in the case of the reinforcement of all responses above the required value.

THE DIFFERENTIAL REINFORCEMENT OF WEAK RESPONSES

One other problem in the differentiation of the intensity of a response may be mentioned here. Can a rat be trained to press the lever with a force *below* a given value? The question is difficult to answer experimentally for several reasons. The normal force is so

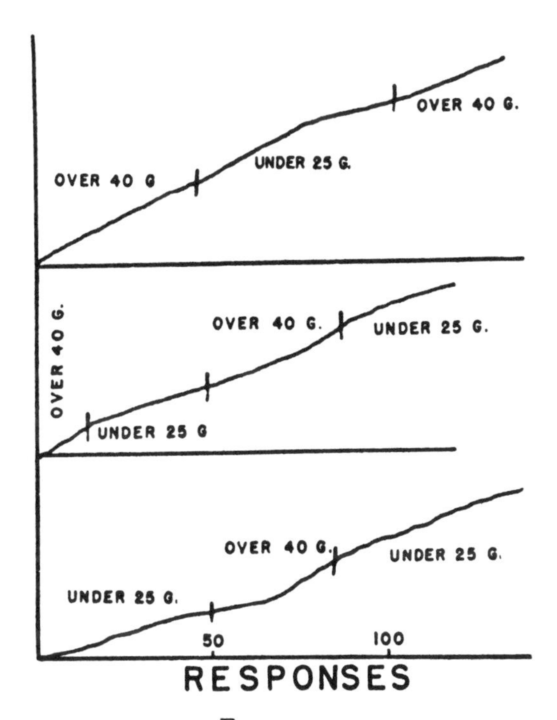

FIGURE III

DIFFERENTIAL REINFORCEMENT OF RESPONSES ALTERNATELY ABOVE AND BELOW CRITICAL VALUES

near the minimal value required to depress the lever that little change in a downward direction is possible. The present apparatus is too crude to cope with the problem within so narrow a range. The only other possibility is to differentiate downward after a differentiation upward has been carried out. Several factors operate against a simple result. When the differential reinforcement of weak responses is first established, an unreinforced response automatically increases the strength, with the result that subsequent responses are less likely to be reinforced. Considerable extinction ensues, and it is only when the extinction force-curve has begun to flatten out that responses again occur weak enough to be reinforced. Such reinforcement strengthens the response slightly and the whole process must be gone through again. When the upward differentiation is slight, a quicker shift to a low mean force is obtained, but it is again difficult to say whether a true differentiation has been established. Representative records for three rats are given in Figure 111 (page 327). The reinforcement is frequently changed from above 40 grams to below 25 grams as marked. There is little difference between the upward and downward adjustments in force. Either may be rapid or slow, presumably according to the accidental order of reinforcement in each case.

Differentiation of the Duration

The process of differentiating the duration of a response is in many respects similar to that of differentiating the intensity. It is only at the longer durations that a fresh problem emerges. In pressing the lever with varying intensities only one reflex is presumably involved. There are doubtless a few topographical differences between strong and weak responses, but for purposes of description (especially with regard to the reserve) we are dealing with essentially one unit of behavior. In pressing the lever and holding it down for a given length of time, however, there are two responses, which are quite different topographically. At the longer durations they come into conflict with each other and produce an effect not encountered in the case of intensity.

The original development of the differentiation follows the course already described for intensity. In Figure 112 the two records at A are typical of the low mean duration when all responses are reinforced. When reinforcement is withheld, subsequent responses are occasionally of longer duration. Extinction curves comparable with those of intensity have not been taken, but at B in Figure 112 two typical records are given for the duration when groups of four responses are alternately reinforced and unreinforced. The emergence of longer responses shown in the record permits the differential reinforcement of responses three to five seconds long. The effect of the differential reinforcement in developing a greater mean duration is shown in the figure at C, where records for three rats are given. Responses were reinforced at durations of four, four, and five seconds as marked. All reinforced responses are indicated with dots. The rate of increase in mean duration is comparable with that already described for intensity.

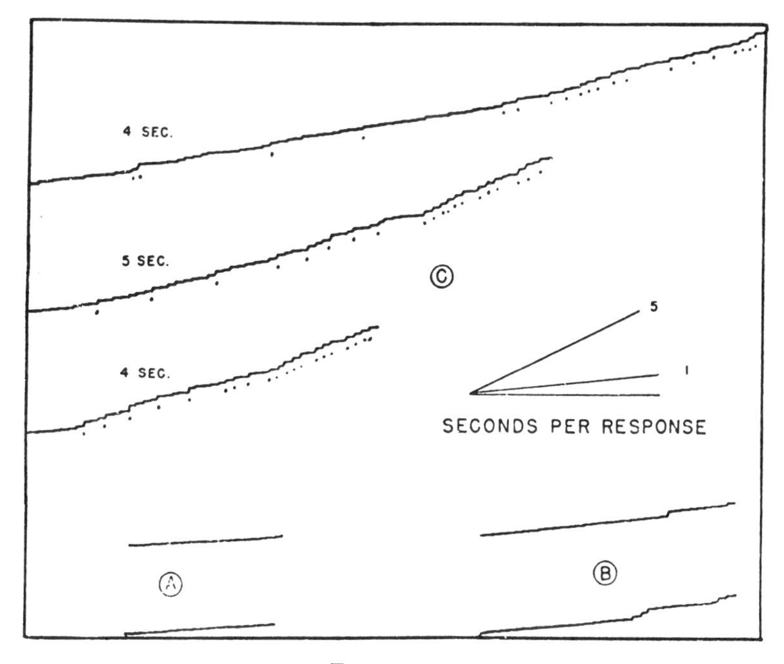

FIGURE 112

DIFFERENTIATION OF THE DURATION OF THE RESPONSE

A: normal durations, all responses reinforced. B: long responses appear when alternate groups of four responses go unreinforced. C: increase in mean duration when responses above a critical duration are differentially reinforced. Records for three rats.

It is when differentiations at extreme values of the duration are attempted that a special effect arises. The differential reinforcement of responses of considerable force yields responses the intensities of which are distributed closely about the required value. In the case of duration there tend to be two classes of responses: of long and short durations. The short responses occur in groups and usually at the beginning of an experiment. Responses of intermediate durations are rare and are scattered throughout the record.

A series of consecutive records under differential reinforcement at 12 seconds is shown in Figure 113. The rat had advanced from two to twelve seconds on seven previous days, twenty-five responses being reinforced each day. The first record in the figure

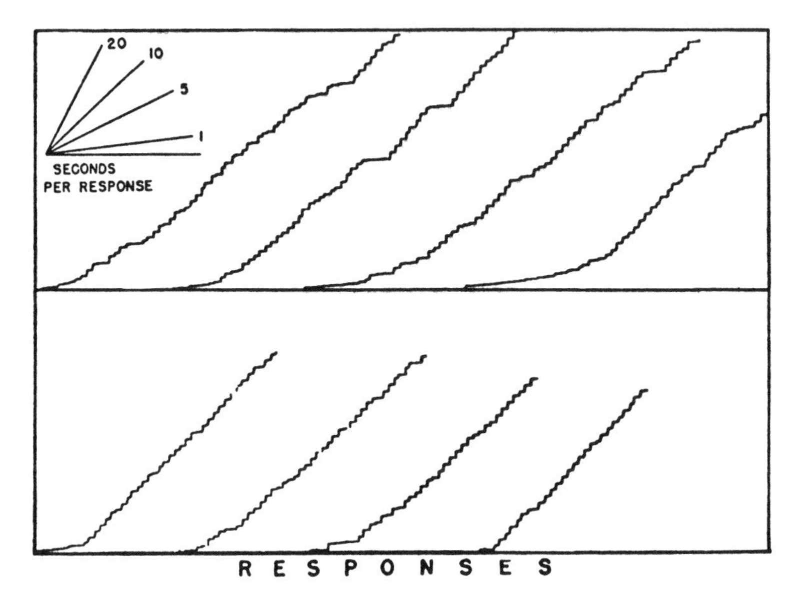

FIGURE 113

DURATION UNDER DIFFERENTIAL REINFORCEMENT OF RESPONSES TWELVE SECONDS LONG

Note the tendency to begin each daily period with short responses.

resembles an intensity-record in the distribution of adequate and inadequate responses. As the differentiation becomes more efficient, the short responses become shorter and are grouped together, usually near the beginning of the period. At the end of this series of records, where the differentiation is well established, very few short responses occur.

After the series recorded in Figure 113, the differentiation was advanced by steps of three seconds every three days. The groups of short responses became more pronounced as the duration increased. In Figure 114 the three records at 27 seconds are given. There are some intermediate short responses mixed with the long, but the principal characteristic of the record is the group of very short responses at or near the beginning of each record. In this figure the rate-curves are also given. The rat begins each period by responding very slowly. One or two of these slow responses may be long. There is then a period of very rapid responding, when all responses are very short. Finally long responses begin to appear

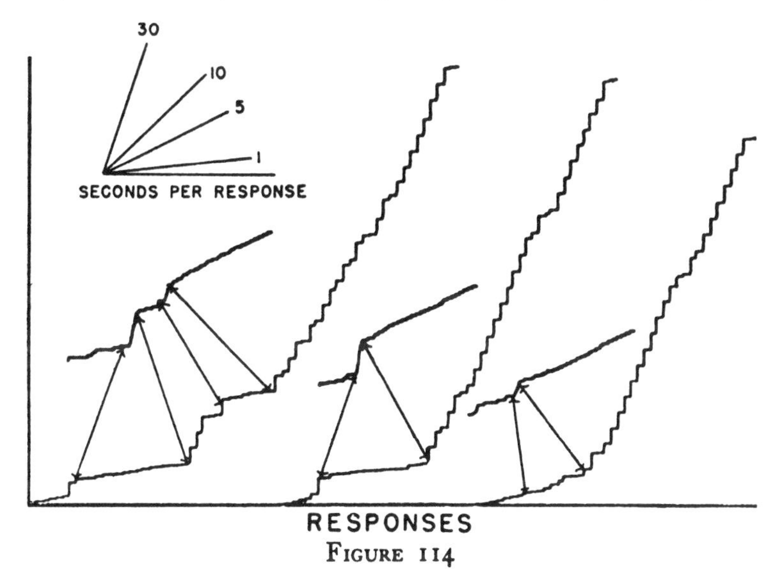

FIGURE 114

DURATION UNDER DIFFERENTIAL REINFORCEMENT OF RESPONSES TWENTY-SEVEN SECONDS LONG

A later part of the series in Figure 113. Rate-curves have been added and corresponding points connected with the duration-curves by arrows. Note further development of groups of short responses at beginning of records.

regularly. The rate at this time is, of course, limited by the duration of each response. After the three days shown in Figure 114 the required duration was increased to 30 seconds. Although the relative increase was slight, the behavior of the rat was greatly disturbed. Five days are shown in Figure 115 (page 332). The groups of short responses still appear, but they begin to occur in the body of the curve. The relation between the rate and the duration may be seen from the two rate-curves included in the figure and connected by arrows to the duration curves.

On only one occasion has a force-curve given any suggestion of a similar grouping of weak responses. In general the typical curve for an extreme value of the intensity (see Figures 103 and 104) is free of this effect. The explanation of the difference seems to be that in coming to hold the lever down for a given length of time the rat encounters a conflict between the response of holding it down and the response of pressing again. Continuing to hold the

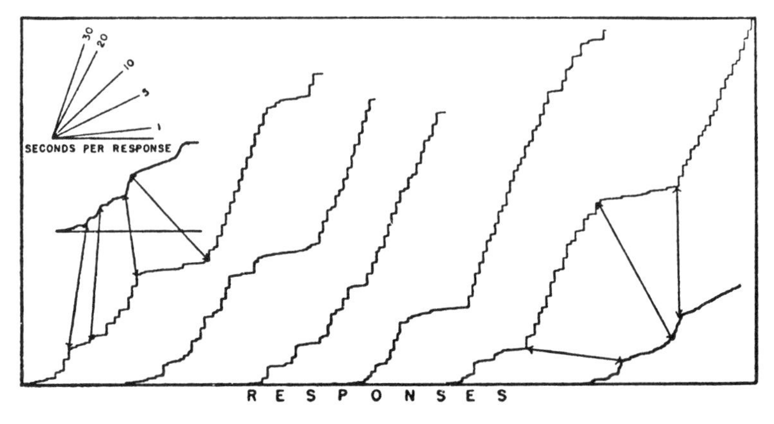

FIGURE 115

DURATION UNDER DIFFERENTIAL REINFORCEMENT OF RESPONSES THIRTY SECONDS LONG

Continuing the series in Figures 113 and 114. Two rate-curves are also given. Groups of short responses now appear in the body of the curves.

lever down is incompatible with making a second response, but both holding and pressing are reinforced and exist side by side with considerable strength. The response of simply pressing exists at a strength which calls for its elicitation more frequently than the required duration will allow. According to this interpretation a long response can be executed only when the response of pressing is weak enough not to conflict. The required weakness is obtained in the rapid elicitation of short responses, which are not reinforced. It is only when the rat has 'got rid of a certain number of pressings' that it can continue to hold the lever down long enough to secure reinforcement. The situation probably also involves the reinforcement of 'not-responding,' a problem which requires investigation in its own right but on which I have nothing at the present time to offer. Such reinforcement may account for the initial low rate of responding on each day.

If the interpretation of the typical result in differentiating the duration as the effect of a conflict between pressing the lever and holding it down is correct, it should be possible to alter the balance of the two forces by changing the drive. The response of holding the lever down involves a temporal discrimination. But a measure of any discrimination which depends upon a ratio of 'correct' to 'incorrect' responses or upon the prepotency of one of two conflicting responses will be affected by the state of the drive existing at the time. Pavlov's demonstration that a 'trace reflex' (interpreted as a temporal discrimination in Chapter Seven) is sensitive to the drive has already been mentioned. In the present case, if the drive is lowered, the tendency to begin a new response should be weakened and the rat should thereby be allowed to hold the lever down for longer times. A very minor experiment on this point offers some suggestion that this is the case. Two rats which had been differentially reinforced at eight seconds were tested. The extent of the differentiation was estimated by counting the number of too-short responses made in the course of completing 20 sufficiently long responses. Calculations were made for thirteen days. Six of these were at the normal drive, the mean number of too-short responses made by each rat being 26 and 21 respectively. Interspersed among these were days when the drive was reduced by feeding either two or four grams of food (see Chapter Ten). The first day in the case of each rat was with two grams and the effect was chiefly a disturbance at the new procedure. The remaining days were with four grams. The mean numbers of too-short responses on the four-gram days were reduced along with the drive to 13 and 12 respectively.

This experiment is too slight to yield more than the suggestion that the appearance of short responses can be affected by the drive in a way that agrees with the interpretation of the process as involving a conflict between the response of holding and the response of pressing anew.

The interpretation of a conflict between holding the lever down and making another response is more adequately supported by the course of the duration when the response is extinguished, to which we may now turn.

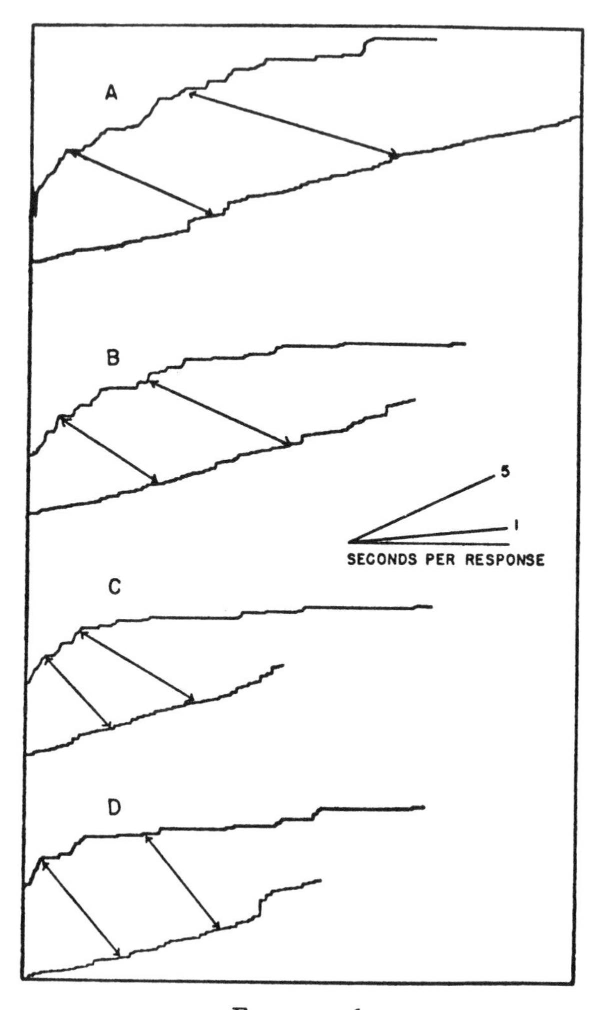

FIGURE 116

EXTINCTION AFTER INCREASING DEGREES OF DIFFERENTIATION OF THE DURATION

The upper curve in each pair is one hour long and is given by the rate of responding (same coordinates as heretofore). The lower curves give the duration. Three days of differential reinforcement at four seconds intervened between each day of extinction.

EXTINCTION AFTER DIFFERENTIATION OF THE DURATION

A series of extinction curves for a single rat after more and more extensive differentiation is reproduced in Figures 116 and 117.

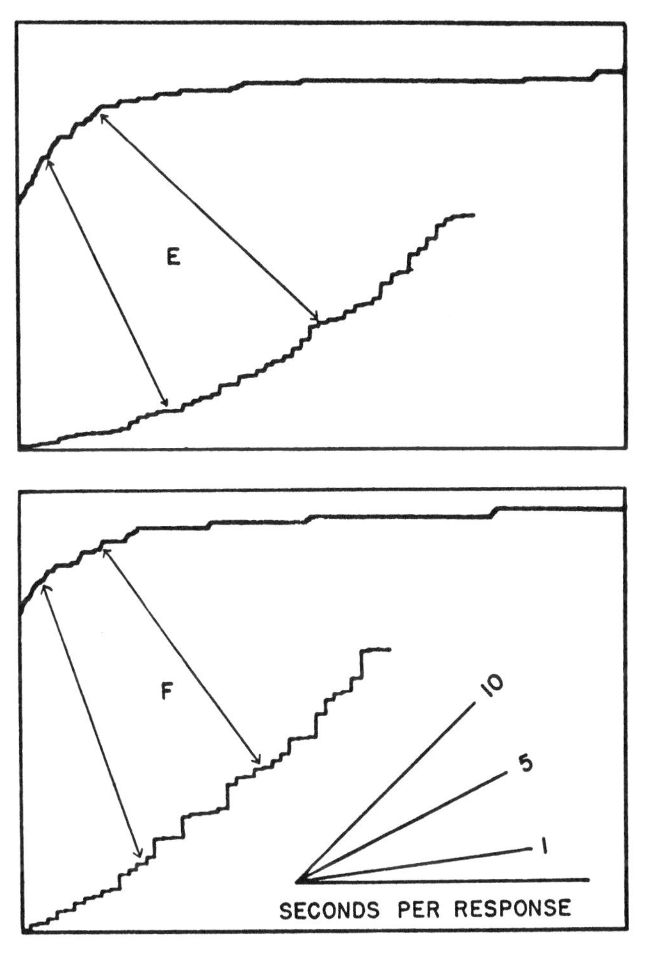

FIGURE 117

EXTINCTION AFTER ADVANCED DIFFERENTIATION OF THE DURATION

These curves continue the series in Figure 116. Prolonged differentiation at eight seconds preceded each extinction. Note the positive acceleration in the duration curves and the normal character of the curves given by the rate.

The records at A were obtained after one day of differentiation, upon which 20 responses three seconds long were reinforced. The responses are recorded in the upper curve, duration in the lower. A few long responses appear during the extinction. Between the curves at A and those at B three days of differentiation intervened, when 20 responses were reinforced each day at a duration of four seconds. The same amount of differentiation also intervened between B and C and C and D. The curves given by the responses plotted against time show a progressive reduction in area, as is generally the case in repeated extinction, and the duration curves

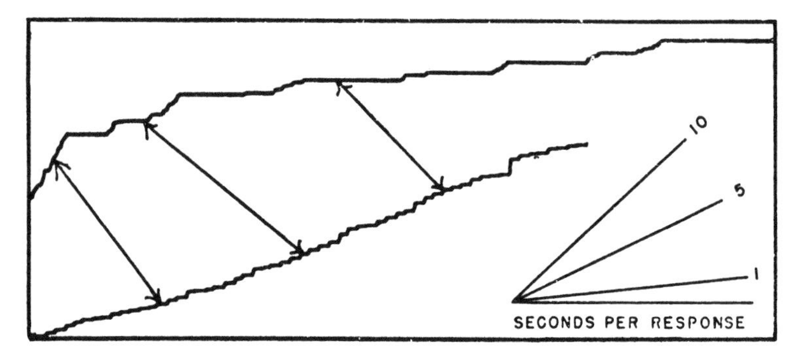

FIGURE 118

EXTINCTION AFTER DIFFERENTIATION OF THE DURATION

All responses produced the sound of the empty magazine at the end of four minutes.

show a slight increase in mean duration. The effect of this amount of differentiation is, however, slight. The extinction recorded at D was then followed by two days at four seconds and three days at eight. Subsequent extinction is shown at E in Figure 117. The rate-curve is of the same order, but the duration-curve shows a considerably greater mean slope and a marked positive acceleration. The curves at F were taken after 16 days of further differentiation at eight seconds. The mean duration is further increased, while the rate-curve keeps the same general form and size.

The series of experiments recorded in these two figures shows first of all that the differentiation has no effect upon the reserve, since the shape and size of the extinction curve are not significantly affected. They show also that the mean duration during extinction is a function of the degree to which the differentiation has been carried. A differentiation of eight-second responses produces a greater mean duration during extinction than one of four-second responses. Finally, in confirmation of the notion of a conflict advanced above, the positive acceleration in the duration-curves indicates that the rat is capable of holding the lever down for longer

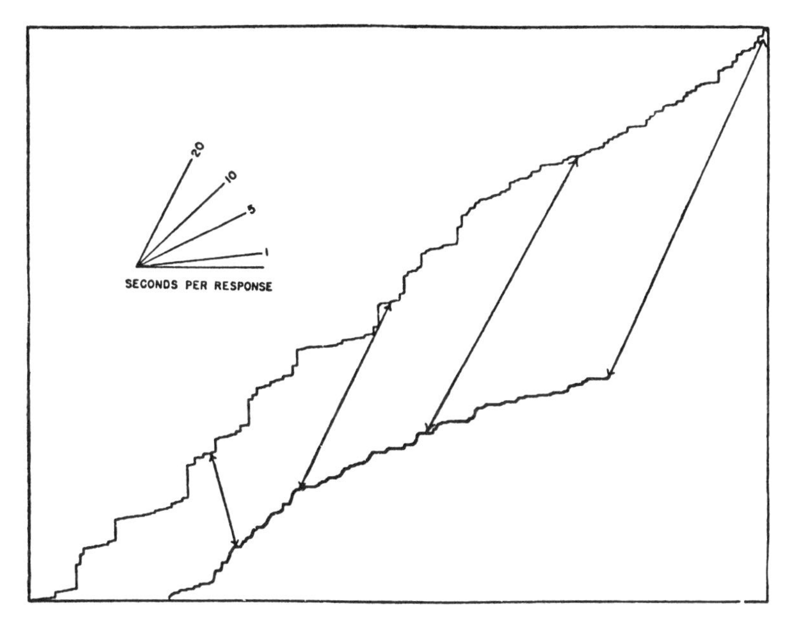

FIGURE 119

EXTINCTION AFTER ADVANCED DIFFERENTIATION OF LONG DURATIONS

The curves are 85 minutes long. The extinction was obtained following the long series concluded in Figure 115. Note the presence of responses more than a minute long and groups of very short responses.

periods when the response of pressing the lever is weakest, *i.e.*, toward the end of extinction.

Another way to extinguish the reflex after differentiation is to allow the magazine to sound at the end of a given interval but to deliver no food. The sound usually terminates the response at the selected time, and there is no tendency to increase the duration as the reflex weakens. A typical curve is given in Figure 118. Differential reinforcement had been carried out for one day at two seconds and three days at four seconds. On the day recorded in the figure each response four seconds long produced the sound of the magazine, but no food was given. If a response is made at all, it is generally prolonged until the magazine sounds.

After extremely long durations have been attained, the extinction curve is quite irregular. Groups of short rapid responses occur, as when differential reinforcement is being provided, and exceptionally long durations are also obtained. In Figure 119 (page 337) the curve following the differentiation shown in Figure 115 is reproduced. Several responses more than one minute long may be observed, although the maximal duration reinforced was only thirty seconds. The groups of very short responses characteristic of Figure 115 are also obvious. The initial slow responding developed during the differentiating is apparent at the beginning of the extinction curve for responses against time (B), although the latter part of the curve is normal for the amount and frequency of reinforcement that have preceded it. The curve for the duration does not show the positive acceleration characteristic of extinction after less differentiation, but in spite of the very long responses in the early part of the curve, there is no flattening off comparable with that observed in the case of intensity.

Combined Discrimination and Differentiation

In saying that the differentiation of the response takes place independently of discriminative stimulation it is not implied that a relation may not be established between these two events. In the ordinary life of the organism the determination of the intensity of a response often rests with discriminative stimuli. We lift an object with a force roughly proportional to its size before the weight of the object has stimulated us, as is readily apparent in the size-weight illusion. The behavior shows a double discrimination. In the presence of a given discriminative stimulus a response above a critical intensity is reinforced. In the presence of another stimulus the response may be governed by another critical intensity. An experimental parallel would be the case in which in the presence of a light only responses above a given force would be reinforced while in the absence of the light any response or only responses below a given force would be reinforced. This case has not yet been tested. Such a discriminated and differentiated operant should be written $sS^D . R^D$ where the reinforcement is conditional upon both *D*'s.

Conclusion

Topographical and quantitative differentiations of the response are both clearly distinct from sensory discrimination. Even if we assume that for each identifiable form of response there is a specific internal or external stimulus exclusively correlated with it, it is still impossible to force the facts described in this chapter into the formulae of Chapter Five. The difference becomes clearer when the non-elicitative character of operant behavior is recalled. Between topographical and quantitative differentiation there are also several important differences. In either the intensive or durational case the various forms from which a differential selection may be made lie on a continuum. The interaction between them is relatively simple, and only one reserve is involved. But it is difficult to say how far a topographical differentiation could be carried before additional reserves would be encountered. The interaction of operants possessing much in common topographically has been studied by Youtz (82). When a rat is conditioned to press either a vertical or a horizontal lever by reinforcing with food, Youtz finds that the extinction of the response to one lever decreases the number of responses in the extinction curve for the other by 63%. This topographical case obviously involves two reserves.

Both topographical and quantitative differentiation follow the rule of original operant conditioning that a response of the required form must be available prior to reinforcement if differentiation or conditioning is to take place. Extreme forms or values are obtainable only through successive approximations. When a single property such as intensity is the basis of the differentiation, the process may be represented in the following way. Originally the responses occur with their intensities distributed (say, normally) about a low value. Reinforcement of members in the upper part of the range shifts the mean upward and with it the whole curve. Responses in the upper part of the new range may then be reinforced, and so on.

The topographical case is not so simply represented, but extreme forms of response are again reached only through a series of steps. Animal trainers are well versed in this method. As a sort of *tour de force* I have trained a rat to execute an elaborate series of responses suggested by recent work on anthropoid apes. The behavior consists of pulling a string to obtain a marble from a rack, picking the marble up with the fore-paws, carrying it to a tube projecting two inches above the floor of the cage, lifting it to the top of the tube, and dropping it inside. Every step in the process had to be worked out through a series of approximations, since the component responses were not in the original repertoire of the rat.[1]

The data on the intensity of the response illuminate one of the distinctions between operant and respondent behavior. In a respondent the intensity of the response is determined by the intensity of the stimulus, and no differentiation of the response is possible. The intensity of the response to a constant stimulus is a direct measure of the strength of the reflex. In operant behavior there are very slight, if any, changes in intensity with changes in the strength of the reflex. In Figure 102 the slight increase in intensity during rather complete extinction is probably due to the incidental differentiation of the intensity that results from the initial tension of the lever. In any event the change is much narrower than the change in strength given by the rate. The intensity of the response in an operant is significant only in relation to the differentiative history of the organism, and this may account for the failure of many attempts to extend respondent techniques to operant behavior.

[1] A popular account of this rat with photographs appeared in *Life*, May 31, 1937.

Chapter Nine

DRIVE

The characteristic changes in the strength of a reflex hitherto described have been associated with the operation called reinforcement. The processes of conditioning, extinction, discrimination, and differentiation, in their many forms, arise from the various ways in which a reinforcing stimulus may be related to behavior. It is obvious that reinforcement is one of the important operations that modify reflex strength. Another perhaps equally important kind of operation is associated with the traditional problem of drive or motivation, to which this and the following chapter will be devoted.

The notion of drive to be advanced here is rather more restricted in its range than the traditional conception, which has embraced a variety of phenomena. At one extreme, 'drive' is regarded as simply the basic energy available for the responses of an organism; at another it is identified with 'purpose' or some internal representation of a goal. A survey of many current views may be found in the work of Young (81). Fortunately our preliminary orientation yields a simpler statement of the problem and a sharper delineation of the field to be covered.

The problem of drive arises because much of the behavior of an organism shows an apparent variability. A rat does not always respond to food placed before it, and a factor called its 'hunger' is invoked by way of explanation. The rat is said to eat only when it is hungry. It is because eating is not inevitable that we are led to hypothesize an internal state to which we may assign the variability. Where there is no variability, no state is needed. Since the rat usually responds to a shock to its foot by flexing its leg, no 'flexing drive' comparable to hunger is felt to be required. Traditional solutions of the problem of hunger and other drives are not our present concern. As in any case of variability in reflex strength, the problem here is to find the variable or variables of which the strength is a function and to express the relationship in a set of laws.

In the example just mentioned, then, we may turn our attention from a 'state of hunger' to typical behavior that is said to be dependent upon it—for example, to the behavior of a rat in approaching, seizing, chewing, and swallowing a bit of food. The elicitation of this chain of reflexes cannot be predicted merely from a knowledge of the stimulating energies arising from the situation. One of the other variables to which we may appeal is time. If a rat is placed under experimental control with a supply of food, its normal eating schedule may be ascertained. Richter (67) has shown that under certain conditions eating occurs with considerable regularity at 3–4 hour periods. The demonstration of a periodicity extends our predictive power, and similar findings have done the same with other kinds of behavior, *e.g.*, thirst, sex, activity, etc.

At least two further advances in this direction may be made. The demonstration of a 'cycle' or 'periodicity' is an inadequate supplement to a knowledge of the stimulative situation because the variable of which the behavior is a function is not simply time but the ingestion of food and the lapse of time after ingestion. The periodicity is incidental: a rat does not eat continuously and hence it must eat periodically. It is true that the orderliness of the period suggests a lawful process, but it does not of itself tell us much about the actual relationship of the strength of the behavior to the operation of which it is a function. We cannot improve upon the predictive value of a hunger cycle by refining our measurements of time, but we may attempt to gain greater control of the important variable, which is the amount of food eaten. Experiments similar to Richter's in which the amounts eaten at each period and the rate of eating were measured, have been performed by Keller, to which I shall refer again later.

The second possible improvement upon the mere recording of a cycle is to recognize not merely the presence or absence of eating behavior but the various degrees of strength in which the behavior may exist. The simple observation of whether or not a rat eats is an all-or-none measure, but our common use of the term hunger indicates that a measure which takes account of degree is necessary.

The problem may be restated as follows: in dealing with the kind of behavior that gives rise to the concept of hunger, we are concerned with the strength of a certain class of reflexes and with the two principal operations that affect it—feeding and fasting. Because the elicitation of eating behavior usually involves ingestion, the strength and the operation affecting strength are easily confused. There is no especially difficult technical problem involved, however. In measuring the strength of the behavior, the techniques already used in the case of conditioning are available,[1] and the operations of feeding and fasting suggest their own methods of measurement.

A reflex concerned with ingestion may vary in strength between two extremes: it is strongest after prolonged fasting and weakest after extensive ingestion. If an organism is deprived of food for some time and then allowed to eat freely, the strength changes from a high to a low value. In the experiments now to be described this change was followed by recording the rate at which food was eaten. The measure serves two purposes, since the rate is at once an indication of the strength of the chain of reflexes of which eating is composed and also a measure of the ingestion of food, which influences the strength.

Pellets of food of uniform size ($1/15$ gram) were prepared from a mixture of whole wheat, corn meal, rolled oats, flaxseed meal, bone meal, and salt. The rate of eating could then be expressed as the rate at which such pellets were taken up and eaten by the rat. Such a rate may be recorded in the following way. The rat stands on a platform and obtains pellets by pushing inward a light door hanging in the opening to a pocket at one edge. The door is counterbalanced and moves with ease. The food is placed below the level of the platform so that the rat must withdraw from the tray before eating. Each time the door is opened, a contact is made and recorded in the usual way. In the model used, two pockets stand at the end of a common platform. One contains water.

At the same hour daily a rat was taken from the animal room and placed in the experimental box with an ample supply of food and water. Ordinarily the rat was left in the box until at least 30 minutes had elapsed during which no eating had taken place. The rat was then removed from the box, returned to the animal room, and given about five grams of surface-dry lettuce leaf. No other food was given to it.

[1] The aspect of behavior that changes during a change in hunger is the same as in the case of conditioning. On this point the reader should refer to Chapter One, page 25.

In Figure 120 a typical record obtained with this procedure is reproduced. It will be seen that the rate of eating varies in an orderly fashion, beginning at a maximum and decreasing regularly throughout the period. The typical curve closely approximates a parabola. The equation is $N = Kt^n$ where $N =$ amount of food eaten at time t measured from the beginning of the period, and K

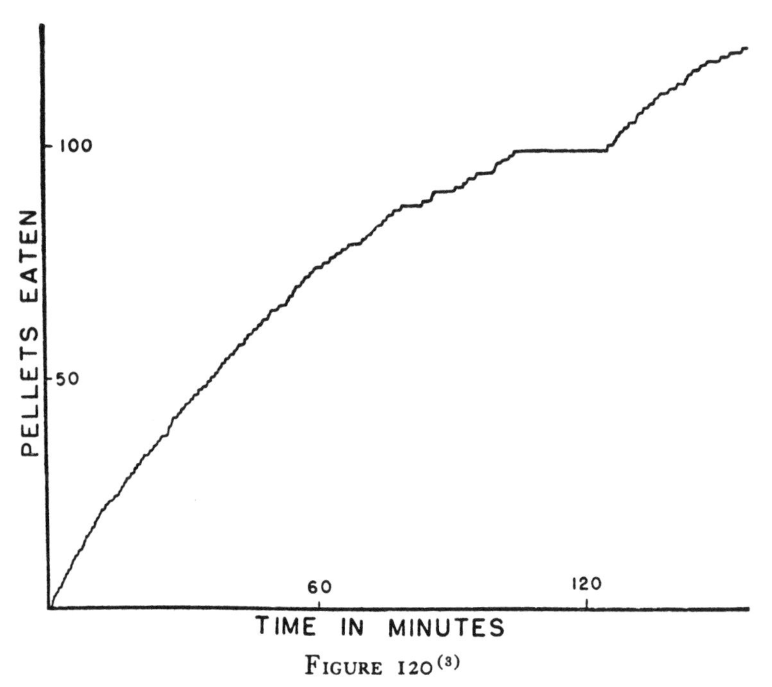

FIGURE 120[3]

CHANGE IN THE RATE OF INGESTION DURING A DAILY EATING PERIOD

At each elevation of the writing point the rat obtained and ate a pellet of food. Note the delay and subsequent recovery toward the end of the period.

and n are constants. The value of n is approximately 0.68. An experimental curve fitted with the curve for this equation is given in Figure 121. The value of n in this figure is 0.70.

These figures are not exceptional. Curves of equal regularity may be obtained daily, if the experiments are conducted with due care. The minor deviations from a smooth curve are attributable to several factors. The recording apparatus may fail to follow the behavior of the animal, especially when this is atypical, as, for example, when the rat takes a piece of food from the tray, drops it, does not recover it, and returns again to the tray. In such a case two pieces will be recorded for one eaten. Accidents of this sort are fortunately infrequent. There are also irregularities in the behavior of the animal which are correctly reported in the records. It will be seen in Figure 120 that near the end of the second hour the rat stops eating, to begin again only after an interval of some 20

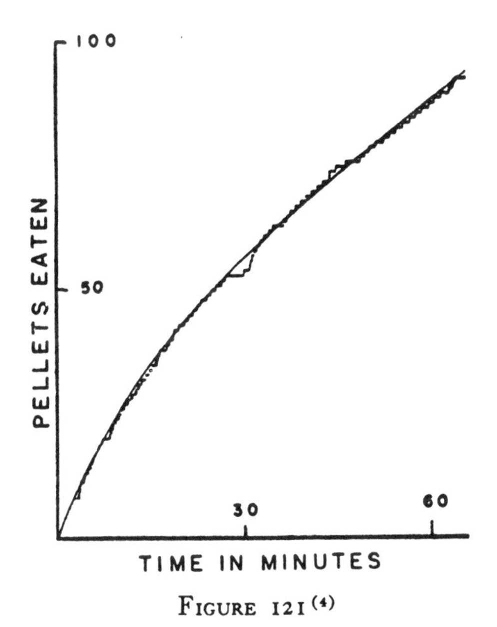

FIGURE 121 (4)

CURVE GIVEN BY THE CHANGE IN RATE OF EATING FITTED BY AN EQUATION DISCUSSED IN THE TEXT

minutes. At the end of the interval the rat is, so to speak, considerably behind the schedule set for it by the earlier part of the curve. When it begins again, however, it eats at a greater rate, so that within half an hour the record has caught up with the extrapolation of the body of the curve. The rate is not constant during the period of recovery, but falls off in much the same way as in the original curve. I have never observed a record to go beyond the extrapolation of the main curve. On the contrary, the extrapolation is not quite reached in most cases. Other minor examples of the same phenomenon may be discovered in Figures 120 and 121.

The interruptions shown in such records may be accounted for in several ways. During some of them the rat drinks, as has been determined with a signal device attached to the water tray. The longer intervals in the latter part of the records cannot be explained in this way, and must be due to conflicting behavior of another sort. They are a function of the strength of the eating reflexes (increasing as the strength decreases), and they may therefore plausibly be regarded as the effect of competing stimuli in taking prepotency over the stimulation from the food. The compensating increase that follows an interruption resembles the compensation already described in preceding chapters.

In some experiments that bear upon the process of recovery a small bellows was arranged to press outward when inflated against the door of the food pocket. The bellows was controlled from the adjoining room and permitted the experimenter to lock the door to the pocket without otherwise disturbing the rat. Intervals of any desired length could thus be introduced into a record, during which the rat could not eat. In the following experiments the intervals were 10–15 minutes long and were begun as soon as the general trend of the curve had been established or when about the thirtieth piece of food had been eaten.

Figure 122 shows a typical record obtained when the pneumatic lock was used for the first time. The curve was interrupted, of course, and following the enforced interval the rat ate at an augmented rate. The extrapolation of the original curve was reached within 20 minutes. This was not always the case, as may be seen in Figure 123, in which the interruption lasted 13 minutes and the extrapolation was never reached. Recovery curves obtained in this way differ significantly from that of Figure 120, as may be most clearly seen by foreshortening the records and sighting along the curves. In Figures 121 and 123 the recovery takes place at an approximately constant rate and the record shows a break where the original curve is reached. In Figure 120, on the other hand, the recovery shows a progressively diminishing rate and the original curve is approached asymptotically.

When the lock has been used several times, the curves begin to assume the character of the normal recovery. Enforced intervals had been introduced on two previous occasions when the record in Figure 124 was taken. Although a break is still discernible where the original curve is reached and the recovery is relatively slow, the curve has become convex and is obviously similar to the normal

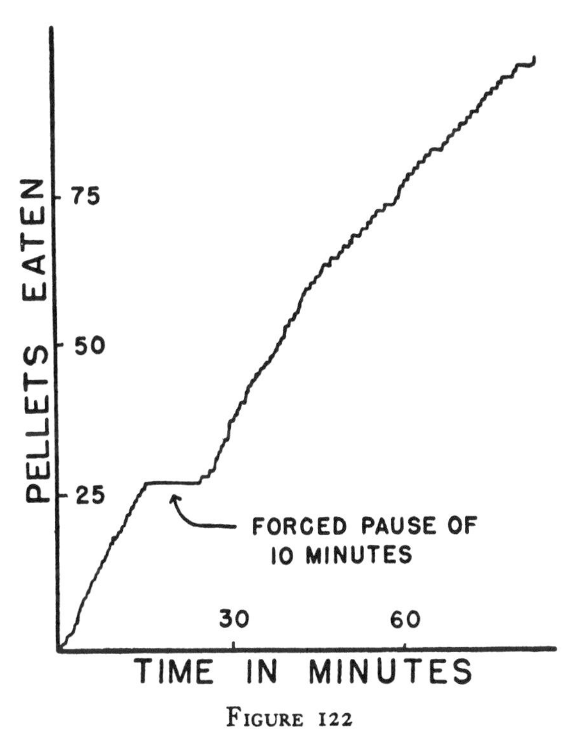

FIGURE 122

COMPENSATORY INCREASE IN RATE FOLLOWING FORCED INTERRUPTION

At the horizontal break in the curve the rat was kept away from the food for ten minutes. The recovery is roughly linear, and there is a definite break when the extrapolation of the original curve is reached.

recovery curve. A tentative explanation of this difference may be offered as follows. The straight-line recovery of Figure 122 is probably the resultant of a normal recovery and an emotional effect set up by the exclusion from the food tray. On subsequent occasions the emotional effect adapts out, leaving the normal recovery curve.

The tendency toward recovery after any deviation gives added weight to the envelop that may be drawn through the top points in a curve and hence to an equation based upon single records.

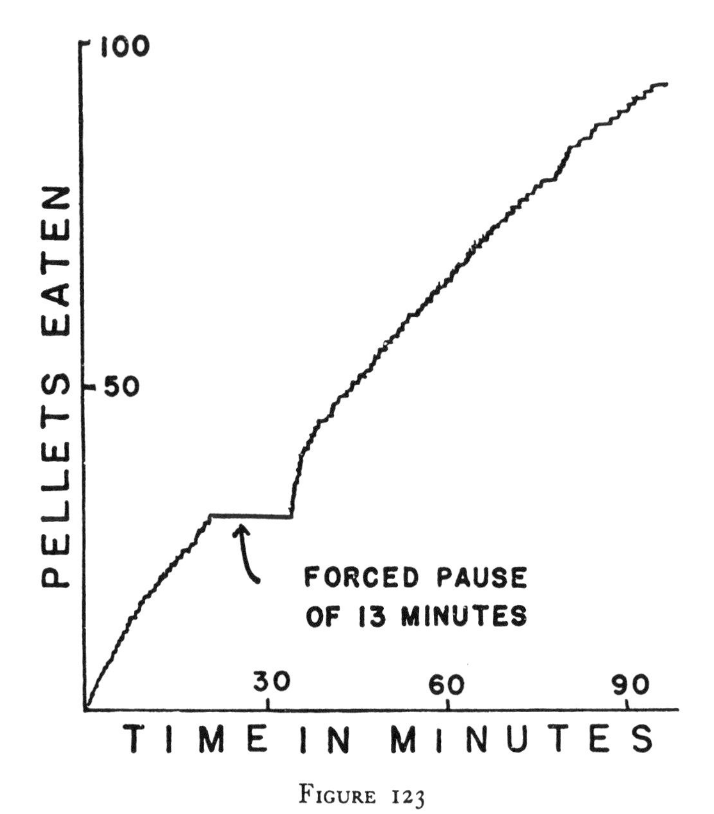

FIGURE 123

COMPENSATORY INCREASE IN RATE FAILING TO REACH EXTRAPOLATION OF ORIGINAL CURVE

This is fortunate, since the averaging of curves of this sort is objectionable. Deviations are apparently always *below* the main course of the record, and although an average would yield a smoother curve, it would not give the curve actually approximated by the rat. Since deviations are, for obvious reasons, more frequent at lower strengths, the shape of the curve would be seriously modified by averaging.

Such a curve as that in Figure 120 is arbitrarily brought to an end when the rat has not eaten for, say, 30 minutes. The final flat section of the curve representing this time is not reproduced in the figures. The satiated rat may not eat again for several hours and after a pause as long as this, compensation is not, of course, adequate to bring the curve back to the extrapolation of its first part.

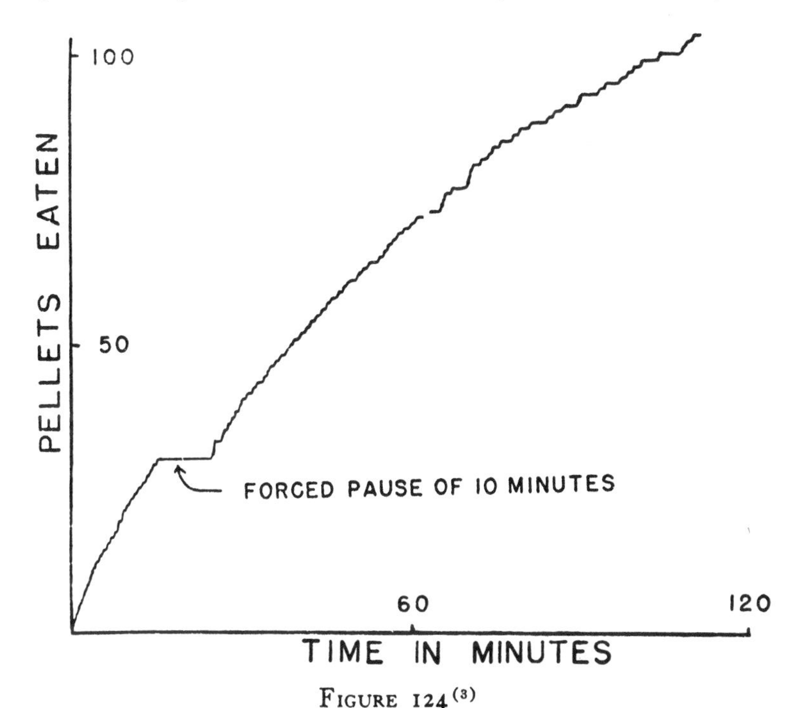

FIGURE 124[3]

COMPENSATORY RECOVERY AFTER FORCED INTERRUPTION

The rat has now adapted to being kept from food and the recovery curve more closely resembles the 'natural' case in Figure 120.

In considering the eventual fate of such a curve we are brought to the question of its relation to the normal periodic eating cycles described by Richter.

Throughout a period of time of the order of weeks or months the mean rate in ingestion may be said to be constant, if we neglect any change due to aging. But the eating actually takes place in a number of daily periods, during which the rate of eating is no nearer zero than the final rate in the eating curve just discussed and perhaps not so near. In a series of records kindly sent to me by F. S. Keller the total amount eaten by a rat during 24 hours is shown to be divided about equally among ten or twelve periods. (This periodicity does not closely correspond to that reported by Richter, but there are many differences to be taken into account—composition of the food, temperature, humidity, strain of rats, and so on.) The rate of eating during these brief periods is practically constant and fairly low. The rat presumably begins to eat before any great hunger has developed and stops soon.

If, now, a period of enforced abstinence is introduced into the twenty-four hour schedule, the 'density' of the eating behavior during the remaining time increases, particularly just after the period of fasting. As the length of fast is increased, the rat comes to eat more and more nearly continuously during the remaining time. The degree of hunger developed during the fast is, of course, increased, and the rate at which the rat begins to eat is therefore increased as well. It is this technique of 'compression' that is responsible for the typical eating curve, when the hunger of the rat passes from a relatively high to a low value. In the present experiment the daily eating behavior had been compressed until it all occurred within two or two and one-half hours.

The particular curves described above depend upon many experimental conditions—not only upon the technique of compressing the behavior but also upon the form and composition of the food and upon many factors entering into the situation. The amount of food eaten in two and one-half hours in Figure 120 could have been eaten in 15 or 20 minutes if it had been moistened with water. Hence, it would probably be idle to look for any universal 'eating curve.' The value of the present demonstration lies, I think, in its bearing upon the lawfulness of behavior. For one fairly typical set of conditions a high degree of regularity may be demonstrated, and we may conclude that the kind of variability in behavior that gives rise to the problem of drive is probably in general susceptible to similar treatment. Under other experimental conditions it should be possible to give a similar quantitative treatment of variations in reflex strength by appeal to the variables that are responsible for the change.

That the change in the strength of eating reflexes during ingestion is an orderly process capable of being reasonably exactly described has been confirmed by experiments on other organisms by Bousfield. In a study with cats (29) a mixture of milk and fish or other food was placed in a dish on a scale. As a cat ate from the dish, the upward movement of the scale was recorded on a kymograph. Curves comparable in regularity with those given above were obtained. Extraneous stimuli were found to arrest the eating behavior (as had been inferred in the case of the rat), and a similar tendency to compensate by a subsequent increase in rate was reported.

Bousfield found that his curves were more accurately described with the exponential equation $f = c(1 - e^{-mt})$. This equation was also shown by him to describe the rate at which chickens ate grains of wheat (31), and he has argued that an exponential equation is fundamentally reasonable as a description of behavior toward food. The constant c is the horizontal asymptote approached by the curve and may be identified with an ideal physiological limit of food consumption. Bousfield calls m the 'coefficient of voracity.' From his equation he is able to show that the rate of eating varies directly with the amount of food still to be eaten before the physiological limit is reached.

Changes in Strength During Deprivation

The preceding experiments on hunger have dealt almost exclusively with the decrease in reflex strength that accompanies the ingestion of food. An additional question of equal importance is how the strength changes in the other direction during deprivation. That it does so change is obvious from the inspection of a number of daily eating curves, but the course of the change has not yet been determined. So far as the first twenty-four hours are concerned there should be no great difficulty in examining the rate at which a rat begins to eat following varying periods of deprivation. Beyond twenty-four hours the problem grows complex. Bousfield and Elliott (30) have described experiments which suggest that a rate of eating cannot be taken as proportional to hunger during prolonged fasts. They have shown that when a rat is allowed to eat a standard food for one hour daily, an additional delay in feeding produces a decrease in the average amount eaten. In an experiment upon twenty-nine rats they report that if the rats are fed, not after the usual deprivation of twenty-three hours, but three and one-half hours still later, the average amount eaten is decreased two per cent. After an additional delay of 12 hours, the decrease is 15 per cent; after 24 hours, 14 per cent. The total amounts eaten reflect a uniform reduction of the whole eating curve. Figure 125, from a paper by Bousfield (32) shows the normal curve obtained during a regular eating period and, below it, the curve obtained after a relatively severe fast of four days. A general reduction in the rate is obvious throughout. Bousfield and Elliott have explained the reductions in rate and in amount eaten in a limited time as due to changes in the capacity of the organism

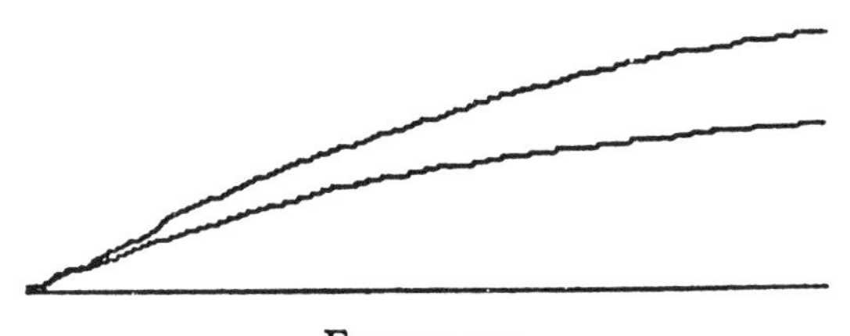

FIGURE 125

COMPARISON OF CONTROL AND DELAYED EATING-CURVES

Curve from Bousfield (32). The upper curve was obtained during a regular daily period of feeding, and the lower curve after a fast of four days. Each step represents the removal of a pellet of food (approximately ¼ gram) from the recording apparatus. The curves are approximately 2¼ hours long.

to assimilate food—for example, to changes in the size or tonicity of the stomach.

It is clear from these experiments that a rat does not always eat more rapidly the more depleted it becomes. But we cannot assert the contrary—that it eats more slowly—since if this relation held during the first twenty-four hours, a rat that was not hungry at all would eat at a maximal rate. The relation between rate and hunger is complex and it is incorrect to take the one as proportional to the other over periods of time during which important factors affecting the relation are permitted to vary. The rate of eating is an imperfect measure of the strength of the initial reflexes entering into the act of eating because the complete act of ingestion involves other factors. Whether Bousfield and Elliott have correctly located these factors is not important here. Presumably no single factor is responsible for the intricate relation between the rate of eating and deprivation over a long period. In the record in

Figure 125 the mere weakness of the rat at the end of a four-day fast might account for the reduced rate of eating, especially where a dry hard food was used as in this experiment.

The use of the rate of ingestion as an index of hunger during a single feeding period is not put in question by the experiments of Bousfield and Elliott, since the factors presumably responsible for the decline in rate observed by them are constant during so short a period. A method which avoids the factors which they note and provides data up to the point of the complete debilitation of the rat will be described in the following chapter. A consistent *increase* in the rate of responding with fasting is demonstrated, because the strength of the initial member of the chain of eating reflexes is only infrequently obscured by the completion of the chain.

Conditioned Reflexes and Their Relation to Drive

The behavior of approaching and picking up a bit of food is so common in the behavior of an adult rat and so uniform from one rat to another that it is likely to be looked upon (mistakenly) as unconditioned. In the adult rat the behavior is so well established that its conditioned status is practically beyond experimental manipulation. In order to investigate the effect of ingestion upon a controlled conditioned reflex reinforced with food it is necessary to attach a new and arbitrary initial reflex to the ingestive chain to achieve this end. The reflex of pressing a lever is suitable. If we reinforce every response with food and observe the rate, we may follow the course of the change in strength exactly as in the preceding experiments.

A typical record from the rat that produced Figure 121 is shown in Figure 126 (page 354). The record represents a period of an hour and a half, during which 124 responses were made and an equal number of pellets of food weighing in all about eight grams were eaten. The rate begins at a maximum and declines steadily throughout the period. The curve follows the theoretical curve already described in Figure 121. The smooth curve in the figure has the same value of n as in Figure 121.

The similarity of these curves suggests that the change in strength following ingestion is independent of the nature of the first member of the chain. In other respects the two kinds of curves are also similar. Both show delays and subsequent compensation, and both end abruptly before the rate has closely approached zero. An obvious exception will arise if the added member consumes so much time that the maintenance of the high initial rate is impossible. This may be the case when the initial member must be

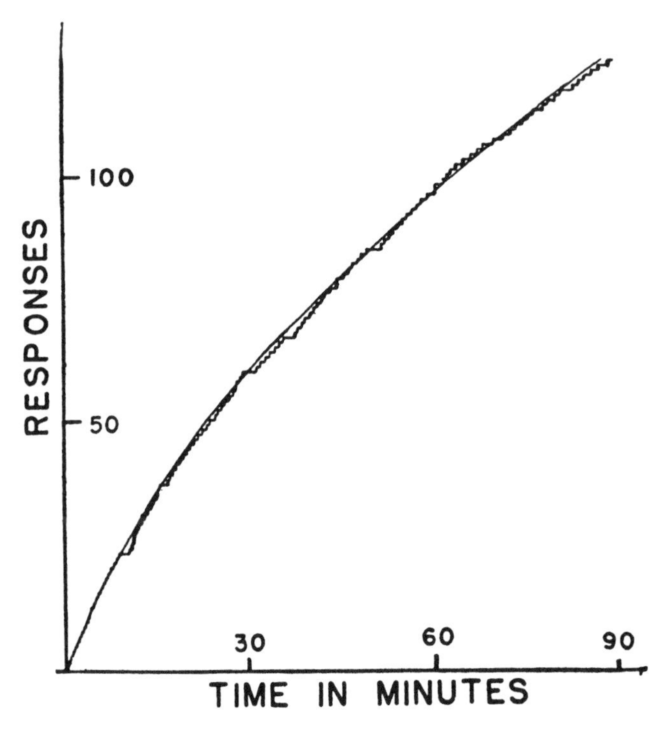

FIGURE 126[4]

CHANGE IN RATE DURING INGESTION WHEN PELLETS ARE OBTAINED BY PRESSING THE LEVER

The smooth curve is for the same equation as in Figure 121.

repeated, as under reinforcement according to a fixed ratio. The ratios in the experiment described in Chapter Seven were too large to permit of any great change in hunger through ingestion, but some of the rats in that experiment were later put on a ratio of 16:1 to test this point, as has already been noted. Four records for one rat were given above in Figure 99. The first record shows the initial positive acceleration which obtains when the ratio is suddenly reduced from a much higher value. Although the pellets in this experiment weighed approximately twice as much as usual, the actual rate of ingestion was reduced. All four records nevertheless show a fairly orderly negative acceleration as the hunger increases. The curves are not so regular as when every response is reinforced.

Another case of this sort was investigated with the collaboration

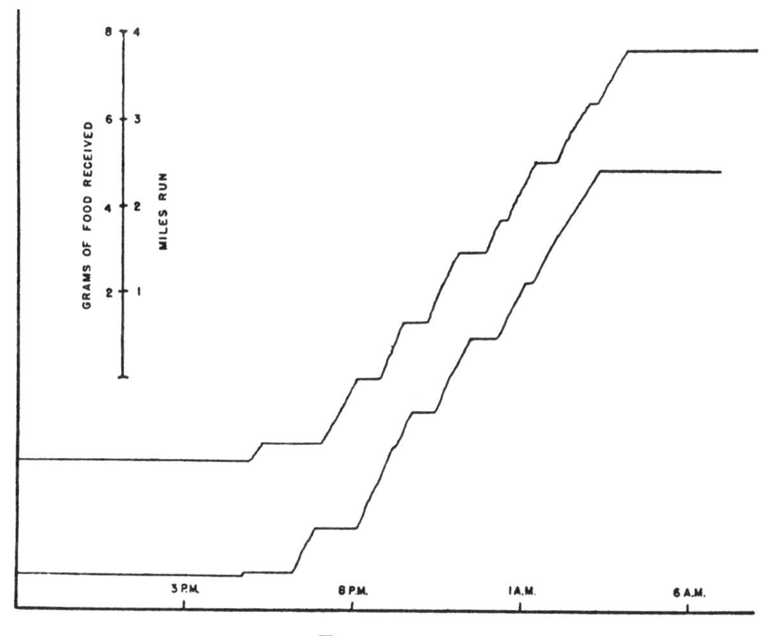

FIGURE 127

RECORD OF THE INGESTION OF FOOD WHEN A RAT MUST RUN APPROXIMATELY ONE-HALF MILE TO OBTAIN ONE GRAM OF FOOD

of Miss M. F. Stevens, where a rat was required to run approximately sixty feet in a running wheel to obtain each pellet of food, delivered automatically at the side of the wheel. Naturally the rat was unable to reach its normal initial rate of ingestion in this case although a more or less orderly decline was observed. Such a procedure also introduces an element of fatigue which may be expected to affect the curve.

In experiments in which rats have continuous access to a wheel and obtain their daily ration entirely by running a given distance for each pellet, I have not been able to parallel Richter's (67) finding of a relatively constant eating cycle. The difference may perhaps be due to the fact that in the present case each rat was isolated in a sound-proofed room and was disturbed only once per day in order to tend the apparatus. In general the rats ate periodically during part of each day but remained inactive during the remaining part. Typical records for two successive days from the same rat are given in Figure 127 (page 355), where the rat ran about one twentieth of a mile to obtain one-tenth of a gram of food. The apparatus was reset at about 9 A. M. daily, at the beginning of the records in Figure 127. It will be seen that the rat ate during the late afternoon and evening and showed the usual periodicity at that time, but that there was a long daily period of quiescence.

On such a schedule the rat can gain weight, but if the required distance is made great enough, more energy will be expended in the running than can be replaced by the food obtained. The rat will then gradually lose weight, although it will continue to run. The distance run per day eventually declines, and the time of running is extended more uniformly throughout the twenty-four hours. In Figure 128 eight consecutive records are reproduced to show the change in behavior as the rat continues to 'operate at a loss' in this way. The weight in grams at the end of each day is given at the beginning of each record. The localized period of eating shown in Figure 127 and evident in the early records of the series gradually disappears. This could be accounted for in terms of a growing need for rest and the impossibility of sustaining a single adequate running and eating period. At the end of the series the rat was exhausted, and the experiment could not be continued. A fairly regular periodicity is to be observed before exhaustion, but apparently only because energy is not available for a concentration of activity.

The bearing of these experiments upon typical ingestive behavior may be stated as follows: When the precurrent reflexes leading to ingestion consume time and energy, other variables beside the mere ingestion of food must be taken into account. These do not invalidate the normal relationship and are presumably capable of quantitative description whenever necessary.

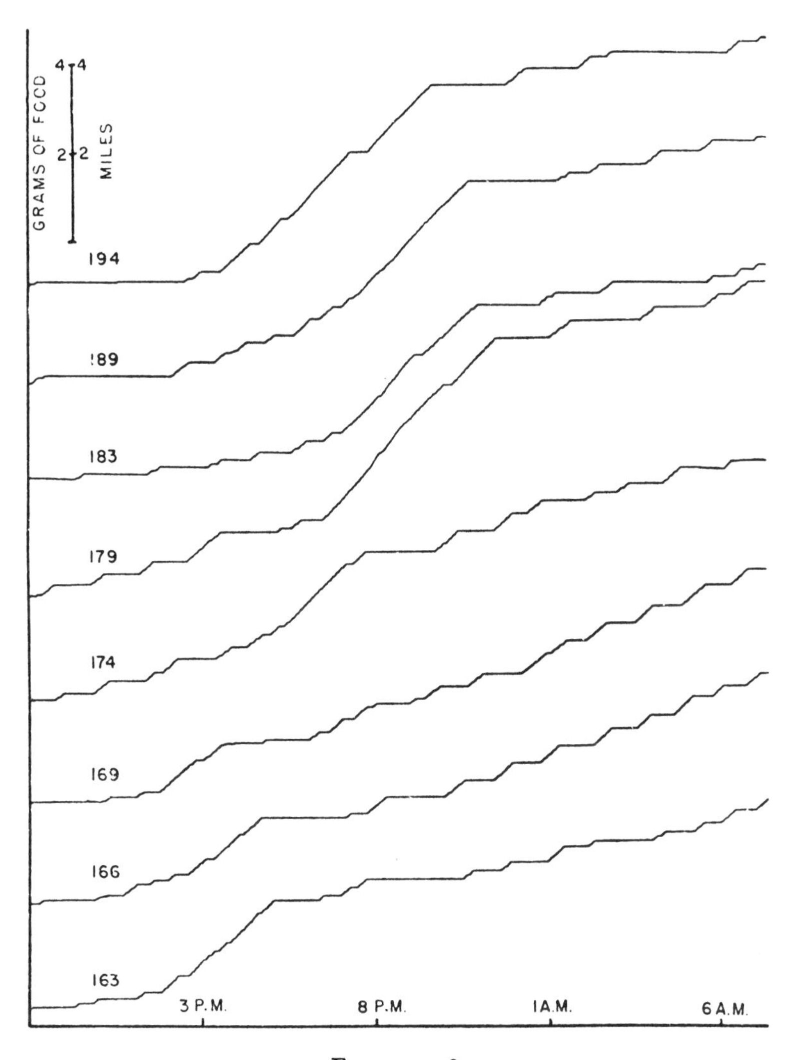

FIGURE 128

CHANGE IN TIME OF RUNNING AND EATING WHEN THE ENERGY REQUIRED IN RUNNING IS GREATER THAN THAT OBTAINED FROM THE FOOD

The distance run per pellet was twice that in Figure 127. The rat tends to run in shorter periods and to remain active a greater part of the day. The weight of the rat declines as shown in grams at the left of each daily curve.

Thirst

The formulation applied to hunger in the preceding pages may be extended to other drives. Each drive has its own defining operations, which raise special technical problems, but the same approach may be taken. Some measure of the strength of the behavior must be obtained and the relation between that strength and the various operations that affect it then determined. In the case of thirst, for example, the behavior may be measured very much as in the case of hunger. The operation of special importance here is the ingestion of water or its deprivation, and the available techniques are similar to those of hunger. Some of the preceding experiments have been repeated with thirst by replacing the food magazine with a device which delivers small measured amounts of water.

Essentially maximal states of thirst are produced by prolonged deprivation of water or through any process that accelerates loss of water from the body. The degrees of thirst used in the following experiments were in general considerably less than maximal and with few exceptions were obtained by allowing the rat to drink freely during a given hour each day and experimenting just before that hour. Food was always available except during the experiments.

A considerable change in the strength of the reflex due to drinking may be shown simply by giving the rat free access to the lever and reinforcing every response. The resulting curves either are continuous or show a single discontinuity. Records showing extreme and various intermediate cases are shown in Figure 129. The continuous curve is similar to, though not identical with, the typical curve obtained when the reflex is reinforced with food (see Figure 120). No effect comparable with the discontinuous curve has been observed with hunger although the typical eating curve is in general abruptly brought to an end. In thirst the abrupt change may occur before any significant change in rate has taken place.

Similar curves are obtained when the response to the lever is reinforced according to a fixed ratio. Four curves for a single rat are shown in Figure 130 (page 360). The ratio was 12:1. The curves compare well with those in Figure 129 in showing a range between a smooth negative acceleration and a sharp break.

As in the case of hunger, there is more than one operation that will affect reflexes leading to the ingestion of water. Simple drinking is only one of them, and the preceding description is hence

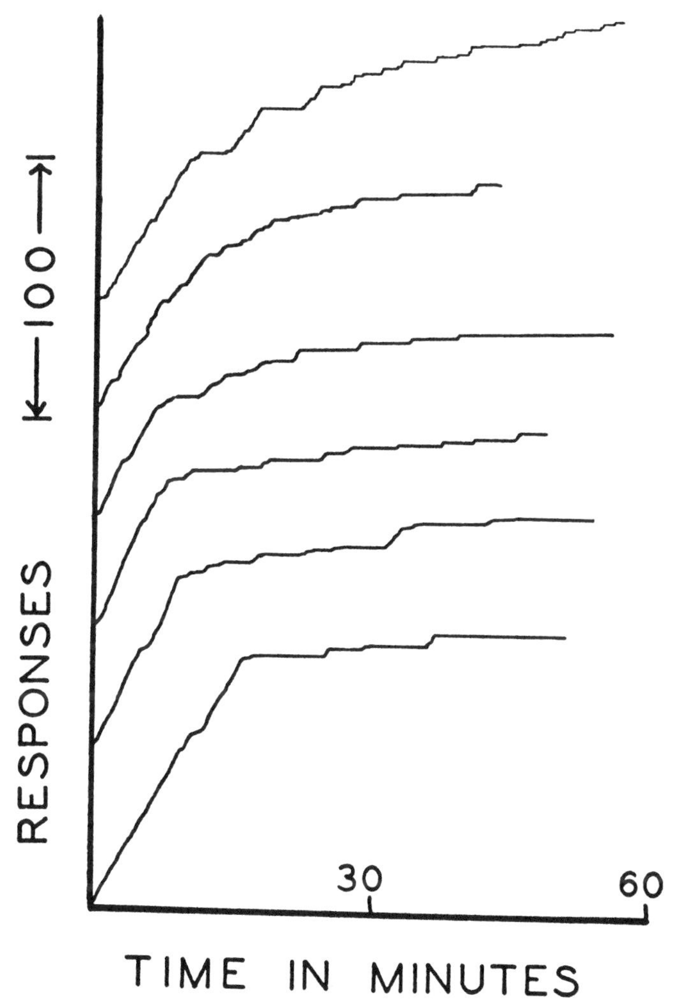

FIGURE 129[20]

CHANGE IN RATE OF DRINKING DURING A DAILY DRINKING PERIOD

Each response was reinforced with a small amount of water. The curves may or may not show a discontinuity. The continuous curve compares well with the curve for ingestion (Figure 120).

incomplete. My present aim is merely to demonstrate the orderliness of a typical case, with the implication that other operations may be treated with comparable success.

Thirst as an Arbitrary Drive

In addition to the different changes in strength that characterize different drives, it is necessary to compare the reinforcing effects of stimuli falling within different drive classes.

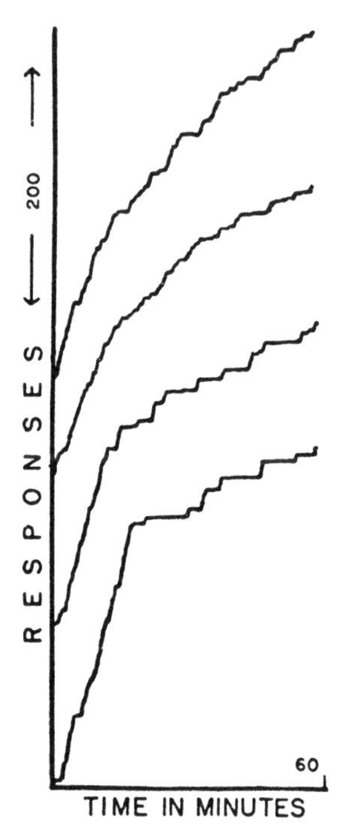

FIGURE 130

CURVES OBTAINED WHEN RESPONSES ARE REINFORCED WITH WATER AT A FIXED RATIO OF 12:1

Whether the results reported in earlier chapters can be generalized to behavior based upon another drive has been tested by repeating a few representative experiments using thirst instead of hunger. In original conditioning one reinforcement followed by extinction showed an effect comparable with that in the case of hunger. In an experiment upon four rats one curve showed an even greater effect of a single reinforcement than has been obtained with food. The curve is given in Figure 131 A. During the first 33 minutes three responses were made but not reinforced.

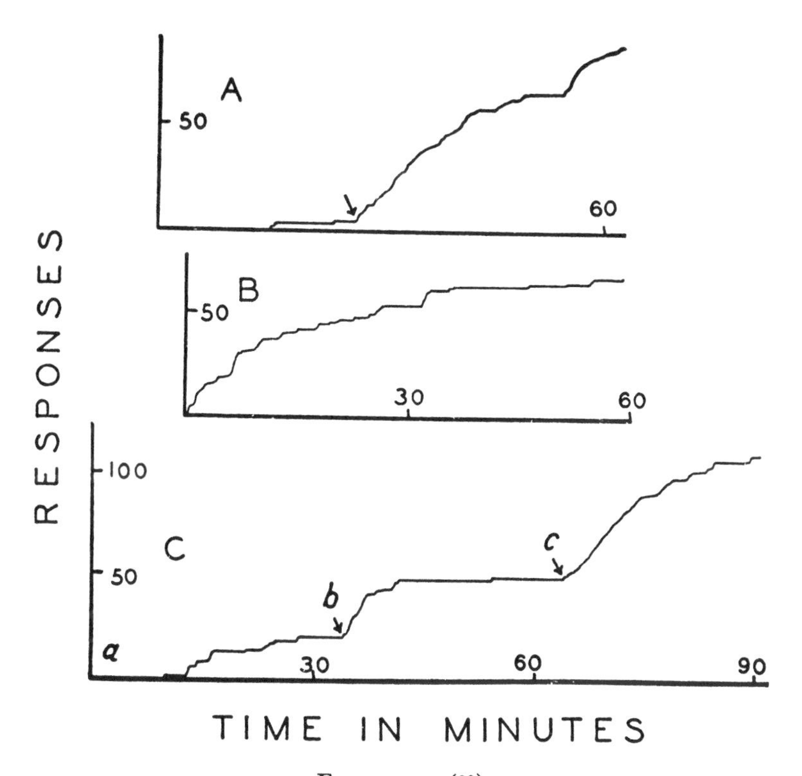

FIGURE 131 [20]

THIRST AS AN ARBITRARY DRIVE

A: effect of a single reinforcement with a small amount of water (*cf.* Figure 15). B: extinction after reinforcement with water. C: at (a) some responding occurs through transfer from previous conditioning to the lever with food; at (b) one response was reinforced with water; beginning at (c) all reponses were reinforced with water.

The fourth, at the arrow, was the only response in the history of the animal to be reinforced. The extinction curve which followed may be compared with Figure 15. Seven curves for extinction after considerable amounts of conditioning were also recorded. A typical example is given in Figure 131 B.

In several cases rats were taken from experiments on hunger. A typical record obtained in the transfer is shown in Figure 131 C. The rats in this group had been conditioned with food three months earlier and had developed various discriminations based upon hunger. They were then placed upon a thirst regimen and trained to the sound of the magazine by reinforcement with water *in the absence of the lever.* In Figure 131 C the rat was released at *a*. No water had been received by the rat in the presence of the lever prior to this experimental period. Food had been constantly available and in this experiment was present in the box, so that the rat could not have been hungry. The responses of the rat during the first 38 minutes were unreinforced. Since no conditioning with water had as yet taken place, this initial responding may be regarded as a transfer of strength from one drive to another via the common factor of the sound of the magazine. The rate shown in this figure (34 responses per hour) is almost exactly the average for the seven rats tested. The early delay is characteristic of five records. The slight negative acceleration during this transfer period is also characteristic and presumably indicates the extinction of the strength acquired from the other drive. At *b* a single response was reinforced with water, and a subsequent extinction curve was obtained. A clearly observable effect of the reinforcement of a single response appeared in all seven records. This part of the curve may be compared with Figure 131 A. At *c* the reinforcement of every response was begun. A maximal rate was developed, followed by a decline as the drive was weakened through drinking. A marked positive acceleration may be noted at *c* as the newly conditioned reflex increases in strength. This is characteristic of all seven records and is, as has been noted, generally observed whenever reconditioning follows extinction.

Spontaneous Activity

Another drive that may be subjected to the same treatment is 'activity.' The fundamental datum is of the same sort: an organism is sometimes active and sometimes not. Can this variability be described through appeal to an operation?

In the traditional use of the term 'spontaneous activity' no attempt is made to distinguish between the various forms that the activity may take. Since each form should have its own units, a quantitative measure of activity as a whole is practically impossible. An indirect measure, such as heat production or oxygen consumption, will not solve this problem. Even if a separate measure of each form of behavior could be devised, we should still have to provide for the reduction of the resulting data to a common unit of activity. The use of a running wheel supplies a practical solution by selecting one form of activity that can be quantitatively measured and by suppressing all other forms so far as possible. If a rat is confined in a small cage with access to a running wheel, its activity may be read from a counter on the wheel simply as distance run. The validity of this measure depends upon the degree of suppression of other behavior, and the results are significant only when a number of factors in the situation are specified—*e.g.*, sounds, temperature, illumination, and so on.

Another important factor, when any close analysis of the behavior is to be attempted, is the wheel itself. Theoretically, any given set of specifications could be used to define a standard wheel. There is good reason, however, to prefer that set which duplicates as nearly as possible the conditions of running on a level surface. This is not only the most natural form of the behavior, and consequently the least disturbing, but also mechanically the simplest. The running wheel may, in short, be regarded as a substitute for a level straightaway, which would be the ideal instrument if it were not impractical; and the wheel should be constructed accordingly. The characteristics of a level straightaway cannot, of course, be exactly reproduced in a wheel. In order to duplicate the work expended in any acceleration, the wheel must have a moment of inertia equal to the weight of the rat. But this introduces rotational and translational effects, which are lacking in straightaway running. Beside this inherent difficulty there are also technical limitations to be considered in the construction of the wheel which arise chiefly from the curvature. The greatest section of the perimeter of the wheel touched at any instant by the rat should be sensibly flat. But the moment of inertia of the wheel increases rapidly with the radius, and the problem is to determine the greatest curvature (the smallest radius) that will still remain negligible in comparing the mechanics of running in a wheel with those of running on a flat surface. A simple compromise is to set a radius at least twice the greatest length of surface touched by the running rat at any time. A moment of inertia equal to twice the weight of the rat might be set as a reasonable value.

Such a wheel can be greatly improved by the addition of a small amount of friction, which damps the oscillation and thus to some extent corrects for the undesirable effect of the necessary weight of the wheel. Assuming the weight of the rat to be 175 grams and the moment of inertia 350 grams, a sliding friction of 10 grams (expressed as the weight acting vertically at the periphery that will just keep the wheel turning) will successfully damp any ordinary oscillation within one or two cycles. Against such a friction the rat runs at a slope of a little over 3°, which is not a very significant deviation from a level path. A wheel having these properties has been described elsewhere (8) together with a device for recording distance run as a function of time.

From this point on the problem lies parallel with that of hunger. There is a similar periodicity of activity and a similar possibility of compression. Records for successive periods of twenty-four hours each are found to be composed of active and quiescent phases, with some relation between the distribution of the activity and the time at which daily care is given to the apparatus and rat. If an interval during which access to the wheel is cut off is now introduced into the twenty-four-hour period, the remaining part of the period shows a greater activity per unit time and the distribution is clearly related to the interval of confinement, in that the latter is followed by a relatively intense burst of activity. If the confinement is then gradually prolonged, the density of the activity during the rest of the period increases, until a length of interval is reached at which the remaining active period shows no quiescent phase. The length of such a period varies with many factors but may be of the order of four to six hours.

If there is no significant diurnal change in the environment, a rat quickly adapts to running at any given hour. Consequently, an experiment may be carried out whenever the conditions of the laboratory are most suitable, especially with respect to noise. The release of the rat is easily accomplished with an ordinary alarm clock. In experiments now to be described the following procedure was adopted. The experimental period ended at 8:30 A. M. At that time the record was removed from the drum and any necessary care given the rat. The cage was then closed, a fresh drum was placed on the kymograph, and the clock set for the release. The length of active period was determined by the time of the release. For example, the alarm was set for 3:30 A. M. if a five-hour period was desired. No other attention was required during the twenty-four hours.

Figures 132 and 133 are reproduced from a series of records obtained with a single female rat aged 125 to 175 days. They are supported, so far as the general comment made upon them is concerned, by several hundred records from many other rats, representing several thousand experimental hours. Most of these records were obtained while the design of the wheel was being evolved through a series of trial models, and differ in various ways which are irrelevant to the present point. The accompanying figures are given mainly to show the sort of result to be expected and the applicability of the method to several problems concerning activity.

The rat was released into the running wheel at 3:00 A. M. and the experiments were terminated between 8:00 and 9:00 A. M. The usual precautions were observed. The temperature of the room varied less than 2°C. during the entire series of experiments and probably less than 0.5°C. during a single period. The average temperature was about 21°C. The room was dark during the running and during most of the period of confinement and was thoroughly sound-proofed. No other animals were present. Drinking water was always available.

Figure 132 is a typical daily record obtained with a friction of about 15 grams. The rate of running begins at a maximum, and during the next three hours the rate shows a gradual but remarkably smooth decline, which is best seen in the figure by foreshortening the record. The variability of this part of the curve appears from inspection to be some function of the slope, increasing as the slope falls off. After three hours the record suddenly shows a greater variability, but the extrapolation of the first part of the curve is later fairly closely approximated.

In order to obtain a record of this length it is necessary to suppress the feeding behavior that would ordinarily appear during the period. This is done by attaching a cover to the food tray, which is closed at the beginning of the period by the release alarm. If this precaution is not taken, the curve is interrupted by delays for which there is no adequate compensation, and the form of the curve as revealed in Figure 132 is obscured. No very considerable hunger is developed during a six-hour period, although feeding behavior is undoubtedly suppressed.

If this procedure is repeated for several days, a conditioned hunger cycle is developed, the peak of which comes at the end of the experiment, when the rat is fed. After the rat has been fully conditioned to eat at, say, the sixth hour after the release, the form of the record changes to that shown in Figures 133 A and 133 B. A well-marked condition of hunger comes to be developed during the latter part of the period, which is reflected in an increase in activ-

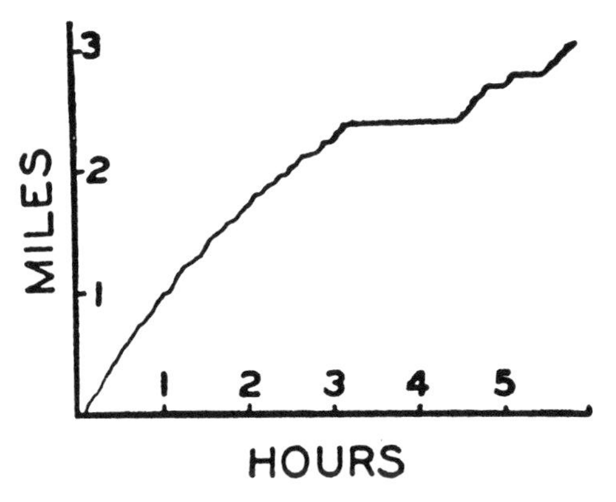

FIGURE 132[8]

CHANGE IN RATE OF RUNNING IN A WHEEL DURING A DAILY RUNNING PERIOD

Compare Figures 120 and 129.

ity. The resultant curve is essentially a straight line, if the removal of the food coincides with the beginning of the period of activity. The difference between the two types of curve is an indication of the way in which the conditioned hunger cycle is built up during the period and, granted the possibility of a quantitative description, is thus a valuable measure of that process. Not all rats give straight lines, but in general each rat has a characteristic curve that may be adequately predicted.

The rat runs in the wheel at a position such that the gravitational component acting tangentially is just equal to the friction. The apparatus, as described, includes a variable friction ranging from practically zero to about 50 grams. Beyond 50 grams technical difficulties are likely to arise in maintaining a constant coefficient of friction and in avoiding chattering or squeaking of the brake. The available range of slopes at which a 175-gram rat can be made to run with this wheel is, therefore, from practically zero to about 16°. Figure 133 demonstrates the effect of friction upon the behavior of the rat. Record A was obtained when the

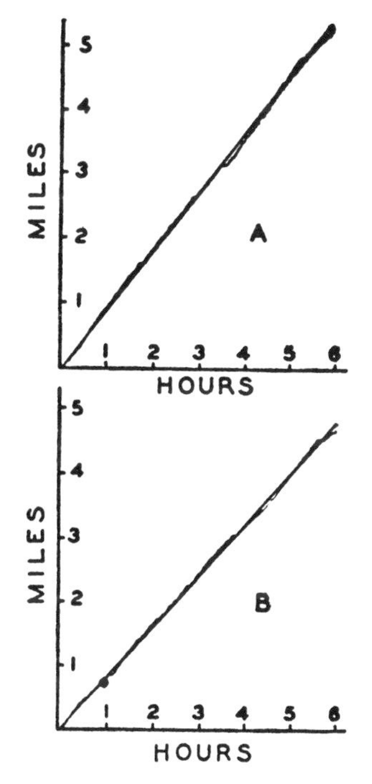

FIGURE 133 [8]

RATE OF RUNNING WHEN THE RAT IS REGULARLY FED AT END OF RUNNING PERIOD

Records A and B were taken with different frictions of the wheel and have different slopes.

friction of the wheel was approximately 14.3 grams. Record B, taken the following day, is for a friction of 19.5 grams. The straight lines drawn through the records have the slopes 52° and 48° respectively. The total record is modified by the change in slope, and to the same degree throughout.

The Nature of Drive

The preceding formulation of drive may be summarized as follows. In measuring the strength of a drive we are in reality only measuring strength of behavior. A complete account of the latter is to be obtained from an examination of the operations that are found to affect it. The 'drive' is a hypothetical state interpolated between operation and behavior and is not actually required in a descriptive system. The concept is useful, however, as a device for expressing the complex relation that obtains between various similarly effective operations and a group of co-varying forms of behavior. The properties assigned to the state are derived from the observations of these relations.

When the operations that affect the drive may be reduced to a single form (as for example, when thirst is said to depend upon the condition of the body with respect to its supply of water, even though the operations affecting this condition are multiple), and when all the forms of behavior affected by the drive vary together (as is apparently the case with thirst), the state of the drive itself may be regarded as simple and unitary. But this is not always true, as may be seen by considering the complex case of hunger. The first fact that suggests the complexity of the hunger drive is that at any moment in the life of the organism different foods will be eaten at different rates. Conditioned reflexes reinforced by different foods will also differ in strength. Young (80) has studied the 'preference' of the white rat for various foods. Not only will one food be eaten when another will not, but two foods will be eaten at different rates. When two foods are present as stimuli at the same time, the stronger reflex takes prepotency as the rat 'chooses the preferred food.' A relation between speed of eating and the gustatory properties of the food has been demonstrated in the new-born cat by Pfaffmann (66).

The notion of 'appetite' has traditionally been advanced to explain differences in eating behavior with respect to different foods, especially by those who are interested in identifying a single mechanism for hunger (36). We are said to eat the main course of a dinner because we are hungry but to eat a dessert merely because of our appetite for delicacies. The observed facts are that at one time we will eat a main course (or a dessert if the main course is lacking) and at another time only a dessert. To say that we eat a delicacy when we are not hungry is to give a restricted meaning to the term hunger. We are certainly hungry for dessert in some degree at such a time. It would be difficult to specify one group of foods the eating of which would indicate hunger and another the eating of which would indicate appetite.

A dual conception of hunger and appetite is a step in the right direction but it does not go far enough. Hunger is not dual but multiple. We must specify the food before we may predict the strength of the behavior of eating it and hence before we may assign a degree of hunger to the organism. At any given moment each form of food commands a certain strength of behavior, and all foods may be ranked in order according to their corresponding strengths. In extreme states of hunger the organism will eat practically anything, although it will still eat different substances at different rates. In complete states of satiation it may eat nothing. In any intermediate state it will eat all foods up to a given point in the order of preference. Such a classification of foods could be based upon gustatory stimulation alone. It is true that the physical state of the food often plays an important rôle, but we may confine ourselves to a basic form or texture varied only in flavor.

So much for the stimulating properties of the food. Before we have finished with the complex structure of hunger, we must also take into account possible differences in the ways of affecting the strength of eating reflexes. In making an organism hungry, or satiated, we may deprive it of, or feed it, different kinds of food. The result is a displacement of certain foods in the rank order previously determined from relative strengths of eating. To take a common example, by feeding large amounts of salt we displace downward the strength of eating all foods in which the flavor of salt is marked. Conversely, by withholding salt we increase the strengths of eating these foods and thus create a salt hunger.

We define a sub-hunger of this sort in terms of the co-variation of relative strengths. We say that an organism is salt-hungry if the strength of behavior in eating all salty foods is relatively high. It is difficult to determine whether a given taste-substance can always be isolated as the basis of covariation. We must then proceed to identify the substance or substances that must be fed or withheld in order to modify the strengths of these reflexes as a group. How many sub-hungers there are, what the sensory stimulation is that is appropriate to each, and what the operations are that affect each, are problems in topography that cannot be taken up here. The technique of isolation is, however, sufficiently clearly indicated in this approach to the problem.

Some experiments on sub-hungers have been conducted by W. A. Bousfield, who has kindly permitted me to give a short report here. Bousfield's work in confirming an orderly change in rate of eating has already been mentioned. In some work on cats he used four different food-mixtures: (A) milk, (B) milk, cooked oatmeal, and raw ground beef, (C) ground fish and milk, (D) ground beef-kidney and milk. He was able to go beyond the relative preferences of his cats for these foods to the question of sub-hungers by showing that when a satiation curve had been obtained with one kind of food, a change to another food might lead again to eating and to production of a second curve. He was sometimes able to get three successive curves in this way. By measuring the amounts of food eaten, he was able to discover the extent to which eating one food reduced the strength of behavior with respect to another. One of Bousfield's cats gave the results shown in Table 9. The figures in parenthesis give the number of grams of the food indicated at the left eaten in the first satiation curve. The other figures give the number of grams of the food indicated at the top eaten in the second satiation curve immediately following. Thus, if

TABLE 9

		Second food in grams			
		A	B	C	D
First food in grams	A	(140)—	(166)52	(128)12	(114)130
	B	(176)—	(128)—	(172)30	(136)—
	C	(180)36	(114)40	(168)—	(192)42
	D	(214)—	(184)—	(280)—	(200)—

Food A or C is given first to satiation, the cat resumed eating when any one of the other foods was presented. If Food D was given first, no other food would be eaten, and Food B reduced all hungers except that for Food C.

Bousfield's method is obviously of great value in studying the question of sub-hungers. The case of prolonged deprivation of a single food and the relative shifts of strength that follow could be treated in the same way. Refinements which suggest themselves are a reduction of the foods to a single texture and the use of more elemental gustatory and metabolic properties.

The Classification of Drives

The multiple character of hunger immediately suggests the broader problem of the identification and classification of drives in general. There is a natural tendency to reduce the drives of an organism to the smallest possible number because of the relative simplicity that is achieved, but we may go only so far in the matter as the behavior itself will allow. Without attempting to make a classification here (it would again fall outside the design of the book to introduce a purely topographical matter) it may be well to consider the ways in which the phenomena of drive are observed and how we may ascertain the number of drives exhibited by an organism.

It is important to make clear that we are not concerned with a classification of a number of *kinds* of behavior but only with behavior which undergoes changes in strength as the result of the kind of operation which defines the field. We do not list a drive for contracting the pupil of the eye because that bit of behavior is relatively invariable. If sexual reflexes were equally as invariable, we should not have a sex drive but only sexual behavior. We are concerned with the number of operations that will modify reflex strength or, to put the matter in a better order, with changes in reflex strength for which we undertake to find variables.

When first observed, each variation in the strength of a reflex controlled by a given operation appears as a separate case and a separate drive must be assumed. If we observe, however, that the strengths of two or more reflexes are the same function of the same operation, a reduction in the number of drives is made possible. In this way we set up hunger, sex, and so on, as fairly inclusive drives relevant to a wide variety of forms of behavior. The unity of the drive depends upon the covariation of the reflexes to which it refers. And whether or not a given reflex belongs to a given drive must be answered by considering covariation rather than any essential property of the behavior itself. Thus, the old question as to whether most behavior is sexual is not a question as to whether it is sexual *in nature* but whether it varies as a function of the operations which define that drive. Does an operation which reduces the sex drive affect the behavior in question in the same way?

Multiple Drives

In a conditioned operant the drive governing the strength is determined by the reinforcement. If pressing the lever has been reinforced with food, the strength varies with hunger; if with water, it varies with thirst; and so on. An operant may be reinforced with stimuli in more than one drive class, and its strength is then determined by two or more drives. In an experiment upon thirst an attempt was made to use some rats from a previous experiment upon hunger. Unlike the experiment described above no food was available in the boxes. Since strong thirst prevents the eating of (dry) food prior to the experiment composite curves were obtained representing responses controlled by both drives as soon as the receipt of a small amount of water had increased the hunger.

The possibility of multiple drives answers an objection frequently made to the use of a rate of eating as a measure of hunger, namely, that eating frequently occurs without hunger and hence cannot be a reliable measure. But this is the case only when the reinforcement has been multiple. Picking up and putting a bit of food into the mouth is ordinarily reinforced by the food itself, but it may be 'artificially' reinforced by presenting water, and a thirsty but non-hungry animal will then put food into its mouth. Such additional reinforcements are numerous in everyday life. The child that must eat its vegetables before getting its dessert does not necessarily eat because its vegetable-hunger is strong but because its dessert-hunger is strong. Eating vegetables in such a case is comparable with any other operant, such as saying 'Please' or practising at the piano. The momentary observation of the behavior would be useless in any exploration of the drive unless the fact of the reinforcement with dessert were known.

All cases of this sort may be identified by the presence of reinforcements in more than one drive-class. They do not invalidate the present formulation nor do they indicate that prediction of behavior cannot be achieved when the relevant factors are known.

Craving and Aversion

It may be objected to a treatment of drive simply in terms of reflex strength that it does not recognize the obviously purposive character of the state of the behavior. A hungry organism may be said to eat, not because it is hungry, but *in order to* reduce its hunger. The relation expressed by 'in order to' is, I believe, irrelevant; but a description of drive would be inadequate if it failed to notice the important fact that the behavior that is strengthened during the heightened state of a drive usually leads to an operation affecting that drive. I cannot see that this involves any teleological principle not equally applicable to, say, digestion. No further use of the relation is made in the description of drive, since the important datum is the strength of the behavior at any given moment. Drive is consequently not to be described as a 'desire' or 'craving' or as any other state directed toward the future.

In the relation between 'craving' and 'aversion' a problem of classification arises which may seem at first sight to require an appeal to the effect of the behavior in modifying its own state. As the strength of an operant approaches zero with the decline of its drive, a point is reached at which the organism may be said to be 'indifferent' to the stimuli in the presence of which it has hitherto responded. With a further decline in drive, which may be obtained in a non-hungry animal through the use of multiple drives, the 'indifference' may give way to 'aversion,' in which an opposing form of response is emitted. Instead of eating, for example, the overfed organism may push food away or turn from it quickly. If we group both eating and pushing away together as 'behavior with respect to food' we may set up a continuous series of states passing from a positive extreme of 'craving' to a negative extreme of 'aversion' through a neutral mid-point. And this is common practice. Such an ambivalent unit of behavior can only be justified by appeal to effect, since eating and pushing food away are quite different topographically. Must some purposive relation be brought in at this point?

The answer dictated by the present formulation is that in no case may we regard topographically different forms of behavior as aspects of the same response, even when one is approximately the reverse of the other. Nor can we assign negative values of strength as such a system might require. Seizing food and pushing it away are dynamically separate responses, and their strengths must be separately described. As Sherrington has pointed out, a response may usually be said either to prolong a stimulus or to cut it short. Organisms frequently possess responses of both kinds with respect to the same stimulus. From a point of view of economy it is not likely that they will be related to the state of the organism in such a way that both will be strong at the same time (although this case may be set up through conditioning). It is also to be expected from considerations of economy that the same operation will affect both their strengths although in different directions. Hence, we may state the case of 'craving and aversion' with respect to food as follows: An organism possesses the two operants of (a) seizing and eating and (b) pushing food away. The operations which weaken (a) strengthen (b) and *vice versa.* If a hungry organism is fed, the strength of (a) decreases while that of (b) increases, until, if the feeding is forced beyond a certain point, the organism pushes food away. The relations between these responses are as follows: (1) the discriminative or eliciting stimuli (*i.e.*, the food) are the same for both, and (2) the operation modifying the strength is the same for both but working in opposite directions. That an organism should possess two responses so closely related is to be explained, as all functions of the organism are explained, by appeal to some form of evolutionary development. If the *effect* of the drive is important anywhere, it is in that development rather than in the description of current behavior.

Drive Not a Stimulus

The attempt to describe behavior as the forced effect of the 'total stimulating situation' is perhaps responsible for the current conception of a drive as a stimulus. In the treatment of hunger, for example, it is commonly stated that a battery of impulses from the empty stomach drives the organism to seek food. On the same model a large number of known or assumed stimuli arising within the organism itself have been worked into stimulus-response formulae in order to account for the variability in behavior that gives rise to the problem of drive. Hunger lends itself most readily to this interpretation because of the interoceptive stimulation responsible for hunger 'pangs.' The analogy with an external goad is strong. The empty stomach, like a tight shoe, supplies that kind of stimulation which in being brought to an end acts as a positive reinforcement. We come eventually, if not originally, to take food or to remove the shoe, as the case may be. But the stimulation originating in the stomach does not accurately parallel the various states of strength of eating reflexes. So long as we hold merely to a rough measure of eating, a crude qualitative explanation may be satisfying; but with a more delicate quantitative measure of the various states of eating behavior something more is required. The temporal and intensive properties of hunger revealed in the experiments described in this and the following chapters can hardly be reconciled with the gross stimulation of hunger pangs. It is difficult to see how an afferent discharge of any sort could be correlated with the nice grading of strength that exists between the wide extremes here exhibited, but in any event the stimulation responsible for pangs certainly cannot, since various degrees of hunger may be demonstrated in its absence (for example, after a small amount of food has been taken).

Stimulation from the empty stomach is not directly concerned with the hunger dealt with in this chapter. It seems to be an emergency mechanism brought into play when the normal variation of hunger has not been efficient in obtaining food. It resembles any noxious stimulus in producing a general activity and in reinforcing any behavior which brings it to an end. Hunger cannot enter into the present formulation with the dimensions of a stimulus and could be regarded as such only with the greatest confusion.

The existence of a special mechanism in the case of hunger has led to a general misunderstanding of the statuses of other drives, the nature of which would otherwise have been clear. In order to parallel the case of hunger it has been necessary to do violence to the term stimulus, as when Watson includes under the term 'any change in the tissues themselves, such as we get when we keep an animal from sex activity, when we keep it from feeding, when we keep it from building a nest' (76). That some change occurs when we keep an animal from sex activity and that it has the effect of heightening the sex drive is obvious; but that the change is a stimulus or provides a stimulus does not follow. The search for stimuli to satisfy the requirements of other drives has not in general been successful. In the present formulation we pass from the operation performed upon the organism to the behavior itself (from deprivation, for example, to an increase in reflex strength).

At no point is it necessary to assume that deprivation involves intermediate stimuli. The effect of drive upon behavior is similar to that of emotion, of certain drugs, fatigue, age, and so on (see Chapter Eleven). The inference of additional stimuli is equally unnecessary in all of these cases.

The conception of a drive as a state rather than as a stimulus is valuable in avoiding arguments about purpose. A drive is not a teleological force nor does the stimulus which acts as an appropriate reinforcement exert an effect before it has occurred. The conception gives little support to a dramatization of the forced character of behavior. 'As a man drives a horse,' says Holt [(45), p. 232], 'so the man himself is driven to action by the moment-to-moment irritation of sense organs, without and within.' The metaphor offers very little help in understanding either the notion of a stimulus or that of a drive. Not only does a drive not *pull* an organism from the future; it does not even *push* it from the past or in the present.

Drive and the Reflex Reserve

I have been anxious to define the concept of the reflex reserve as clearly as possible because so far as I know it has no counterpart in the popular vocabulary of behavior. The reason for this is probably that it is easily confused with motivation. An organism is said to be able to retain over a period of time not only a certain amount of 'knowledge' regarding an act but also an interest or drive. The notion of a reserve is something more than either of these. The question of 'mere knowledge'—whether or not an organism retains the ability to perform a required act—is distinct from the question of the capacity to perform the act in the absence of further reinforcement. Whether or not the organism performs the act is usually regarded as a matter of motivation, but the conditioned status of the behavior may also be involved. Where the single factor of motivation would ordinarily be invoked, the principle of a reserve points to two factors. Not only may the relation between reserve and strength be modified, but the size of the reserve itself may be changed. Two special cases arise in human behavior in which this distinction is important—one concerned with an empty reserve, the other with a full. In certain forms of neurasthenia, where the motivation does not seem to be at fault, the reserve may be empty. The condition could arise from an ineffective reinforcement, either because few or no stimuli possess reinforcing power or because, as is often the case in daily life, the reinforcement is remote from the act. Either of these circumstances would produce a failure to respond without modifying the drive in any way. The opposite case of a full reserve raises the question of sublimation and seems to answer the objection that sublimating behavior is ultimately incapable of modifying ('satisfying') a drive. Sublimating behavior is obviously an example of topographical induction. The effect is not a change in drive but an emptying of the reserve. The state of 'strain' which gives rise to sublimating behavior is not only an intense drive but a full reserve which because of external or internal circumstances cannot be emptied except through topographical induction. These are, however, problems which lie beyond the proper scope of this book.

Drive as 'Inhibition'

Since a change in drive often involves a weakening of a reflex, it is not surprising that examples have occasionally been referred to as 'inhibition.'[2] It scarcely need be repeated that if inhibition refers merely to the fact of a decline in strength, it is too broad a term to be useful. Except for the direction of the change, drive has little in common with extinction or emotion or with any other factor producing decreases in strength.

Drive and the Concept of the Reflex

The present formulation of drive has a bearing upon the general extension of the principle of the reflex to the behavior of the intact organism. In a study of the concept of the reflex (2) I have pointed out that the central defining property that has governed the use of the term is, not the presence of neurological structures, but the lawfulness of the relation of stimulus and response which has made the inference of neurological structure and function plausible. The ever widening field of the reflex (compare Pavlov's extension of the concept to acquired behavior) has proceeded, not with neurological discoveries, but with further demonstrations of lawfulness.

[2] Cf. Anrep, G. V. (25) where 'spontaneous inhibition' seems to refer to a weakening of the salivary reflex either from ingestion or a gastric disturbance.

In no reflex is the elicitation of a response by a stimulus inevitable, but in the case of the spinal reflexes the operations which modify the elicitation are obvious. Thus, it is easy to regard the flexion reflex as lawful, even though the stimulus does not elicit the response after complete fatigue, because the state of fatigue has clearly been induced by the experimenter. The obviousness of the operations affecting spinal reflexes explains why they were first discovered. But according to the present position any response of the organism shows a comparable lawfulness, although the variables responsible for differences in strength may not be so easily identified or controlled. Among the more elusive variables in the behavior of the intact organism, drive stands preeminent. The concept of 'volition,' in its historical opposition to the 'involuntary' reflex, rests largely upon the 'spontaneous' changes in strength to be accounted for through operations falling within the field of drive. In gaining control of these additional variables we are able to extend the term reflex (in its implication of predictability) to behavior in general. The older study of spinal reflexes and the modern study of behavior are logically and to a great extent historically continuous.

Chapter Ten

DRIVE AND CONDITIONING: THE INTERACTION OF TWO VARIABLES

The Problem

The observations upon which the concepts of drive and conditioning are based are essentially of the same kind. So far as any one reflex is concerned, we observe merely a change in its strength occurring as the result of the manipulation of some variable. Such variables may be divided into the classes called 'drive' and 'conditioning' but their effects upon behavior itself offer no useful criteria for differentiation. When we turn from the immediate strength of a reflex to its reserve, a difference is revealed. Conditioning involves the size of the reserve, but drive concerns the relation between the size and the momentary strength. The operation of reinforcement increases the reserve in a definite way, while the operation of feeding or fasting changes the strength without influencing the reserve. These relations were stated more or less dogmatically in Chapter One, and it is time to support the statement with experimental evidence.

We now reach a second stage in the investigation of behavior. Heretofore the account has been confined so far as possible to single variables, but we must now study the manipulation of two variables at once. The interaction of variables is important because there is perhaps nothing in a science of behavior that has actually the status of a completely isolated variable. Although it is possible to hold the drive constant while conditioning or extinguishing, it does not follow that the state at which the drive is held is not significant for the result. Consider the case of the unreinforced elicitation which produces extinction. In operant behavior the process of extinction depends upon the rate of elicitation. But this is controlled by the drive, and the shape of the curve is therefore dependent upon the degree of drive arbitrarily selected. This is the kind of problem to which we must now turn.

The State of the Drive and the Dynamics of Extinction

In the preceding chapter it was argued that the notion of drive rested upon the observation of various changes in the strength of a reflex following a certain kind of operation and that a proper measure of drive was the strength itself. Turning to the notion of a reserve we may assume that in the normal ingestion curve the reserve remains full. Every response is reinforced, and although the rate declines with ingestion, a maximal number of responses could be obtained from the organism by increasing the drive and extinguishing the reflex. Hence it has been argued that the effect of the drive is to change the proportionality of the rate and the reserve. Now, there is no reason to confine this relation to reinforced responses. It should apply as well to the rate of elicitation during extinction and should yield the relation referred to in the preceding paragraph.

A rigorous mathematical formulation of the process of extinction at a given drive is complicated by the fact that extinction does not normally complete itself [1] within a space of time during which the drive and other variables may be held constant. The best case is that of original extinction, which is brief; but such curves are complicated by an emotional factor. Extinction after periodic reconditioning is a more uniform process, but it may require many hours. I have adopted the device of taking measurements for one hour on successive days, when the drive may be adequately controlled; but the process of spontaneous recovery enters here as another complication. In view of these difficulties a theoretical formulation can at present be only a rough approximation.

The notion of a reserve and the shape of the usual extinction curve suggest the hypothesis that the rate of responding at any given time is proportional to the remaining reserve. Theoretical curves based upon this hypothesis may be set up to approximate experimental curves very closely. In Figure 134 a total reserve has been assumed as indicated by the horizontal line at the top of the graph. Curve A is obtained by assuming a definite relation between the rate of responding (the slope of the curve) and the remaining reserve. This is, of course, only one of the possible cases,

[1] It would appear from the shape of the curve (as well as from the present interpretation) that the process is never wholly completed. For practical purposes an arbitrary near-zero rate of responding may be adopted as an end-point.

which depend upon the size of the reserve and the proportionality between rate and reserve assumed. The curve compares favorably with actual experimental curves, such as those in Figure 26.

The principal qualification which must be imposed upon this scheme is the fact, deducible from the process of spontaneous recovery, that the reserve is not fully available at any one time in influencing the rate. The qualification is a minor one, with respect to the degree of approximation here attempted. In spite of it the

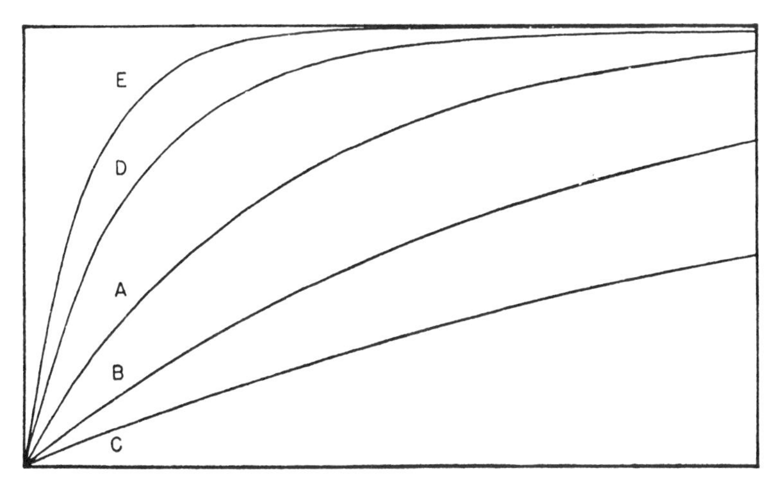

FIGURE 134

THEORETICAL EXTINCTION CURVES AT DIFFERENT DRIVES

The curves are based upon the assumption that the rate of responding is proportional to the responses still remaining in the reserve and that the effect of the drive is to change the proportionality.

hypothesis permits us to consider to advantage the effect of a change in the proportionality of rate and reserve upon the curve of extinction.

Two cases obtained when the proportionality is reduced are represented by Curves B and C. The curves begin at lower slopes because of the reduced drive, but since the reserve is drained less rapidly the rate falls off more slowly. At the end of the period represented in the figure the lower drives exhibit higher rates, although that would not, of course, be the case if the initial rate were too low. The limiting case approached by the reduction in drive is that of complete satiation at which no responding occurs and hence at which no extinction can take place.

In investigating the effect of a lower drive some device must be available to reduce the 'normal' drive to other definite and known degrees. For the purpose of this experiment the following technique was found satisfactory. The normal drive was established through the usual method of feeding the rats once per day for a limited time (in this case, one hour). With this stable value of the drive as a base it was possible to obtain lower values by feeding definite amounts of food or feeding for definite lengths of time just before an experiment. The two methods come to the same thing, since with a constant initial hunger the amount eaten is a function of the time. Practically, it is better to use amount rather than time, because of possible temporary variations in the rate. It is only necessary to decide upon the amounts that will produce the desired degrees of drive in each case.

In some of the following experiments the amount to be eaten was placed in the apparatus. When the experiment began, the rat first ate the food, then went on with the experiment. The apparatuses were used for successive groups of animals, and this method consumed a great deal of extra time, especially when large amounts of food were eaten. An alternative method was therefore devised. At a length of time prior to the experiment determined by the amount to be eaten, each rat was placed in a separate cage with food. By starting the members of a group at different times it was possible to have them finish their different rations at approximately the same time, when the experiment proper was begun. This second method avoids the disadvantage of the first where, if a part of the food is accidentally dropped and not recovered, the rat may start the experiment prematurely.

Extinction curves following periodic reinforcement were used, since they are rarely marked by the emotional disturbances that characterize original extinction. Fifteen rats were taken from other experiments which had involved periodic reconditioning for various lengths of time.[2] The reconditioning interval was five minutes

[2] As follows: four rats from an experiment on the effect of an interval of time before reinforcement, involving 23 days of periodic reconditioning; three from a similar experiment for 19 days (a fourth in this group had been accidentally killed); four from a similar experiment for 22 days; and four from an experiment on periodic reconditioning, where various numbers of successive responses were reinforced at each period, for ten days.

in each case. Immediately before the first experimental period in which extinction was observed the drives were varied by feeding zero, two, four, and six grams. The effect of these amounts upon the curve is clearly indicated in experiments to be reported later in this chapter. One animal in each group was assigned to each drive (no case for six grams in the group of three). A record of one day of extinction at a reduced drive was obtained. On the second and third days of extinction no food was given prior to the experiment. The effect of the return to a normal drive supplies additional information concerning the main hypothesis.

The resulting averages for the 15 rats are plotted as the lower set of curves in Figure 135. On the first day of the graph the average rate under periodic reconditioning for two days prior to extinction is given, and on the second, the averaged extinction curves at the different degrees of drive. Four measurements of the heights were made in order to follow the change during the hour. The third and fourth days of the graph show a continuation of the curves at maximal drive, only the end-points being measured. In their original form the data are difficult to interpret because the effect of the drive is to some extent obscured by the influence of the previous periodic slopes, which differed slightly. As has already been shown and as may be inferred from the relation of the rate to the reserve, the slope of the extinction curve is a function of the periodic slope. A rough correction may be made by multiplying the data for each group by a factor chosen to bring the periodic rate to some arbitrary value. In this case the value for the group at no grams was used, the necessary factors being 0.00, 1.32, 1.15, and 1.14 for the groups at zero, two, four, and six grams respectively. The resulting curves are given in the upper group in Figure 135 (page 384), where the effect predicted in Figure 134 is clearly shown. (It may be noted that the factors chosen for this correction are based upon data available prior to the beginning of extinction.)

The curve at no grams in this figure is typical of the extinction curve described in Chapter Four and is represented by Curve A in Figure 134. One characteristic is that it begins at a rate considerably higher than that observed under the preceding periodic reconditioning. The effect of the reduction in drive is to reduce the slope of the curve. The initial rate shows a consistent decline corresponding to increases in the amount of food previously eaten.

The decrease is not, however, a linear function of the amount but is most severe at the first step. The curves at the lower drives lack the curvature of those in Figure 134 but show the slower negative acceleration.

On the following day, at normal drive, the curve at no grams declines as usual but the others show an increase in rate, not only

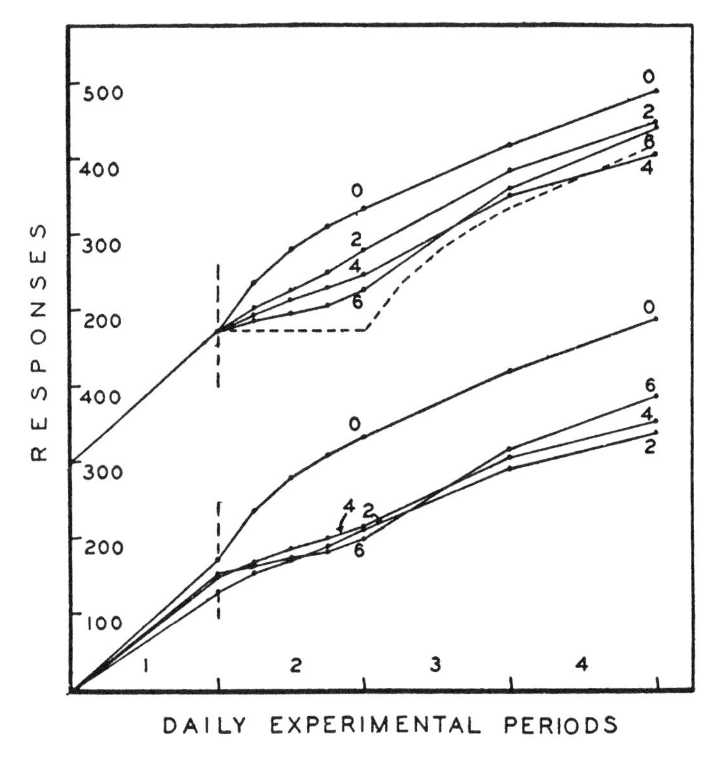

FIGURE 135 (19)

EFFECT OF LOWERED DRIVE UPON EXTINCTION

The lower curves give the original data; the upper curves have been corrected for differences between the rates under periodic reinforcement.

On the first day of extinction (Day 2 in the figure) the drives were reduced by feeding amounts of food just before the experiment as indicated in grams at each curve. The last two days were at normal drives. The dotted curve is the assumed case in which the drive is so low that no responding occurs.

above their previous values at the reduced drives, but also above the contemporary record for the group at no grams. Such an increase, which is the more pronounced the lower the drive, is to be expected from the hypothesis.

In these experiments the extinction curves were not continued at reduced drives long enough to test for the eventual appearance of a constant maximal number of responses in each case. But by returning to a full drive on the second day the same effect is shown in an accelerated form. The effect of a change in drive upon the family of curves in Figure 135 is to postpone the appearance of some of the responses that would have been elicited on the first day of extinction and thus to shift the body of the curve to the right. A limiting case which illustrates this shift has been entered in the graph in broken lines. Here the drive on the first day of extinction is assumed to be so low that no responding occurs. Consequently no extinction occurs, and on the following day (the drive being now normal) the curve begins *ab initio* exactly as at zero grams on the first day. Here the whole curve has been moved one day to the right. In the case of six grams a similar effect is clearly shown. The low drive on the first day of extinction allows only a small number of responses to appear. The process of extinction is consequently not advanced very far, and on the second day (at full drive) the rate is relatively high. A similar effect is detectable in the other two groups at reduced drive. The group at four grams, however, is obviously approaching a lower asymptote, which may mean that the correction is not wholly adequate or that there is a difference due to sampling.

It will be shown later that the normal drive maintained by daily feeding for a limited time is perhaps only one-fourth or one-fifth of the maximal drive exhibited during starvation. We must therefore consider what effect a further increase in drive would have upon the curve for the normal drive shown in the preceding experiments. Figure 134 gives the expected effects of an increased drive in Curves D and E. If the hypothesis holds, there should be an even more rapid expenditure at the beginning of the curve, followed by a necessarily abrupt drop to a low rate of emission. No deliberate experiments have been performed on this subject, but a few extinction curves at maximal or near-maximal drives have been obtained accidentally [3] in experiments similar to those to be reported later.

Two extinction curves at near maximal drives are shown in Figure 136. The ordinates are reduced as in the experiment to be described later but a dotted line has been added to Curve B to indicate its position on the usual coordinates. It is clear that the expected result is actually obtained. The reserve is quickly emptied, and a very low rate of responding follows. Doubtless some additional responding would be observed on a second day, since the strength should spontaneously recover to some extent. A trace of further responding appears toward the end of Curve A.

If future responding after recovery may be supposed to be slight, the curves in Figure 136 should be compared with those in Figure 39. The total number of responses made by the rat in the latter case during five days of extinction is approximately equal to that in Figure 136. The difference in the shapes of the curves for the first day of extinction demonstrates the expected effect of the maximal drive. Approximately the same number of responses exist in the two reserves but they are called out much more rapidly in the case of Figure 136.

The rapid emptying of the reserve resembles that which takes place after the establishment of a temporal discrimination under reinforcement at a fixed ratio, as may be seen by comparing the present curves with those of Figure 95. The reasons for the very rapid initial rate are different in the two cases, but the result is approximately the same.

The effect of drive upon original extinction curves is more difficult to follow because of the relatively severe deviations that characterize them. Nevertheless some indication of the effect of hunger may be obtained; and it is in good accord with the interpretation advanced above. Twelve rats approximately 130 days old were conditioned in the usual way and 20 responses were reinforced. On each of two successive days 20 other responses were reinforced. After two days on which the rats were fed in their cages at the experimental hour, extinction curves were obtained. Immediately before the experiment food was given to each rat in a separate cage as follows: to four rats no grams, to two rats two grams,

[3] An accidental curve is obtained when the magazine jams or the reset apparatus fails, as is sometimes the case.

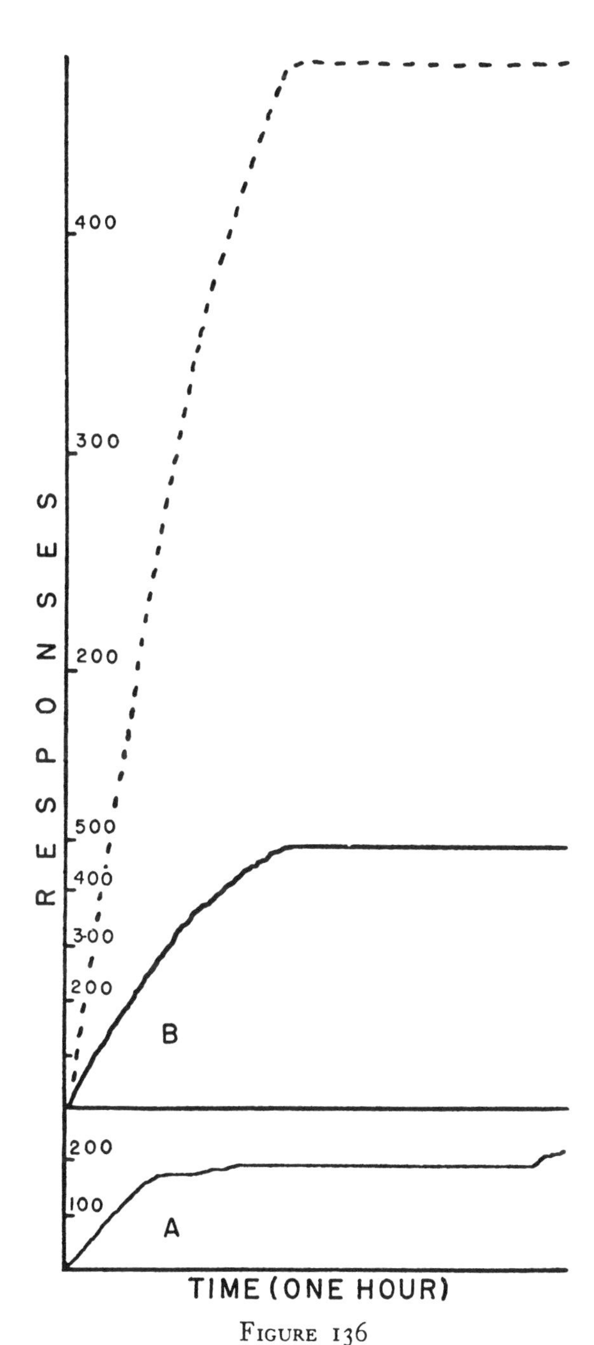

Figure 136

Two Extinction Curves at Nearly Maximal Drives

Coordinates had been reduced to accommodate the high rate. Dotted curve indicates the position of Curve B on the usual coordinates.

to four rats four grams, and to two rats six grams. A representative record at each degree of hunger is given in Figure 137. The course of each curve is indicated with a broken line, which is offered not as a theoretical curve, but merely to show the character of the experimental curve more clearly.

The most important of the characteristics of original extinction which distinguish it from extinction after periodic reconditioning

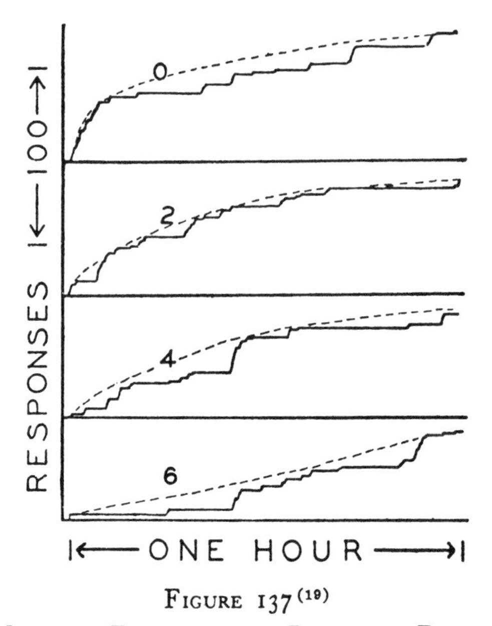

FIGURE 137[19]

ORIGINAL EXTINCTION AT DIFFERENT DRIVES

Different amounts of food were fed prior to the experiment as indicated in grams at each curve.

is speed. Under the conditions of this experiment the greater part of the change in strength is normally completed by the end of one hour. Thus, the curve at no grams in Figure 137 has closely approached a zero rate of responding by the end of the period. Unless the postponement of responses by the reduction in drive is too great, we should be able to obtain in a complete form and within one hour a family of curves similar to that which in Figure 135 was interrupted by the return to maximal drive. With the exception of the curve at six grams, this is the case. To the degree of approximation remaining in spite of the irregularity it is apparent in Figure 137 that the animals reach a low rate of responding by the end of the hour. It is also apparent that this is accomplished in spite of a progressive reduction in the initial rate by the reduction in drive. The curves are thus in rough agreement with the assumption that the final height is not affected by the drive, and that the only effect is a modification of the relation between

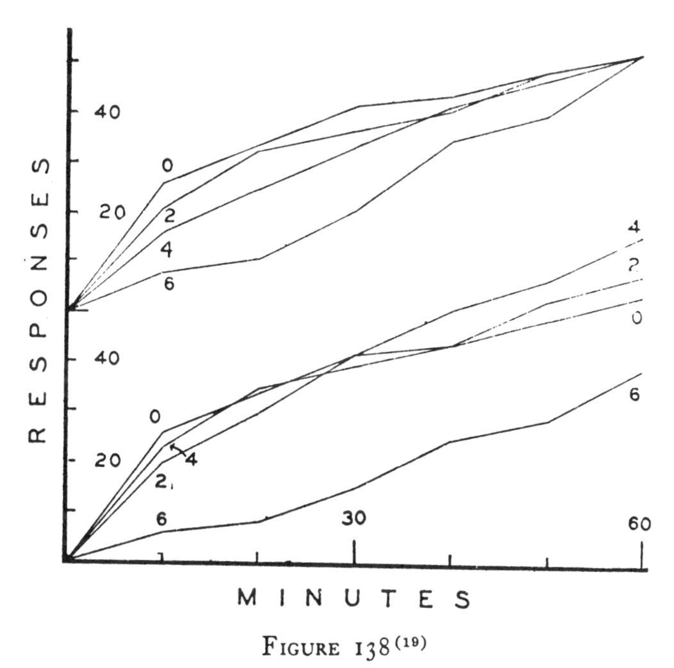

FIGURE 138 (19)

AVERAGED CURVES FOR ORIGINAL EXTINCTION AT DIFFERENT DRIVES

The lower curves are the original data. The upper curves have been obtained by multiplying the values in the lower by factors chosen to bring the ends together. The drive was reduced by feeding amounts of food as marked in grams at each curve.

the rate of responding and the number of responses still to be elicited. In the case of six grams part of the curve has obviously been postponed too long to appear during the experimental period.

The averages for all records in the group are given in Figure 138. Since the deviations in curves of this sort are all in one direction (*i.e.*, below an envelop) an average will not correct for them. An average is scarcely more useful than a single curve, except that it serves to express the result for the whole group. In the case of the raw data in Figure 138 (lower curves) the effect of the change in drive is again not very clearly shown because of the different slopes for the four groups, and unfortunately it is not possible to make a simple correction as in the case of extinction after periodic reconditioning, because we have no preceding data to work from. The number of responses to be elicited as the result of conditioning is not a simple function of the number of the preceding reinforcements when these occur grouped together. Organisms differ rather widely in their extinction ratios and in the heights of original curves following comparable amounts of conditioning. It is therefore at present impossible to predict in advance what the height of an original extinction curve is going to be. If, however, we make the assumption that the same height is to be reached at the end of the hour (noting the improbability of the assumption in the case of the six grams), we may multiply all values by factors chosen to bring the ends to some arbitrary point (say, the value for no grams). This has been done in the upper group of curves in Figure 138, where the required factors are 0.00, 0.93, 1.82, and 1.38. The characteristic effects of the change in drive that I have already noted in connection with the individual records may be observed.

A more elaborate experiment should obviously be performed to determine whether the number of responses obtained in extinction is wholly independent of the drive. The preceding experiments lack the controls necessary to decide the question definitely, although they indicate that there is no very great difference in the total number of responses obtained with high or low drives. It may be tentatively concluded that the effect of the non-reinforcement of a response is independent of the drive. In common terms 'failure to receive a pellet means as much to a slightly hungry rat as to a very hungry one.' In both cases a single response is subtracted from the reserve. But this is not the case for the effect of reinforcement, as will be seen from the experiments that follow.

Drive and the Extinction Ratio

It was shown in Chapter Four that the constant rate of responding observed under periodic reinforcement varied with the frequency of reinforcement. But this rate, like any other, should also vary with the drive, and the interaction of drive and conditioning should therefore be significantly revealed in an investigation of the relation of the rate under a given frequency of reinforcement to the operations of feeding and fasting. The rates of responding described in Chapter Four existed under the normal drive maintained by feeding once each day for one and one-half hours. As in the case of the extinction curves the degrees of hunger lying both above and below this normal value should be investigated.

In the first experiment to be described the hunger was reduced by feeding given amounts of food immediately before experimenting as in the experiment already described. The results for three groups of rats were as follows.

Group A. Four rats, 105 days old at the beginning of the experiment, were tested at four drives resulting from the feeding of zero, two, four, and six grams. The food was placed in the apparatus and was eaten before responses to the lever were elicited. Thereafter responses were periodically reinforced for one hour. Each rat was tested several times at each drive in random order during the 16 days of the experimental period. Of the 64 records obtained, two were lost through technical faults in the procedure or apparatus. The remaining 62 were distributed as follows: 16 at zero grams, 15 at two grams, 15 at four grams, and 16 at six grams, each rat contributing three or four records to each group.

The average rates at the four drives in responses per hour are given as open circles in Figure 139 (page 392). The relation between the rate and the amount eaten is roughly linear. The best fit with a straight line extrapolates to an excessively high value for the amount of food necessary to bring the rate to zero. The line drawn through the points in the figure takes a more reasonable extrapolation into account.

Group B. Four rats (age not known but probably about five months) were tested at similar drives. The food was placed in the apparatus, and the rats ate it before responding to the lever. In this group each amount of food was fed to one rat on two successive days; but otherwise the amounts were shifted at random, each rat contributing a number of records at each drive. Forty records for periods of one hour each were obtained from the group. Four of these were lost through technical mistakes. The remaining 36 were distributed as follows: 10 at zero grams, 11 at two grams,

12 at four grams and 3 at six grams. With this group six grams of food reduced the rate to so low a value that considerable irregularity was encountered, and the experiment was therefore confined chiefly to the higher values of the drive. The result with six grams should obviously not have full weight. The averages for all drives are given as the lower solid circles in Figure 139. It will be seen that the rate declines as a linear function of the amount

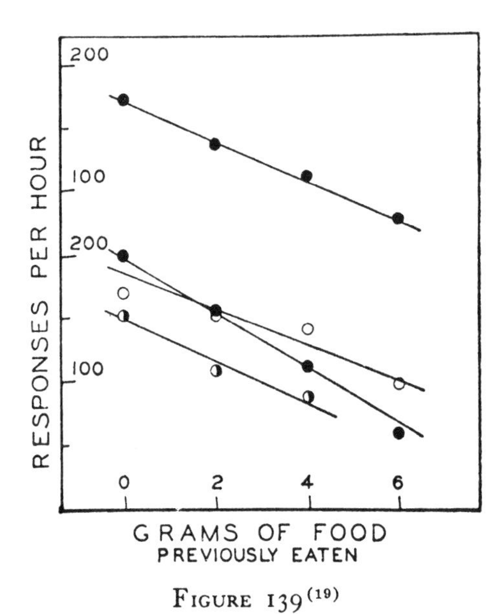

FIGURE 139[19]

RATE OF RESPONDING UNDER PERIODIC REINFORCEMENT AS A FUNCTION OF THE AMOUNT OF FOOD EATEN PRIOR TO THE EXPERIMENT

The lower curves are for groups of four rats each. The upper curve represents the averages for all three groups.

fed, and in this case a better fit is obtained. Sample daily records for one rat at the four drives are given in Figure 140 (page 393).

Group C. Three rats of the same age as in Group B were tested in the same way. Thirty records of one hour each were obtained from this group. None of these was lost for technical reasons, but three were at odd drives for the sake of exploration and are here omitted. The remaining 27 records were distributed as follows: nine at zero grams, eight at two grams, and ten at four grams. Two exploratory records at six grams showed that at the resulting drive the rate was too low to yield a satisfactory result. The averages for all drives are given as shaded circles in Figure 139. The rates for this group are considerably below those of Groups A and B but show the same approximately linear relation to the amount previously eaten. Sample records for one rat at the three degrees of drive are given in Figure 141 (page 394).

The averages for the three groups are given in the upper curve in Figure 139. The point at six grams is not to be taken as of equal

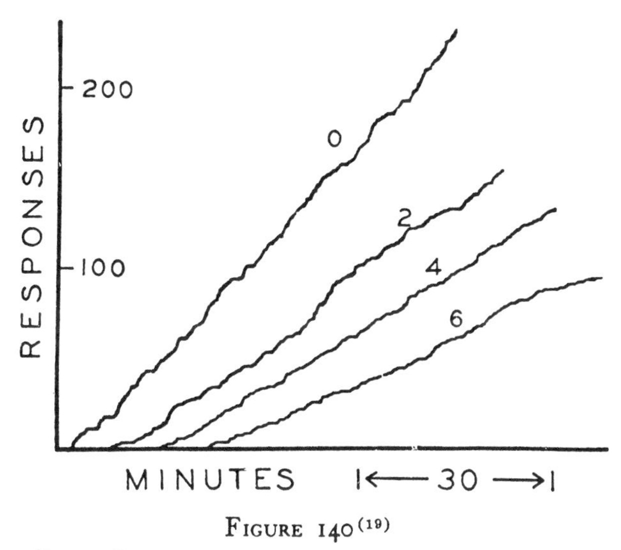

FIGURE 140[19]

FOUR DAILY RECORDS FOR ONE RAT IN FIGURE 139, GROUP B

The amounts of food eaten just before the experiments are indicated in grams. Note that the curves remain essentially linear.

weight because of the absence of a value for Group C, which raises the average because the extrapolation for the group passes below the average for the other two. At the higher drives, however, an average serves to reduce the scatter resulting from the smallness of the samples and to reveal the relationship more clearly. In evaluating the graph it may be noted that it includes every record made during the experiment with the few exceptions already mentioned.

This experiment permits us to test whether or not the degree of drive reached by feeding part of a daily ration remains constant for some time thereafter. It might be expected that the digestion of the food already taken would bring about a later change, but this is not the case. Such a change would produce a curvature in the records during the hour. If the height of a daily record is measured at the middle and the end of the hour, the former height should be one-half the latter if the curve is a straight line. Deviations from the expected middle value may be expressed as per-

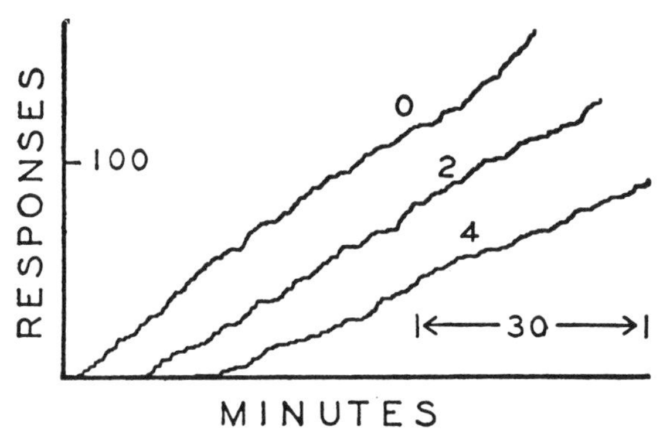

FIGURE 141 (19)

THREE DAILY RECORDS FOR ONE RAT IN FIGURE 139, GROUP C

The amounts of food eaten just before the experiments are indicated in grams.

centages. The average curvatures for all records grouped together according to drives are given in Table 10. The average values are either zero or slightly positive, and in no case are they significant. Even at the low rates obtaining under six grams, the records are straight lines. It is, therefore, shown that if a rat is interrupted

TABLE 10

CURVATURE AT DIFFERENT DRIVES (IN PERCENTAGES)

Grams fed	0	2	4	6
Group A	0.0	+ 8.5	+ 7.7	0.0
Group B	+ 2.2	− 4.7	− 8.6	—
Group C	− 0.5	+ 0.9	+ 4.5	—
Average *	+ 0.6	+ 1.6	+ 1.2	0.0

* Signs taken into account.

during the ingestion of a daily ration of food, the degree of hunger existing at that moment will persist without significant change for at least one hour. We are omitting the possible effect of 12 pellets of food administered periodically during the hour, having a total weight of a little over one-half gram.

At first sight, of course, we seem to be lifting ourselves by our own bootstraps. We start out to discover whether the degree of drive has any effect upon the periodic slope; we then use the periodic slope to show that the drive is constant during an hour. But the circularity is only apparent. We show that we are able to produce either a decrease in rate by permitting the ingestion of food or, from one day to the next, an increase in rate by withholding food. We then show that no change takes place during the hour similar to that which would be produced by further feeding or fasting.

The rate of responding at degrees of hunger above the normal may be investigated by following the change during starvation to death. Experiments on this subject have been performed in collaboration with Professor W. T. Heron. Data were obtained on thirteen rats which were 150 days old at the beginning of the experiment, with the exception of four rats about 100 days old. The apparatus and method were essentially as already described. The experimental periods were one hour long. Except during the experimental period the rats were kept in a constant temperature cabinet at 25°C. Water was available at all times.

With Professor Heron's permission I quote with slight changes from the published report of these experiments (22).

> On the day previous to the initiation of the starvation period, the rats were allowed continuous access to food for 24 hours. From that time on they were allowed no food except that which was necessary to recondition them. Since the interval of reconditioning was four minutes and the daily test period one hour, each animal received a daily ration of about 15 pellets or a total mass of approximately 0.7 grams. Under these conditions the animals were tested daily until death by starvation. It was not originally intended to carry the experiment so far as this, but by the time the course of the change had been clearly worked out the animals could not be saved.
>
> In general terms, the results may be stated as follows: the number of responses per hour increases with the period of starvation until a maximal rate is reached. After this point there is a

relatively rapid decline in the rate until death ensues from inanition.

Figure 142 is a graph showing the daily mean number of responses. Since there are individual differences in regard to the day on which the maximal rate is reached, the interpretation is somewhat difficult. The rat which reached its maximal rate first did so on the fourth day, while at the other extreme one rat prolonged its rise to the thirteenth day after the beginning of the starvation period. If this difficulty is disregarded and if we assume that there is a direct relationship between the rate of re-

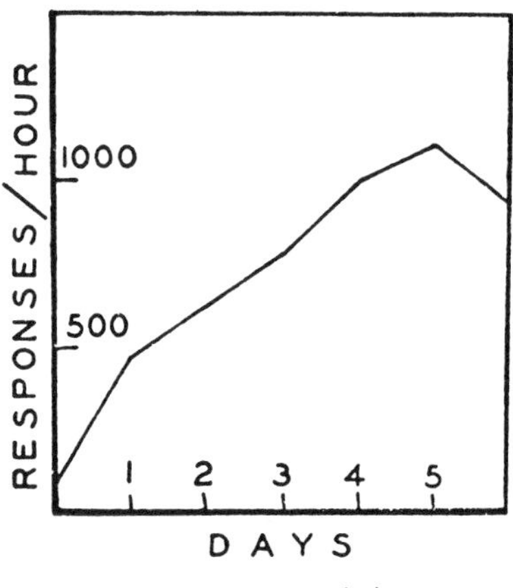

FIGURE 142 [22]

CHANGE IN THE MEAN RATE OF RESPONDING UNDER PERIODIC RECONDITIONING DURING STARVATION

The point at zero days shows the rate after twenty-four hours of continuous access to food. The rate rises rapidly during the first twenty-four hours and continues in a roughly linear fashion to a peak at the fifth day. Experiments in collaboration with W. T. Heron.

sponding and the strength of drive, the mean maximal drive for the group occurred on the fifth day after the beginning of the starvation period.

Figure 142 also indicates that the relationship between the increase in mean drive and the progress of inanition is approximately linear until the peak is reached. The greatest deviation from linearity is the relatively abrupt rise between the first and second points on the curve, but this is an artifact due to the fact that the first period was preceded by a 24-hour period of continuous access to food. The curve in Figure 142 has been plotted only to the sixth day after the beginning of starvation. A number of

animals remained in the experiment after this point but several also died before the seventh record was taken. It would be misleading to continue the curve since it would no longer be representative of the group as a whole.

Because of the individual differences with respect to the time at which the peak was reached, it was thought desirable to superimpose the individual records at their maximal rates. This was

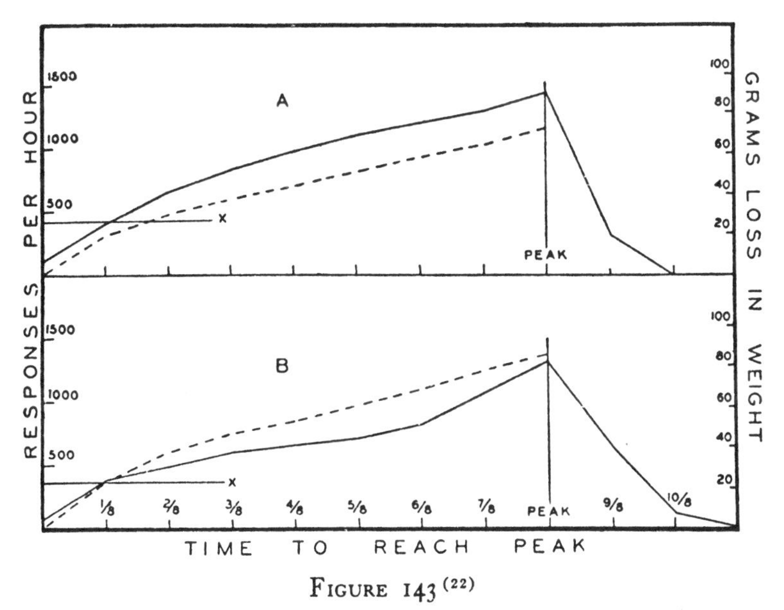

FIGURE 143 (22)

CHANGE IN MEAN RATE AND BODY-WEIGHT WHEN INDIVIDUAL RECORDS ARE SUPERIMPOSED IN SUCH A WAY THAT THE PEAKS ARE BROUGHT TOGETHER

Solid lines represent the rate, dashed lines the absolute loss in weight. Group A shows a more rapid decline after the peak than Group B. The horizontal lines marked X indicate the rate prevailing under the normal feeding schedule. Experiments with the collaboration of W. T. Heron.

done in the following way. The point of the beginning of the experiment and the point of reaching a peak were indicated on a sheet of graph paper. The distance between them was divided into a number of equal parts corresponding to the number of days taken to reach the peak by one rat. The data for this rat were then plotted and the points connected by straight lines. The data for each rat were treated in the same way. Ordinates were then erected to divide each of the individual curves into eight equal parts. The intersections of the ordinates with the individual curves were read off on each vertical and averaged. The curves in Figure 143 (solid lines) are for the averages thus obtained. The parts of the curves beyond the peaks were also spaced out on the coordinates assigned to each rat and averages obtained in the same way. Since there was some individual variation in the amount of time elapsing between the attainment of the peak and death, the sections of the curves to the right of the peaks are not representative of the whole group throughout their entire length.

Before group curves were made, the daily records were plotted for each rat. An inspection of the individual curves indicated that it would be convenient to deal with them in two groups. The first group (A in Figure 143) is composed of the eight animals which rose to their maximal rates and then dropped precipitously toward zero. The second group is composed of the remaining five animals, the curve for which is shown as B in Figure 143. They maintained a lower mean rate for the first six-eighths of the time. Their rate from that time on increased more rapidly until they reached a peak, which was not as high as that reached by the first group. After the peak their rate does not decline so rapidly, and their mean survival time is longer.

The differences in these two curves may or may not indicate a fundamental difference in the animals. The difference after the peak may be an artifact caused by the fact that a continuous process was sampled at relatively gross intervals of twenty-four hours. For example, in the rats which are represented in Curve B the drive may have reached its maximal rate in the 24-hour interval elapsing between the test period showing the highest rate and the next test period. In other words, the peak shown on the curve is possibly misplaced to the left of its true position. This difficulty is inherent in the present technique, but it could be minimized by using shorter test periods spaced at closer intervals.

After each animal had passed its peak, it was obvious that it was in an extremely impoverished condition. It was cold to the touch (bodily temperatures were not taken), its hair was erect and shaggy, and in many cases a normal posture could not be maintained. From these observations and from the early death of the animal after its peak was reached, it may be concluded that the decline in rate was due to physical weakness, rather than to any independent decrease in the state of the drive. The experiments do not confirm the human report of an early decrease in hunger during prolonged fasting.

The two lines drawn parallel with the base-line and marked X in Figure 143 indicate the rate of response under the usual feeding method as determined for the respective groups before the beginning of the starvation period. It is obvious that the level of drive maintained by the usual feeding method is far below the maximum.

The dashed curves in Figure 143 represent the absolute loss in weight for the two groups. Each rat was weighed daily after each test period and the loss of weight calculated. The data thus secured were plotted in the same way as the rates of responding, letting the end point for each rat's weight curve be determined by

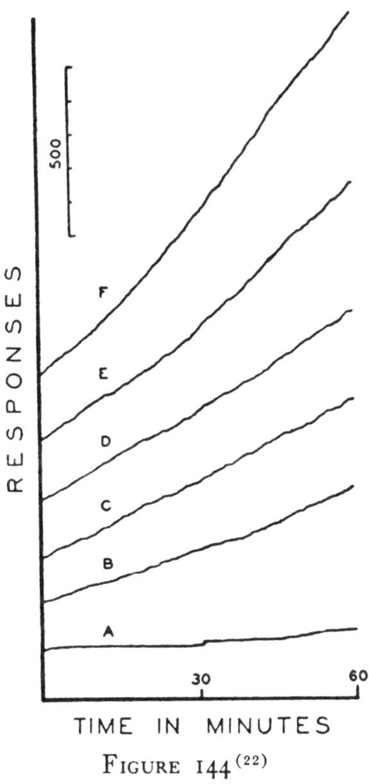

FIGURE 144 [22]

A SET OF RECORDS FOR ONE RAT FROM GROUP A IN FIGURE 143

The record at A was taken after twenty-four hours of access to food. Records B, C, . . . follow at twenty-four hour intervals. The slight acceleration during the hour in Records E and F is not typical.

the day on which he reached his maximal rate of responding. An inspection of Figure 143 shows that the correspondence between the mean loss of weight and the mean rate of responding is close.

Figure 144 shows the type of record obtained on successive days. In order to accommodate the high rates obtained in this experiment, the ordinate scale was reduced four times. The curve at A is the record made after 24 hours of continuous access to food. Since the rat was almost satiated, its rate was very low. The record at B is that made after one day of starvation and is approximately the rate maintained under the usual feeding method. The remaining records are for successive days of starvation until the rat reached its peak on the fifth day (Curve F). This rat belonged to the first group and did not survive to give a record on the day following the peak.

These experiments on starvation together with the preceding experiments on subnormal degrees of hunger apparently cover the entire range of available degrees. The behavior of the rat is consistent with the assumption that the primary effect of hunger is upon the rate of responding. No other change in the behavior has been revealed. The daily records remain approximately linear, and the momentary variations (the 'grain' of the records) are apparently unchanged. Hence we may generalize the relation discussed in the preceding chapter still further. The effect of the drive upon the rate is the same not only when every response is reinforced (as in the normal ingestion curve) or when it is not reinforced (as in extinction), but also when a constant rate is maintained by periodic reinforcement.

If the change in rate is the only effect of the change in drive, three inferences may be drawn. (1) The behavior of responding for one hour at a low drive should increase the reserve, since responses are being reinforced just as often as at the normal drive, but the reserve is not being drained at the same rate. (2) Conversely, responding for one hour at a higher drive should drain the reserve, since no more responses are being reinforced than at the normal rate, while many more responses are being given out. (3) The normal drive obtained by daily feeding for a limited time must be regarded as luckily just the drive needed to obtain a balance between input and output. The third inference is not likely to hold, since the procedure of controlling the drive is quite arbitrary. The first and second inferences are quite definitely wrong.

A strain upon the reserve or a contribution to it should appear as a curvature in the record after a change has been made from a low to a high drive or *vice versa*. But this is not the case. We want to know first whether there is any curvature as the result of passing from a higher to a lower drive. Measurements of all records in the preceding experiments on subnormal drive at a drive lower than that of the preceding day were grouped together and the average curvatures expressed as heretofore were found to be of the following directions and magnitudes: Group A: + 1.1 per cent; Group B: — 3.2 per cent; Group C: — 6.0 per cent. The average is — 2.7 per cent. Although the curvature is in the right direction to indicate some drain upon the reserve at the preceding higher rate, the value is probably insignificant. The case is similar for records at drives higher than that of the preceding day. When these are grouped together, the resulting curvatures are: Group A: + 4.1 per cent; Group B: + 0.6 per cent; Group C: + 2.2 per cent. The average is + 2.3 per cent. Here again there is some sign that the reserve has gained from the reinforcements at a low drive but the value is again scarcely significant.

In the case of Groups B and C in the experiment on subnormal drives the days were paired in an attempt to disclose the effect of the periodic reinforcement upon the reserve. The averages for all pairs of days which followed a lower slope are 183 and 153 for the first and second days respectively. This confirms the tendency just indicated. The first day at the higher drive is especially high because the reserve has profited from reinforcement at a lower rate of expenditure. But the averages for all pairs following a higher slope are 107 and 102, which shows a trend in the same direction and contradicts the foregoing evidence. Some drift in this direction is to be expected from the spontaneous decline of the rate during periodic reconditioning described in Chapter Four. A similar inconsistent result is obtained from Group A by averaging all values at any one drive in groups according to the preceding drive. Not all such groups are represented, as the random changes in drive were not well distributed; but the following determinations (and no others) can be obtained from the data: (a) (contradictory) the values for all records at zero grams following a day at six grams are 11 per cent higher than those at zero grams following a day at four grams; the values at four grams following a day at six grams are 11 per cent higher than those at four grams following a day at two grams each; (b) (confirming) the average value of all records at two grams following days at zero grams are 11 per cent higher than those following a day at four grams; and the values at six grams following zero grams are 12 per cent higher than those following four grams. These irregularities are probably due to sampling; and for the present degree of approximation at least, it may be concluded that no effect upon the reserve is felt.

The additional fact contributed by these experiments is that the reconditioning effect of a single reinforcement is a function of the drive. (It will be remembered that the extinguishing effect of a failure to reinforce is probably not.) The change in the rate of responding with a change in drive is an affair of the extinction ratio. It is only on such an assumption that the maintenance of very high rates during starvation can be explained since no reserve exists of a size that would carry a rat through the starvation period at that rate without additional contributions from a more effective reinforcement.

The Measurement and Comparison of Drives

The question how we are to measure the strength of drive is practically answered when we have once decided upon a definition. We are to measure, not the drive as such, but behavior. This statement does not identify the behavior and the drive. It has been shown that a drive is most accurately regarded as a state of proportionality between the reserve and the momentary strength. But if no changes occur in the size of the reserve and if other variables are held constant, the strength of a drive is proportional to the strength of any reflex associated with it. Hence any technique for measuring drive will be a technique for measuring reflex strength. It is a poor substitute to measure the operation responsible for the state of a drive until the relation between the state and the operation is accurately known. A 'twenty-four-hour hunger' is perhaps a relatively constant degree of hunger, but the 'twenty-four' is not a convenient representation of the value since hunger does not vary quite linearly with the time of fasting.

A well-known way of estimating the strength of behavior as a function of a drive-operation is to observe the relative prepotency of the behavior over incompatible behavior assumed to be of relatively constant strength. In the Columbia 'obstruction method' (74) a response away from an electric grid is opposed to a response toward the grid conditioned by placing food or some other reinforcing stimuli on the other side. The strength of the response across the grid is estimated by counting the number of occasions upon which it takes prepotency over the response away from it in a given period. Several objections may be raised to such a method. (1) The response away from the grid is affected by variations in the intensity of the shock (which are hard to avoid), upon the development of methods of walking which reduce the intensity, upon adaptation to the shock, and so on. (2) The conditioning of the response across the grid may not begin or remain maximal. (3) The conflict of responses and the shock itself may be additional (emotional) variables modifying the strength during the experiment.

The use of a standard reflex against which other reflexes are compared is unnecessary. The same methods of measuring strength are available in the study of drive as in any other field in which the strength varies. But it may be argued that the obstruction method excels the direct measurement of behavior by providing a means of comparing different drives. Drives A and B stand in a relation to each other given by the relative prepotency of reflexes typical of each when opposed to the same standard reflex. But the third term of the standard reflex is unnecessary. The simplest way to determine whether hunger is stronger than sex is to place appropriate stimuli for each drive before the organism at the same time. If it is argued that drives mutually influence each other, so that maximal degrees of two drives cannot exist at the same time, then the question of comparison is academic and may be passed by.

Any method of measuring drive is handicapped by the fact that if more than one measurement is to be taken, either the drive must be modified by a consummatory reflex after the first measurement or the degree of conditioning must be affected by withholding the reinforcing stimulus. Sex is particularly hard to measure for that reason. Even in the case of hunger, where one consummatory response may have a very slight effect, some weakening of the strength must occur. Because of different solutions to this problem the curves for the change in hunger during starvation given above (Figures 142 and 143) differ significantly from those obtained with the obstruction technique. In order to avoid the effect of consummatory reflexes Warden and Warner (74) used groups of rats for each period of starvation and took one measurement only with each individual. In the experiment described above some reinforcement with food was supplied each day, and although this delayed starvation to some extent it was then possible to follow each individual throughout the process. When the individual curves are averaged, the resulting curve closely resembles that of Warden and Warner, except that the peak comes slightly later. But this averaged curve differs in two important respects from the individual case: the increase in hunger is slower and the final drop sharper in the individual curve, and the height at the peak is greater. Both results follow from the fact that different rats reached peaks at different times. If this was also true, as is probable, in the Columbia experiments, the curve there obtained does not correctly represent either the course of the change in hunger or the maximal value attained. Some of the rats that were starved for four days had probably reached their maximal drive before the test period, and some certainly reached it afterwards. The fourth day happened to catch more rats at or near their maximal drive than any other period used. In the present experiment the animals reached peaks anywhere from the fourth to the thirteenth day of starvation. The mean is at 7.3 days and the median is at 7. By averaging the rates for all rats a peak at *five* days is obtained. A comparison of the height of the peak in Figure 142 with the heights in Figure 143 will show the depressing effect which the averaging of the group without respect to individual curves has upon the maximal value.

This difficulty would not be serious if the measurements of all drives were affected in the same way, but unfortunately this is not the case. There are certain conditions under which the depressing effect will be minimized. (1) If the drive rises very rapidly to a peak (*cf.* thirst) the method of averaging groups will catch a great many more animals at or near their peak than would be the case if the drive rose slowly as in the case of hunger. (2) If the drive is maintained at the peak for a relatively long time, the chances of finding all animals at or near a peak value are very good. An example is the maternal drive, which is probably maintained at its maximal level for a number of days while a litter is young. It may be that the hunger drive is actually weaker than the thirst drive; if so the method has exaggerated the difference. Or it may be that the hunger drive is actually stronger than the thirst drive, but because the mean for the hunger drive has been depressed, it appears to be weaker. If maximal drives are to be compared in strength, the comparison should be based upon the mean individual peak.

It is doubtful whether the use of groups of rats is a satisfactory way of avoiding the slight change in drive that ensues when repeated measurements are made on individuals.

Chapter Eleven

OTHER VARIABLES AFFECTING REFLEX STRENGTH

Emotion

The remaining important kind of change in reflex strength commonly observed in normal behavior may, in spite of certain current definitions, be called 'emotional.' Perhaps the commonest conception of emotion is that it is a form of response. Even when the primary datum is said to be an 'experience,' some sort of response is generally appealed to, either as the expression of the emotion or as an antecedent or concomitant activity. It is tempting to accept this well-established formulation, since it could be incorporated into the present system with very little trouble. If a child weeps when hurt, and if weeping is the emotion (or at least the only behavioral datum to be taken into account), then we might establish a correlation between the injurious stimulus and the flow of tears exactly as in the case of any other reflex. But this disposition of emotional behavior is not without its difficulties. Why should a certain part of the reactions of an organism be set aside in a special class? Why should we classify weeping in response to a bruised shin as emotional but weeping in response to a cinder in the eye as not? Most persons readily and consistently separate many of the responses of an organism into these groups but with what justification?

No satisfactory criterion, I believe, has been advanced in answer to these questions. The definition of an emotion as a response involving certain effectors—principally those under the control of the autonomic nervous system—is by no means rigorous, since there are probably no effectors involved in emotion which are not also involved in non-emotional behavior. The criterion of an identifiable characteristic of experience is available in a study of behavior only in translation, and it is there probably reduced to identifiable proprioceptive or interoceptive stimulation, but responses which provide a unique form of stimulation should also provide for a topographical classification, so that the preceding criterion is implicated. A third criterion is the diffuseness of the emotional reaction, a fourth its disorganization or ineffectiveness, and so on.

Attempts to define emotion as a special form of response need not concern us, since a quite different conception is necessary here. According to the present formulation emotion is not primarily a kind of response at all but rather a state of strength comparable in many respects with a drive. In so far as a response to an emotional stimulus occurs, it is to be dealt with like any other response, but the response does not define the stimulus as emotional and is only the accompaniment of the central emotional change. The changes in strength induced by the same stimulus provide practical criteria, and they are the commonest data in the field. I know that a man is angry, not because he is secreting adrenalin or because his blood pressure is increasing, but because he greets me dully, shakes hands slowly and weakly, responds to my remarks curtly, and avoids me if possible. All the responses which he is accustomed to make in my presence have undergone a significant change, and that change is the primary datum upon which I base the statement that he is angry. Similarly, I know that a companion on a dark road is afraid, not because I see that his palms are sweating or that his pulse is rapid, but because he starts easily, speaks in a whisper if at all, keeps his eye on his surroundings, and so on. It is true that upon closer examination in either case I may discover *responses* (rather than changes in the strength in reflexes) which are to some extent characteristic of, though not peculiar to, each state. If these responses are to be called emotional, it is not because of any essentially emotional character which they possess, but because they are elicited by stimuli which typically induce changes in reflex strength.

A simple experiment by Bousfield and Sherif (33) provides a good example of this aspect of emotion. These investigators recorded the effect of a loud sound upon the rate at which chickens and guinea pigs ate. Hungry animals were given food, and shortly after they had begun to eat, a shot was fired. The time elapsing before the animal resumed eating was measured in each case. One property of the effect shown by Bousfield and Sherif was its relatively rapid adaptation. In these experiments it could also have been shown that the shot produced at least two other kinds of effects: (1) a startle response involving the skeletal musculature, and (2) autonomic responses affecting blood pressure, pulse rate, and so on. Neither of these changes has been shown to be typical of emotion. The change in the state of the ingestive behavior, on the other hand, requires a separate classification of the shot as a stimulus. So far as we are concerned here, this change in strength is not only the defining characteristic of emotion but the only effect that needs special treatment.

This conception of emotion is, of course, not new. Many theories have been based upon changes in normal behavior rather than upon the production of specific responses, and most of these are compatible with the present definition. The central process in emotion has been largely overlooked because so many investigations start in search of the correlates of experience or of specific expressions. In cataloguing the phenomena of emotion, one is naturally tempted to seize upon manifestations which are common to most organisms. Changes in strength lack this generality because they depend largely upon individual repertoires. But although this makes the identification and classification of emotional effects difficult, it should not bring about a misinterpretation of the central characteristic.

The problem is similar in many ways to that of drive. Specific responses in emotion are much like specific stimuli in drives. Just as the hypothesis of a particular afferent stimulation cannot circumvent the study of all the behavioral changes due to a change in the drive, so the appeal to specific responses will not supply a simple substitute for the analysis of all the possible changes in reflex strength consequent upon the presentation of an emotional stimulus. In both cases we must describe the covariation of the strengths of a number of reflexes as functions of a particular operation. Drive and emotion are separate fields only because the appropriate operations can be separated into different classes. In many cases, this distinction is thin. A female rat is said to care for its young because of a maternal drive; when the same rat kills its young as food, it is said to be acting according to another drive; but when, after being moved into new and disturbing quarters, it kills its young and does not eat them, it is said to act emotionally. It is only because the operation responsible for killing differs from the operation of hunger or the maternal drive that the distinction is made. The effect upon behavior is of the same sort in both cases.

It is not essential to this formulation that drive and emotion constitute two distinct classes. The important thing is the recognition of a change in strength as a primary datum and the determination of the functional relationship between the strength and some operation. The terms drive and emotion may easily be dispensed with whenever they lose their convenience. Indeed, it seems to me one of the virtues of this conception of emotion that it so closely resembles that of drive. An emotion is a dynamic process rather than a static relation of stimulus and response.

The term emotional has been used in the preceding chapters according to this definition. Some more or less temporary state of reduced strength (an increase in strength would fit into the same formulation) has been related to a disturbing stimulus or to some other emotional operation, such as withholding a reinforcement. In several cases the property of rapid adaptation has been noted. I have, however, no special experiments to report under this heading and it is beyond the scope of the book to attempt a topographical classification either of the various kinds of emotional operations or of the reflexes varying together in different emotions. It may be noted that one problem of classification, which seems so hopeless in terms of emotional responses, is much less difficult in terms of variations in reflex strength. Anger and fear clearly involve different states of strength in different reflexes, no matter how similar the responses of gland and smooth-muscle may be. The mild emotions, for which corresponding responses are difficult to isolate, are at no disadvantage; although variations may be less intense, the distinctions are topographical and easily made so long as the changes are at all observable.

Drugs

The effects of drugs upon reflex strength have been fairly intensively studied in the case of spinal and postural reflexes [*cf.*, for example, Magnus (61)], but so far as the behavior of the whole organism is concerned, much less information is at hand. Most of the available information has been collected and described by pharmacologists in the terms of a popular vocabulary, and because of the difficulty of talking about behavior quantitatively the effects described are often confined to relatively isolated neural systems which lend themselves to simple description. Just how drugs affect the structure or the dynamics of behavior as a whole is at present difficult to say. It would appear that various reflex systems are differentially affected, following lines of demarcation which are sometimes close to those given by various drives and emotions, and that a drug may affect the topographical relation of stimulus and response. Whether the effects are upon the reserve or the relation between the reserve and immediate strength is not clear.

As in the case of emotion I have very little to report here. Some experiments upon the effects of caffeine and benzedrine made in collaboration with Professor Heron using the present technique may, however, be described as an example of the kind of analysis needed to bring the subject into line with a system of behavior. The caffeine or benzedrine was given in 0.5 cc. of physiological salt solution by subcutaneous injection. On control days an equal amount of the salt solution was injected. In all cases the drug was given immediately before putting the animal into the apparatus and beginning the experiment. I shall quote with slight changes from a report of these experiments (23).

> Figure 145 is a graph showing the daily and mean number of responses per hour of four rats which were given 10 mgm. of caffeine sodiobenzoate on the days indicated. The rats had had several weeks of periodic reconditioning at four-minute intervals before the caffeine was administered. Without exception the caffeine increased the mean rate of responding, although on several occasions an individual rat did not show an increase. (On the fourth day the increase in mean rate was not large. The solution of caffeine was several days old, and it was thought that some deterioration might have taken place. On the last day a new solution was used, and the mean rate returned to its former position. Whether deterioration actually occurred cannot, of course, be decided by this single case.)
>
> After the caffeine had been given twice, it occurred to us that the increase in rate might be caused indirectly through an increase in hunger (see Chapter Ten). As a check on this possibility the amount of food eaten by each rat following the experimentation each day was determined by weight. The mean food consumption is plotted at the top of Figure 145, and there is a close correspondence between the variations in the amount of food consumed and the rate of responding for that day. It is apparent

that the caffeine does increase the food consumption and that presumably the rat is hungrier while it is in the apparatus. This would account for some of the effect upon behavior but probably not all (see below).

Since caffeine is given immediately preceding the experiment, it might be supposed that the cumulative curve would be positively accelerated, since some time might be required for the drug

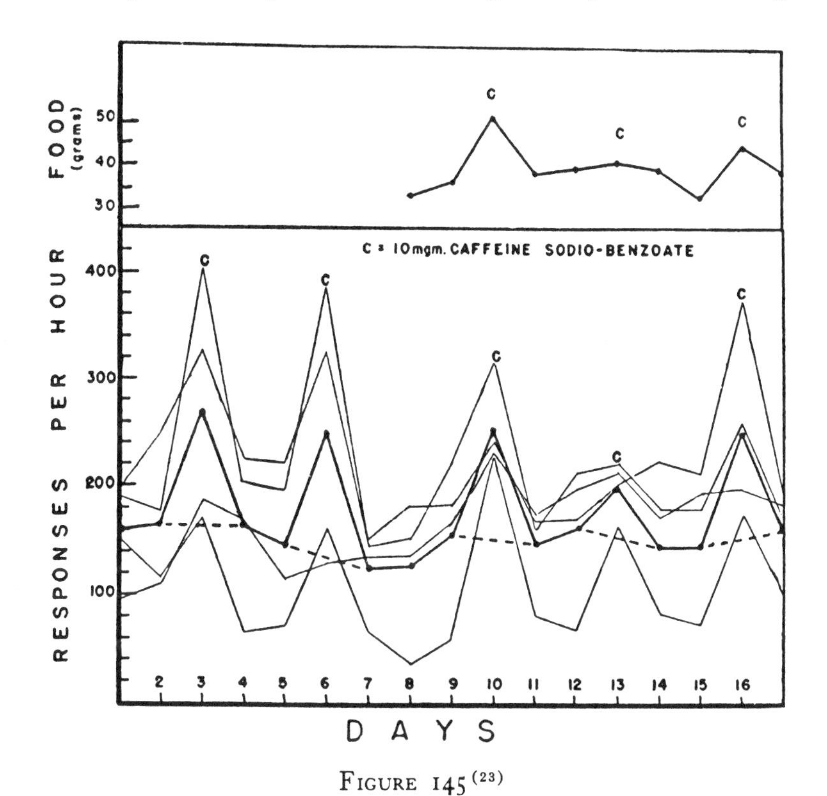

FIGURE 145 [23]

EFFECT OF CAFFEINE UPON THE RATE UNDER PERIODIC REINFORCEMENT AND UPON FOOD CONSUMPTION

The lower curves give the individual rates (lighter lines) and the mean (heavier lines). The upper curve shows the mean food consumption. Caffeine was given before experimenting as marked. Experiments with the collaboration of W. T. Heron.

to act. Figure 146 (page 412) gives the mean cumulative curve for the first day. The curve is not significantly accelerated and the action of the drug must therefore begin immediately. The mean curve for the preceding control day is also given.

In Figure 147 B the extinction curve for the four rats in Figure 145 is shown. On the fifth day caffeine was given and the rate was restored almost to the level prevailing during periodic reinforcement. On the second day following the administration of caffeine there is another rise in the extinction curve as a sort of rebound. Both of these increases are accompanied by corresponding in-

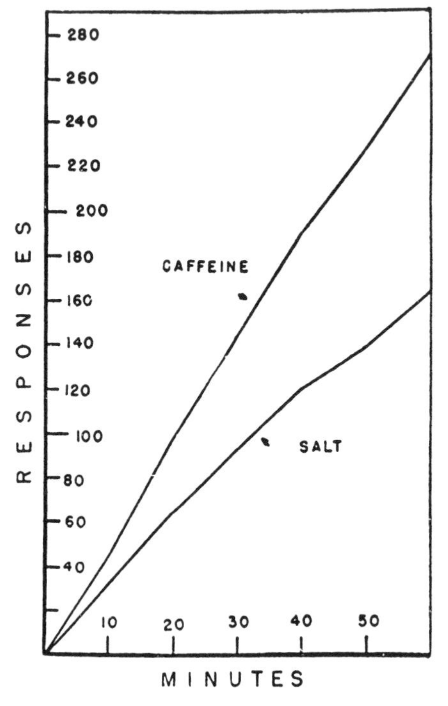

FIGURE 146 (23)

RATE DURING ONE HOUR UNDER CAFFEINE AND UNDER A CONTROL INJECTION OF SALT SOLUTION

The effect of the caffeine is felt immediately.

creases in the food consumption as shown in Curve A of Figure 147.

Curve C in Figure 147 shows a repetition of the experiment with a group of eight rats, which had also had considerable experience in the experimental situation before the extinction was started and gave a much higher constant rate under periodic reconditioning. Here again the caffeine restored the rate almost to its original level and a 'rebound' occurred on the second day following the caffeine. Unfortunately and inexcusably the food consumption of this group was not measured.

The only explanation that we can give for the secondary rise is that the rat overeats on the caffeine day and, therefore, eats a sub-standard ration on the following day. The result is that on the third day it is hungrier than normal. However, the same phe-

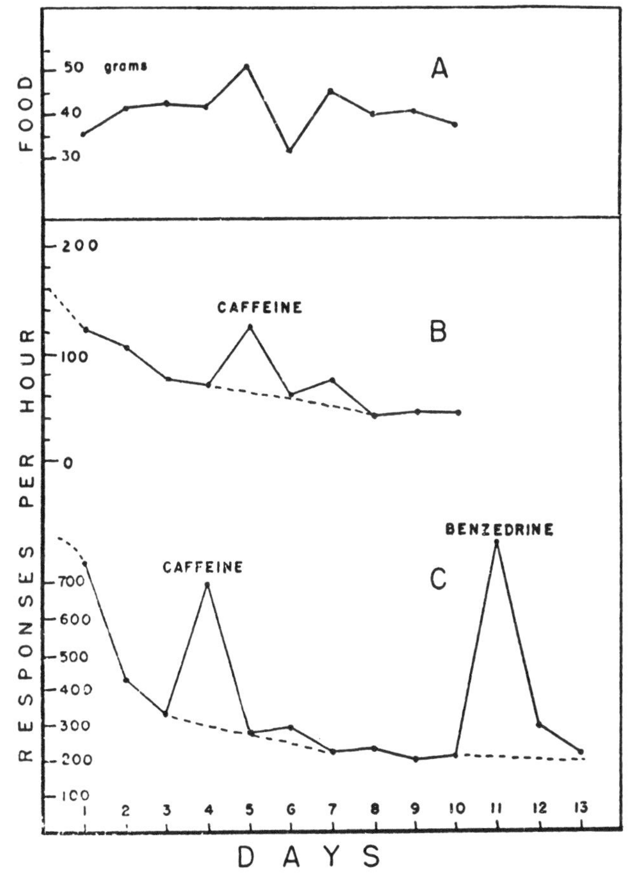

FIGURE 147 [23]

EFFECTS OF CAFFEINE AND BENZEDRINE UPON THE RATE DURING EXTINCTION

Curve A gives the mean food consumption for the four rats of Curve B. Curve C gives the result with a group of eight rats. Note the 'rebound' on the second day following the administration of caffeine. Experiments with the collaboration of W. T. Heron.

nomenon does not seem to occur under conditions of periodic reinforcement (see Figure 145). The rate on the second day following caffeine is not consistently higher in this case, nor does the food curve show the expected rebound.

The attempt to explain the effects of caffeine in terms of hunger is weakened by the results of the administration of benzedrine. The animals in the second group just mentioned were given 0.5 mgm. of benzedrine sulphate (a relatively large dose according to human standards) on the 11th day of extinction. Curve C of Figure 147 shows that the drug produced a complete restoration of the rate of responding to the level of periodic reinforcement. Nor does the rate return to its former extinction level on the day following. The reason for this, if it is significant, is not clear, as the drug is presumably entirely eliminated in the course of a few hours. The results from benzedrine are not consonant with the idea that changes in hunger account for the effects of caffeine, since benzedrine decreases hunger, as has been shown by Wentink (78).

Some further light on the effect of benzedrine and its relation to hunger has been obtained from an experiment in which the influence of hunger was eliminated. Four rats at an advanced state of extinction were given injections of benzedrine approximately one-half hour after the beginning of an experimental period. Two typical records (A_1 and A_2) are shown in Figure 148. Each curve begins with the usual rate at that stage of extinction. At the vertical marks a 0.5 mgm. dose of benzedrine sulphate was given, with the usual effect of an almost immediate rise in rate. The records were carried out for about two hours, during which the effect of the benzedrine was only slightly diminished. Immediately after this experiment the rats were given a generous supply of food, which remained in their cages until the following day. Additional food was also placed in the experimental boxes, together with a supply of pellets in each tray. On the following experimental day, therefore, each rat had not only been given continuous access to food for more than twenty hours but was surrounded by food in the apparatus. Further extinction was then carried out as shown at the beginning of Curves B_1 and B_2 in Figure 148 (for the same rats respectively as at A). That the rats responded at all under these circumstances was unusual. The twenty or thirty responses made by each rat during the first half-hour may be due to the intense activity of the preceding day and may thus represent a conditioned effect of the drug. It will be noted that a similar carry-over was revealed in Figure 147. At the vertical lines the same dose of benzedrine was again given, and it produced an unmistakable increase in rate, which in the case of Rat 2 is almost as great as that occurring on the preceding day when the rat was hungry.

It is obvious from these experiments that the effect of the ben-

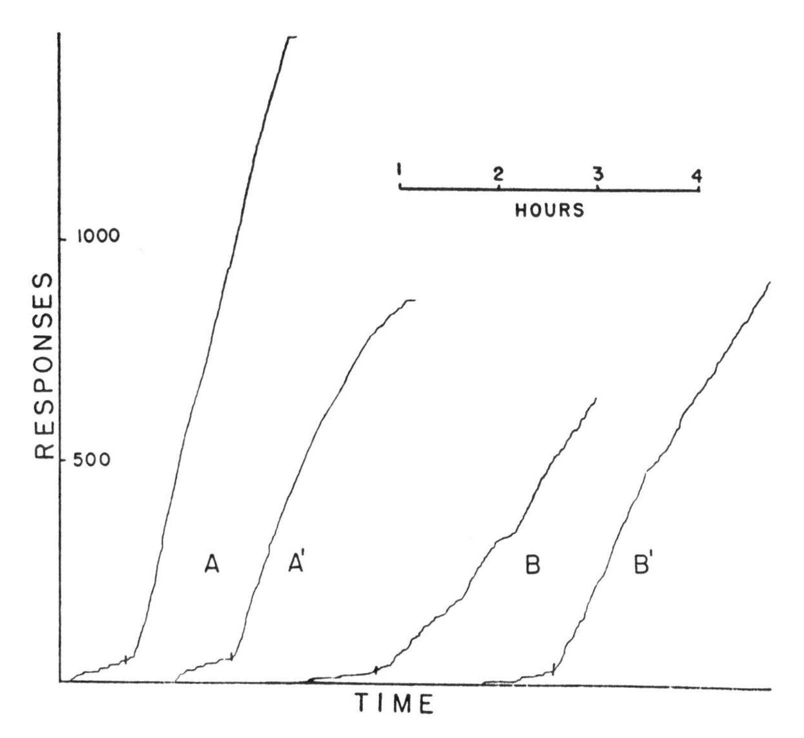

FIGURE 148

THE EFFECT OF BENZEDRINE UPON THE RATE IN LATE STAGES OF EXTINCTION

Benzedrine was administered at the vertical dashes. The curves at A were obtained under normal drive. The curves at B for the same rats respectively were obtained after twenty-two hours of access to food and with food in the experimental boxes. In spite of the lack of drive the benzedrine produces an extensive increase in rate. Note that the curves have been greatly reduced in the figure.

zedrine is to a considerable extent independent of the drive. The drug seems to produce a state of general excitability in which a response characteristic of the situation is emitted at a high rate.

The reflex reserve, which in the undrugged animal controls the emission, seems not to be directly involved. It might be said that the drug merely multiplies the responses existing in the reserve, although that is little more than a figure of speech. Obviously further work is necessary. A study of the heightening of spontaneous activity under benzedrine has been published by Zieve (83).

The effect of benzedrine upon the rate under periodic reconditioning has been confirmed by Wentink using the same apparatus and technique. Miss Wentink has also studied the effects of a number of other drugs and hormones. Sodium bromide had no effect upon the rate in the dosages given (0.018 gm., 0.0216 gm., and 0.036 gm., on successive days in that order), while 0.022 gm. of phenobarbital almost completely suppressed responding within ten minutes. Adrenalin (0.025 cc.) produced a decrease in the rate after a latent period of several minutes. The mean rate for one hour was reduced by 25 to 30 per cent. Insulin increased the rate when given alone, but when insulin was combined with benzedrine, which also increased the rate alone, an unexpected and severe decrease in rate was observed in three out of four rats. The details of these experiments may be obtained from the published report (78).

Other States of Strength

There are also certain metabolic or pathological states of the organism to be taken into account in completing a description of reflex strength, such as general fatigue, asphyxiation, and disease. Operative and other lesions, especially of the nervous system, may be included here. As in the case of drugs most of the information in this broad field is at present available only in the unanalyzed terms of a popular vocabulary and it can be translated only very crudely into the terms of a scientific formulation of behavior.

The commonest examples of illness that I have observed in the course of this work were confined to 'colds,' where a general decline in strength followed a decline in hunger and seemed to show no direct effect. On one occasion, however, a rat developed a vestibular infection, the first symptom of which was a disturbance of posture. The head was held twisted to one side, and incipient movements of rolling over were made during progression. The rat was not discarded from the experiments; and when the infection became severe, a marked effect upon the behavior in the apparatus was observed. Shortly after the postural difficulty began to show itself the strength of the response to the lever under periodic reconditioning began to drop. At the end of two days it reached practically zero and remained there for two more days. It then began to rise rapidly and five days later had become abnormally high in comparison with the other rats in the experiment. Until the end of the experiment four days later a rate of about 600 responses per hour was maintained. This was exceptionally high for the conditions of feeding then in force. The factors responsible for such a change could not be isolated in a casual observation of this sort, but the possibility of following the course of a disease, particularly one attacking the nervous system, is clearly suggested.

Another factor producing a change in strength is, of course, sleep. It is observed when records are taken for several hours and appears as an abrupt change in the rate of responding from any value then in force to a value of zero. Whether or not the rat falls asleep at such a time has not been determined by direct observation. The present fact is simply that these abrupt changes will be observed if an experiment extends over a considerable period of time.

The age of the organism is also a factor influencing strength that should ultimately be taken into consideration. The problem of maturation could be stated in terms of the changes in strength due to this variable, particularly as opposed to conditioning.

This list of additional variables of which reflex strength is a function is no doubt incomplete. My purpose in giving it is simply to remind the reader of the many variables that have been held reasonably constant in the preceding chapters of the book and to indicate the extent of the territory still unexplored. Not all of these problems seem to me to be the concern of the student of behavior, but, since behavior is involved, he should at least be able to supply techniques for the use of others. That the general formulation of behavior here proposed is fully adequate to so great a task cannot be asserted until a great deal of further investigation has been carried out.

Chapter Twelve

BEHAVIOR AND THE NERVOUS SYSTEM

If the reader has accepted the formulation of behavior given in Chapter One without too many reservations, and if he has been reasonably successful in excluding extraneous points of view urged upon him by other formulations with which he is familiar, he has probably not felt the lack of any mention of the nervous system in the preceding pages. In regarding behavior as a scientific datum in its own right and in proceeding to examine it in accordance with established scientific practices, one naturally does not expect to encounter neurones, synapses, or any other aspect of the internal economy of the organism. Entities of that sort lie outside the field of behavior as here defined. If it were not for the weight of tradition to the contrary, there would be no reason to mention the nervous system at this point; but an analysis of behavior is rarely offered without some account of the neurological facts and theories supposedly related to it. Although I have no intention of dealing with such facts or theories in detail, I can scarcely avoid some discussion of the all but universal belief that a science of behavior must be neurological in nature.

The various forms of neurological approach are too diverse to be considered exhaustively. I have already mentioned (in Chapter One) the primitive and yet not altogether outworn view that the phenomena of behavior are essentially chaotic but that they may be reduced to a kind of order through a demonstration that they depend upon an internal fundamentally determined system. This is the view which most naturally presents itself as a materialistic alternative to a psychic or mentalistic conception of behavior. The sort of neural homunculus that is postulated as a controlling force bears an unmistakable resemblance to the mental or spiritual homunculi of older systems, and it functions in the same way to introduce a kind of hypothetical order into a disordered world. The argument rests historically (and depends logically) upon a demonstration that neurological phenomena are intrinsically more lawful than behavior. It is only recently that this could not be appealed to as an obvious fact. The science of neurology achieved a degree of experimental rigor long before a science of behavior could do so. Its subject matter was chiefly 'physical' (in a somewhat naïve sense) while the data of behavior were evanescent; it could adopt the methods and concepts of its relatives in the biological sciences; and it could more easily confine itself to isolated parts of its subject matter. But the historical advantage has not been conserved. It is now possible to apply scientific techniques to the behavior of a representative organism in such a way that behavior appears to be as lawful as the nervous system. I know of no experimental material, for example, concerning the central nervous system which consists of smoother or more easily reproducible curves than are illustrated in many of the figures of this book. Accordingly, if we are to avoid historical influences in arriving at a modern verdict, we must discount the priority of the science of neurology; and in recognizing that the two sciences are of, let us say, equal validity, we may no longer subscribe to a point of view which regards a chaos of behavior as reducible to order through appeal to an internal ordered system.

A more sophisticated neurological view, which acknowledges the orderliness of behavior, is based upon the contention that a law of behavior cannot be fully validated until the neural events which participate in the observed phenomena have been accounted for. In particular, the discontinuity of the relation of stimulus and response is cited as an objection to dealing exclusively with the observed correlations of terminal events. Neurology is regarded as *explaining* the laws of behavior, and it is held that in ignoring neurological facts a science of behavior abandons its only hope of explanation. This conception of the aim of a science is, of course, far-reaching, but the problem may be dealt with for our present purpose by considering the relatively small point at issue here. In order to put a representative case before the reader I shall review the procedure of investigating the neural events underlying the kind of reflex that I have called a respondent. This is not the only case to which the argument applies, but it will serve as an illustration.

The neurologist begins, as does the student of behavior, with the observation that a given stimulus elicits a given response. His first step is to discover conducting tissue between the loci of these events, first as gross structure but eventually as a chain of specialized cells. Such a chain is a 'reflex arc,' a neurological entity which has no counterpart in behavior. The arc serves to account first of all for the mere connection between a stimulus and a response demanded by their approximate simultaneity of occurrence, but it must also account for the differences between their forms. A stimulus and its response differ in time of inception, in duration and in form and amount of energy. By various procedures, which we shall not need to consider, the steps in the conversion of a stimulus into a response are assigned to parts of the arc. The gross conversions of energy are, of course, referred to end-organ and effector, part of the elapsed time to afferent and efferent nerve, and so on. By processes of logical and surgical isolation a certain group of properties are shown to be independent of the activity of end-organ, effector, and nerve-trunk. They are properties of the central nervous system and presumably of the points of contact between nerve-cells called synapses. In Sherrington's (68) classical treatment the properties attributable to the synapse are expressed as differences between nerve-trunk and synaptic conduction, but they may be restated in a simpler form, as in the following examples: (a) a period of time elapses between the arrival and departure of a discharge at a synapse, (b) the duration of the efferent discharge is frequently greater than that of the afferent, (c) the intensity of the efferent discharge does not vary rectilinearly with the intensity of the afferent, (d) a single small afferent discharge is often not effective in producing an efferent but succeeding discharges following closely upon it may be, (e) repetition of an afferent discharge (with certain temporal specifications) evokes progressively weaker efferent discharges, (f) a second afferent discharge may be ineffective or submaximally effective for a short period of time after a first, (g) two discharges arriving at a synapse from separate sources may combine in producing an efferent discharge, and so on.

The traditional procedure of the science of reflex physiology in dealing with these facts has been to set up some such basic concept as synaptic 'conductivity,' 'excitability,' or 'resistance' to refer to the state of the synapse and subsidiary concepts of 'latency,' 'after-discharge,' 'refractory phase,' and so on, to refer to its processes. In Case (e), for example, repeated afferent discharges are said to increase the synaptic resistance or lower the excitability in accordance with the special law of 'reflex fatigue.' There is little difference between this kind of neurology and the system of behavior established in Chapter One. The synaptic processes are not directly observed as such but are inferred from a comparison of input and output very much as in the case of behavior. The basic concept of synaptic conductivity or its congeners is for most purposes identical with reflex strength. The laws of behavior do not exactly match those of the synapse because the parts played by end-organ, nerve-trunk, and effector are eliminated in the latter case, but there is a closer correspondence than might be expected. The events immediately preceding and following the passage of a synapse have the dimensions of nervous impulses, and hence the modifications imposed by the synapse are temporal or intensive. But in stating a correlation at the level of behavior it is assumed that both stimulus and response are measured on scales appropriate to the form of energy involved in each case, and the forms are not important in the principal laws. The description of a reflex at the level of behavior is largely in terms of time and intensity also. In the static laws applying to a specific case a correction must be made for the activities of receptor and effector in comparing the corresponding laws of the synapse; but in the dynamic laws, which are by far the more important, little or no correction is required.

The concepts and laws of reflex physiology at this level differ from those of behavior principally in the local reference implied in the term synapse. If it were not for this reference, the traditional 'C. N. S.' might be said to stand for the Conceptual Nervous System. The data upon which the system is based are very close to those of a science of behavior, and the difference in formulation may certainly be said to be trivial with respect to the status of the observed facts. The same argument applies as well to other concepts—for example, to the connective network which is offered to account for the topographical relations of stimuli and responses. Here there is often a gross local reference, but a single 'reflex arc' is otherwise as inferential as synaptic processes. In dealing with interlocking arcs it is often possible to establish the order of the loci of specific events, up to and including the final common path, but there is only a gross anatomical knowledge of absolute position. In any one of these examples the essential advance from a description of behavior at its own level is, I submit, very slight. An *explanation* of behavior in conceptual terms of this sort would not be highly gratifying. But a Conceptual Nervous System is probably not what the neurologist has in mind when he speaks of the neural correlates of behavior. The correlation demanded as an explanation is with a science of neurology which completes its local references and devises techniques for the *direct* observation of synaptic and other processes. The network is to be carefully traced and its various parts described in physico-chemical terms. The notion of synaptic conductivity is to be translated into terms, say, of permeability or ionic concentration; while the subsidiary processes of latency, fatigue, and so on are to be described in terms at the same level. Hypothetical steps toward such a system have been taken. Sherrington's (69) hypothesis of E and I substances or states is a familiar example. Factual material has also begun to accumulate, and it may be assumed that a science of the nervous system will some day start from the direct observation of neural processes and frame its concepts and laws accordingly. It is with such a science that the neurological point of view must be concerned if it is to offer a convincing 'explanation' of behavior.

What is generally not understood by those interested in establishing neurological bases is that a rigorous description at the level of behavior is necessary for the demonstration of a neurological correlate. The discovery of neurological facts may proceed independently of a science of behavior if the facts are directly observed as structural and functional changes in tissue, but before such a fact may be shown to account for a fact of behavior, both must be quantitatively described and shown to correspond in all their properties. This argument becomes more cogent as independent techniques are developed in neurology and hence applies more directly to a physico-chemical neurology than to a conceptual. That is to say, a demonstration of a correlation comes to demand greater rigor as neurology and a science of behavior begin to deal with different methods and subject matters.

The practical independence of these two kinds of neurology was asserted in an early paper by Forbes (42). 'It may be that inhibition opposes excitation by affecting the permeability or other properties of the synapse and preventing impulses from reaching the moto-neurone; or it may be that it acts by arousing in the cell body processes which oppose those of excitation. It seems to me that no data at hand suffice to determine this question, and further, that its solution is not essential to a consideration of the dynamic properties of reflex inhibition.' Add to this a conception of inhibition as a property of behavior, and the three levels that I am pointing out are obtained. At the level of behavior a law of inhibition was stated in Chapter One in this way: 'The strength of a reflex may be decreased through presentation of a second stimulus which has no other relation to the effector involved.' In terms of a conceptual synapse, strength comes to be stated as 'conductivity' and a second stimulus becomes a second afferent path. The 'dynamic properties' of inhibition are presumably not very different in the two cases. At the level of a structural synapse regarded as a physico-chemical system inhibition becomes (at present hypothetically) an affair of permeability, the inactivation of a conducting substance (adsorption?), or some such process. A correlation between the first two levels is not difficult because the data are largely identical. But the correlation of a physico-chemical process, once it is observed, with inhibition at the level of either behavior or an inferential synapse will require a rigorous quantitative formulation at these latter levels. If the two processes do not match, the 'explanation' will be inadequate.

The very notion of a 'neurological correlate' implies what I am here contending—that there are two independent subject matters (behavior and the nervous system) which must have their own techniques and methods and yield their own respective data. No amount of information about the second will 'explain' the first or bring order into it without the direct analytical treatment represented by a science of behavior. The argument applies equally well to other sciences dealing with internal systems related to behavior. No merely endocrinological information will establish the thesis that personality is a matter of glandular secretion or that thought is chemical. What is required in both cases, if the defense of the thesis is to go beyond mere rhetoric, is a formulation of what is meant by personality and thought and the quantitative measurement of their properties. Only then can a valid correlation between a state of endocrine secretion and a state of behavior be demonstrated. Similarly, in the developmental sciences, no principle of development—part out of whole or whole out of part—will account for an aspect of behavior until that aspect has been independently described.

I am asserting, then, not only that a science of behavior is independent of neurology but that it must be established as a separate discipline whether or not a rapprochement with neurology is ever attempted. The reader may grant this and at the same time object that the neurological side should not be ignored. He may contend that the two fields are admittedly related and that much might be gained from exploring both at the same time, rather than in holding to the strict isolation represented by the present book. The arguments for this view are much less convincing than its general acceptance at the present time would seem to demonstrate.

Much of the tendency to look to the nervous system for an 'explanation' of behavior arises from clinical practices where explanation has a relatively simple meaning. The discovery of a cerebral lesion as the 'neural correlate' of, let us say, aphasia is doubtless an important step in the understanding of the condition of a patient. But the success in this instance of finding 'what is wrong' with behavior by looking into the nervous system depends largely upon the negative nature of the datum. The absence (and in many cases the derangement) of a function is much more easily described than the function itself. 'He speaks' is admittedly an inadequate description of verbal behavior, which demands great amplification. 'He cannot speak' is a fairly complete description of the opposite case, so long as the unanalyzed notion of speaking is accepted. The significance of this difference for the present argument may be pointed out by comparing the correlation of aphasia and a lesion with the correlation of normal speech and the neural processes involved in it. It is not difficult to point to a mere damage to verbal behavior and a corresponding damage to the nervous system, but almost no progress has been made toward describing neurological mechanisms responsible for the positive properties of verbal behavior. This argument is provisional, of course; eventually a correlation of important properties may be reached. The point at issue is not the possibility of successful correlation but its significance. Although the discovery of a lesion may be of first importance for diagnostic or prognostic purposes, a description of the phenomena of aphasia, in their relation to normal verbal behavior, is aided very slightly if at all by this added knowledge. It is wholly a matter of the interests of the investigator, whether he makes this excursion into the nervous system. In general a descriptive science of behavior can make little use of the practices of the clinician, except in so far as they are descriptive. Usually the descriptive side is neglected because of the pursuit of the neural correlate. Thus, to continue with the example of aphasia, the monumental work of Head (43) is of little value to the student of behavior because his analysis of the nature and function of language is antiquated and obscure.

The clinical practice of looking into the organism is carried over in the widespread belief that neurological facts somehow *illuminate* behavior. If my statement of the relation of these two fields is essentially correct, the belief is ill-founded. It obviously springs from the ancient view of behavior as chaotic. If there is any illumination at all, it is in the other direction. Behavior is by far the more easily observed of the two subject matters, and the existence of an intermediate science dealing with a conceptual nervous system testifies to the importance of inferences from behavior in neurology. In any event, I venture to assert that no fact of the nervous system has as yet ever told anyone anything new about behavior, and from the point of view of a descriptive science that is the only criterion to be taken into account.

The same statement of the relation between neurology and behavior will serve to dismiss the claim that neurology offers a *simpler* description of behavioral facts. This view is again reminiscent of the belief that simplicity is not to be sought for in behavior itself; but aside from this it may be contended that different kinds of behavioral facts may eventually be found to spring from a single neurological source and that the number of terms required for description may therefore be reduced by resorting to neurological terms. Perhaps such a view lies behind the interpretation of 'brain waves' as the basis of thought or endocrines as the basis of personality, since the physiological system is apparently simpler than the behavior to be explained. But just what kind of correspondence between behavior and physiology this implies I am not prepared to say. Either it is not a one-to-one correspondence, or there must be a common 'simplifying' property in the behavior itself. If, for example, the discovery of a single kind of synaptic process is some day made to account for the various kinds of 'learning' discussed in previous chapters, it can successfully account for them only if some common property between the several cases may be demonstrated at the level of behavior. It is toward the reduction of seemingly diverse processes to simple laws that a science of behavior naturally directs itself. At the present time I know of no simplification of behavior that can be claimed for a neurological fact. Increasingly greater simplicity is being achieved, but through a systematic treatment of behavior at its own level.

Another objection to the independent development of a science of behavior is that an investigation of the relation to the nervous system may lead to useful hypotheses and hence to fruitful experimentation. It may be true, this objection will run, that a complete description of both terms is necessary in demonstrating a rigorous correlation, but a tentative correlation, based upon meager information, will suggest important experimentation in both fields. In so far as this rests upon what is regarded as important experimentation, it is a matter of tastes and there is no disputing it. But the sort of experimentation suggested by such hypotheses is presumably directed toward establishing a better correlation rather than advancing either field separately, and hence the answer to the objection reduces to the argument already given regarding the attempt at correlation. For research that is not directed toward establishing correlations, the significant points of attack are most expediently determined from a systematization of the data. It is certainly possible to design research in behavior without an eye to neurology and (it may be added) with an expectation that the result will contribute something of greater permanence than the disproof of a hypothesis. The gain to the science of behavior from neurological hypotheses in the past is, I believe, quite certainly outweighed by all the misdirected experimentation and bootless theorizing that have arisen from the same source.

Unless the reader has clearly grasped the conception of an independent science of behavior, it is not likely that he will be convinced by these arguments. A purely descriptive science is never popular. For the man whose curiosity about nature is not equal to his interest in the accuracy of his guesses, the hypothesis is the very life-blood of science. And the opposition to pure description is perhaps nowhere else as strong as in the field of behavior. I cannot expect that a mere demonstration of the independence of a science of behavior will dissuade the reader from his willingness to let the two disciplines proceed together as closely enmeshed as they are at the present time. There are, however, arguments of a more positive sort that he should take into consideration.

The first of these is hygienic. A definition of terms in a science of behavior at its own level offers the tremendous advantage of keeping the investigator aware of what he knows and of what he does not know. The use of terms with neural references when the observations upon which they are based are behavioral is misleading. An entirely erroneous conception of the actual state of knowledge is set up. An outstanding example is the systematic arrangement of data given by Pavlov. The subtitle of his *Conditioned Reflexes* is 'An Investigation of the Physiological Activity of the Cerebral Cortex,' but no direct observations of the cortex are reported. The data given are quite obviously concerned with the behavior of reasonably intact dogs, and the only nervous system of which he speaks is the conceptual one discussed above. This is a legitimate procedure, so long as the laws established are not turned to 'explain' the very observations upon which they are based; but this is commonly done, as for example, by Holt (45). Holt's procedure is especially interesting because he is clearly aware of the kind of fallacy of which he is the victim. In the early pages of the book cited he quotes Molière's 'coup de grâce' to verbalism—

> 'I am asked by the learned doctor for the cause and the reason why opium induces sleep. To which I reply, because there is in it a soporific virtue whose nature it is to lull the senses.'

He then proceeds to explain behavior with a conceptual nervous system! I can see little difference between the use of the term instinct, to which he objects, and his own explanation of learning in terms of 'Pavlov's Law,' except that a neural reference is assigned to the law which is lacking for the instinct. The reference is not at present supported by the data.

A second argument for maintaining the independence of a science of behavior is that it is then free from unnecessary restraining influences. Behavior, as I have said, is far more easily observed as a subject matter than the nervous system, and it is a mistake to tie one science down with the difficulties inherent in another. A single reflex arc, identifiable as such and as the correlate of a reflex, is at present inaccessible. Even gross dynamic properties are equally obdurate. Although the neurologist may speak, for example, of an afferent discharge from the stomach or of some other process as the basis of hunger, no method has to my knowledge been devised to obtain measures of resulting cortical or sub-cortical states of the drive as delicate as the measures of behavior described in Chapter Ten. We shall accept too great a handicap if we are to wait until methods have been devised for the investigation of neural correlates in order to validate laws of behavior. It is especially necessary to avoid restricting the term reflex to correlations for which arcs have been located. The restriction is commonly urged by the neurologist who is perhaps justifiably dismayed by the so-called 'units' of behavior which are featured in psychological work. But the isolation of an arc is not a useful criterion to appeal to in order to exclude the misuse of the notion of a unit. Other criteria are available, as, for example, those considered in Chapter One, which are based upon the lawfulness of the unit during various changes in its state.

The current fashion in proceeding from a behavioral fact to its neural correlates instead of validating the fact as such and then proceeding to deal with other problems in behavior seriously hampers the development of a science of behavior. The first of the experiments described in this book was on the change in the rate of ingestion of food. The 'natural' course would have been to turn to the identification of the physiological processes with which the change was correlated. Various hypotheses suggested themselves: the curve reflected a change in the condition of the stomach, or in the concentration of blood sugar, or the oxidation of a 'hunger hormone,' and so on. Doubtless this would have been a profitable line of research, but it would have meant renouncing an interest in behavior itself. For the purposes of a lawful description of behavior the quantitative change in the strength of ingestive reflexes was enough. No detection of a correlated physiological change would have increased the validity of the law, and by turning instead to other laws at the level of behavior more about behavior could be learned. Meanwhile the physiologist was provided with a method of investigation, whenever he might wish to carry out his side of the correlation.

I am not overlooking the advance that is made in the unification of knowledge when terms at one level of analysis are defined ('explained') at a lower level. Eventually a synthesis of the laws of behavior and of the nervous system may be achieved, although the reduction to lower terms will not, of course, stop at the level of neurology. The final description will be in terms of whatever quasi-ultimate physical units are then in fashion. How important an advance this will be depends upon one's view of science as a whole. One of the objectives of science is presumably the state-

ment of all knowledge in a single 'language' (38). Another is prediction and control within a single field. What I am arguing for here is the advantage to be gained from a rigorous prosecution of a field at its own level.

The intensive cultivation of a single field is to be recommended, not only for its own sake, but for the sake of a more rapid progress toward an ultimate synthesis. Far from thwarting neurology, an independent science of behavior has much to offer it. A careful systemization and investigation of behavior is of value to anyone who takes behavior as his starting point and seeks an explanation in neural terms. The neurologist has till now been able to do without this help because he has confined himself to relatively simple cases. He has not yet reached the point at which a popular conception of behavior breaks down. A current neurological theory of learning, for example, may content itself with the simple notion of sensory-motor connections (*cf.* neurobiotaxis) because both the neurological and (to a lesser extent) the behavioral facts are so few that any more cogent theorizing would be idle. Eventually only the most rigorous formulation of learning will suffice for a neurological starting point, and it may be one in which the mere connection of paths seems trivial. As a general proposition it may be said that the facts of behavior now appealed to by neurology can be satisfying only in the early stages of a science. It is too much to ask that the neurologist refine both fields as he proceeds. Whether or not he must do so depends largely upon the future of the science of behavior.

Perhaps I can best indicate the kind of contribution that a science of behavior may be expected to make by selecting from the preceding chapters a number of properties or aspects of behavior that are already of obvious significance for neurology in its exploration of neural correlates.

1. *The uniformity of changes in the rate of emission of such a relatively complex response as 'pressing a lever' and the practical separation of the total act into the component parts of a chain of reflexes.* In view of current controversies over the analysis of behavior and the possibility of identifying functional units, the orderliness of many of the processes here reported should be reassuring to the neurologist who wishes to preserve the hypothesis of a unitary 'center' in connection with a relatively complex act.

On the other hand the obvious generic nature of the response as measured behaviorally raises an acute problem for neurology which for the most part may be ignored in behavior.

2. *The concept of reflex strength and its usefulness in stating the principal dynamic laws.* The neurological parallel of strength is some form of synaptic conductivity or excitability. A reduction of the principal phenomena of behavior (drive, conditioning, discrimination, emotion, and so on) to states of reflex strength presents a very much simplified problem to neurology.

3. *The notion of operant behavior and its emission rather than elicitation.* The notion of conductivity as the essential function of a center must be supplemented with a state of excitation in which impulses are simply emitted.

4. *The distinction between the various functions of stimuli.* A discriminative stimulus which brings about the emission of a response (which 'sets the occasion' for the response) differs quantitatively in its action from the eliciting stimulus and must be 'explained' by a different neural mechanism.

5. *The conception of drives and emotions as states rather than as stimuli and responses.* The search for afferent stimulation specific to each drive has with few exceptions been futile and even ridiculous. The attempt to define emotions in terms of specific responses has fared only slightly better. Both of these endeavors may well give way to an analysis of states of excitability of covarying reflexes and of the forces which produce changes in them (perhaps including afferent stimulation in part).

6. *The grouping of reflexes according to drives and emotion.* The traditional scheme of sensory-motor organization in the cerebrum pays almost no attention to the covariations of otherwise unrelated reflexes during changes in drive and emotion. Some additional organization must be exposed to account for these obvious facts.

7. *The reflex reserve.* The concept of a reserve demands a neural mechanism different in kind from the momentary excitability or conductivity of a center or the mere connection of pathways. Whatever state or condition is found to correspond to the reserve must have the property of surviving relatively long periods of time without significant change of magnitude and must obey the other laws here established.

8. *The relation between the reserve and the momentary*

strength. The mechanism of (7) must be under the control of another mechanism in order to produce (5). The difference, for example, between an empty reserve and a full reserve with no drive or a full reserve under emotion is clear at the level of behavior and must have some neurological counterpart.

9. *The distinction between conditioning of Type R and of Type S and the formulation of types in terms of the contingencies of the reinforcement.* Schemes for explaining Type S in terms of simultaneously active paths are inadequate for Type R, which presents a special problem in the apparently retroactive action of the reinforcement.

10. *An analysis of the concept of inhibition.* The repeated objection to basing a term solely upon the direction of the change in reflex strength applies to neurological hypotheses as well. A variety of neural processes may be required to account for the instances commonly referred to by one term.

11. *The formulation of a temporal discrimination.* The status of time in a description of behavior has many bearings upon neural processes (*cf.* the 'trace stimulus'). A careful reconsideration of popular conceptions is necessary if experimentation upon the 'biological time' of the organism is to make sense.

12. *The analysis of the differentiation of a response.* As in the case of operant conditioning, a telephone-analogy breaks down when no topographical changes are made. The differentiation of the intensity of the response calls for a modified neurological theory of learning.

The list could be extended and the argument for its significance in neurology could be greatly amplified. Indeed, if we should turn to the history of neurological hypothesizing for our examples of the criteria of proof and for our standards as to the adequacy of facts, a really prodigious amount of speculation could be based upon the present experimental material. Like Pavlov's *Conditional Reflexes* the book might have been put forth as a neurological treatise. I have already stated my belief that an account which is not a mere translation of behavioral data into hypothetical neural terms must be the fruit of independent neurological techniques, which it is not within the province of a science of behavior to develop. Leaving the material in this form will illustrate the relation between a science of behavior and neurology which should prove most fruitful. In the case of most of the items listed, a number of quantitative properties have been fairly well established. It is this quantification, together with a rigorous formulation, which places a science of behavior in a quite different position from casual observation and analysis.

A quantitative science of behavior may be regarded as a sort of thermodynamics of the nervous system. It provides descriptions of the activity of the nervous system of the greatest possible generality. Neurology cannot prove these laws wrong if they are valid at the level of behavior. Not only are laws of behavior independent of neurological support, they actually impose certain limiting conditions upon any science which undertakes to study the internal economy of the organism. The contribution that a science of behavior makes to neurology is a rigorous and quantitative statement of the program before it.

The relation between neurology and a science of behavior that I have been trying to express is somewhat more temperately stated in the following words of Mach (60) at the beginning of a chapter on Physics and Biology:

> 'It often happens that the development of two different fields of science goes on side by side for long periods, without either of them exercising an influence on the other. On occasion, again, they may come into closer contact, when it is noticed that unexpected light is thrown on the doctrines of the one by the doctrines of the other. In that case a natural tendency may even be manifested to allow the first field to be completely absorbed in the second. But the period of buoyant hope, the period of over-estimation of this relation which is supposed to explain everything, is quickly followed by a period of disillusionment, when the two fields in question are once more separated, and each pursues its own aims, putting its own special questions and applying its own methods. But on both of them the temporary contact leaves abiding traces behind. Apart from the positive addition to knowledge, which is not to be despised, the temporary relation between them brings about a transformation of our conceptions, clarifying them and permitting of their application over a wider field than that for which they were originally formed.'

Chapter Thirteen

CONCLUSION

It may be desirable to comment in somewhat more general terms upon the systematization of behavior put forward in the preceding pages. Two important characteristics scarcely need to be pointed out. The work is 'mechanistic' in the sense of implying a fundamental lawfulness or order in the behavior of organisms, and it is frankly analytical. It is not necessarily mechanistic in the sense of reducing the phenomena of behavior ultimately to the movement of particles, since no such reduction is made or considered essential; but it is assumed that behavior is predictable from a knowledge of relevant variables and is free from the intervention of any capricious agent. The use of analysis seems absolutely necessary in a science of this sort, and I know of no case where it has in practice really been avoided. The way in which 'pressing a lever' is defined as a unit of behavior and the way in which the unit is validated experimentally are, I hope, beyond the reach of current criticisms of over-simplified stimulus-response formulae.

What has been called the topographical description of behavior includes the listing and classification of reflexes, of the operations which induce changes in reflex strength, and of the groups of reflexes affected by such operations. But neither a reflex nor a group of reflexes appropriate to, say, a drive or emotion can be identified from topographical features alone. A *part* of behavior, isolable in terms of some classificatory scheme, is not known to be a *unit* of behavior until certain dynamic properties have been demonstrated, nor can the behavior characteristic of a drive or an emotion be identified except in terms of dynamic properties. There is, therefore, no wholly independent physiognomic or taxonomic field in a science of this sort. A successful classificatory description may prove useless when applied to the dynamic side of behavior.

The dynamic properties which are fundamental to a science of behavior can be properly investigated only in the laboratory. Casual or even clinical observation is ill-adapted to the study of *processes,* as distinct from momentary features. A process, which necessarily involves time, can be made available for analysis only through the use of quantitative observations and records. Because it is experimental, a science of behavior may justifiably claim greater validity than popular or philosophical formulations whenever disagreement arises, but the advantage gained from experiment is frequently misunderstood. It is not merely that additional data are supplied by experimentation, or that the data are more reliable, or that by experimenting we are able to check hypotheses against reality. The principal advantage that compensates us for the necessity of subordinating a topographical description to an experimental investigation of dynamic processes lies in obtaining a system of behavior which has a structure determined by the nature of the subject matter itself.

The term *system* when applied to behavior frequently oscillates between two fairly distinct meanings. On the one hand it may refer to a systematic compilation or classification of data by some person or school, often with the use of a special terminology. The opinion is commonly expressed, perhaps with some justification, that we have had enough systems of this sort. Without the check provided by the consideration of dynamic properties a purely topographical description may be carried out in many different ways, and investigators may therefore emerge with different systems in this sense. On topographical grounds alone there may be no way of obtaining agreement. For example, a large part of the behavior of an organism may appropriately be called sexual if it bears the proper hall-mark according to some scheme of classification. There is nothing to prevent anyone from making such a classification if only the features of the behavior are taken into account, nor from making another classification in which the term 'sexual' would be given a much more limited or conflicting application. Disagreements among systems of this sort are largely verbal.

A second kind of system, to which the term is intended to refer here, is clearly exemplified by the systems encountered in physical chemistry. Such a system consists of an aggregation of related variables, singled out for the sake of convenient investigation and description from all the various phenomena presented by a given subject matter. In the case of behavior, a system in this sense can be arrived at only through the kind of experimental analysis to which this book is devoted, in which the parts or aspects of behavior which undergo orderly changes are identified and their mutual relations established.

The disturbing differences which now exist among the current systems of behavior which are not merely topographical seem to be the result of differences at this elementary stage of the selection of variables. There is no general agreement as to what the principal variables in behavior are. Very often no attempt is made to define them explicitly: an investigator simply enlarges upon some current popular or philosophical system and brings in new defined terms at a few points. Or, on the other hand, variables are selected which yield a convenient system but are not representative of the behavior as a whole. I shall try to elaborate these points by considering a number of current examples.

A system of behavior based upon the concept of the tropism seems to satisfy the requirements of a system in this sense except on the point of generality. In the extensive experiments of Crozier and Pincus [a convenient account may be found in (40)] variables have been isolated which are capable of being treated quantitatively and which behave in lawful ways. They may also be combined in larger complexes with predictable effects. But any system which takes orientation or oriented progression as the only property of behavior to be accounted for and which regards a stimulus only as a field of force is seriously circumscribed. In the case of the higher organisms at least it is presumably possible to set up an independent descriptive system based upon the concept of the reflex that will yield an equally satisfactory result.[1] Where behavior is largely orientation and where stimuli are fields of force, we may prefer the concept of the tropism on grounds of simplicity while at the same time rejecting it in the case of more complicated organisms.

Much the same objection may be levelled against Lewin's concepts of vectors and valences (58). Such a system applies readily to behavior which can be conceived of as orientation or movement in a field of force and where the stimulus can be said to generate such a field, but it is not an expedient system for handling other kinds of behavior in response to other kinds of stimuli.

[1] Compare Hunter's objection to the work on the tropisms of mammals (50, 51).

Field behavior of this sort may also presumably be treated in terms of stimulus and response as here defined, and greater generality may therefore be claimed for the reflex. Another advantage of the reflex over the vector or valence is that behavior is defined with a sharper reference to the topography of the organism, as will be noted again in a moment.

A third important current system, in which the problem of the isolation of variables becomes acute, has been worked out by Hull (49). Hull begins by defining a number of terms, some of which (*e.g.*, 'extinction,' 'reinforcement') have a more or less technical meaning but the majority of which (*e.g.*, 'discouragement,' 'success,' 'disappointment,' 'frustration') are taken from popular vocabularies or from various psychological systems. With the aid of certain postulates (*e.g.*, 'Each reaction of an organism gives rise to a more or less characteristic internal stimulus'), Hull states and proves a number of theorems (*e.g.*, 'Organisms capable of acquiring competing excitatory tendencies will manifest discouragement'). The demonstration of theorems of this sort is offered as 'specific evidence that such problems, long regarded as the peculiar domain of philosophy, are now susceptible of attack by a strictly orthodox scientific methodology.'

The virtue of Hull's work lies in an insistence upon the experimental validation of statements about behavior and upon the necessity of confining oneself to statements that are internally consistent and may be experimentally verified. But he has failed to set up a system of behavior as distinct from a method of verification. The only terms in his list which might be regarded as fundamental variables are brought in without definition (*e.g.*, 'reaction,' 'stimulus complex,' and 'excitatory tendency,' which is his nearest approach to 'strength'). Most of the terms that he defines are supernumeraries, drawn from various inexhaustible sources. Several hundred acceptable definitions of the same sort could readily be obtained, and a dismaying number of theorems could be derived. No procedure is supplied for reducing the number of necessary terms to a minimum, and there is no guarantee that such a method can by itself ever attain to that ultimate simplicity of formulation that it is reasonable to demand of a scientific system. The terms which Hull selects do not compose a system in the present sense, nor has he actually applied his methodology to the problem of designing such a system.

By beginning with a deductive procedure, Hull has necessarily made the formulation of hypotheses and the design of critical experiments the central activities in an investigation of behavior. A quite different emphasis is to be found in the preceding chapters. Deduction and the testing of hypotheses are actually subordinate processes in a descriptive science, which proceeds largely or wholly without hypotheses to the quantitative determination of the properties of behavior and through *induction* to the establishment of laws. The difficulty seems to lie in the model that Hull has chosen. A science of behavior cannot be closely patterned after geometry or Newtonian mechanics because its problems are not necessarily of the same sort. This is especially true with respect to the problem of isolating fundamental variables. If Hull had chosen experimental physics or chemistry as a model, the place of deduction in his system would have been much less important.

Tolman has presented a system which is in many respects close to that described here (72, 73).[2] Behavior is taken as the dependent variable and shown to be a function of age, heredity, drive-operations, and so on. What has here been treated as a 'state,' as distinguished from the operation responsible for the state, is called by Tolman an 'intervening variable.' For example, his 'demand' for food is equivalent to a 'state of hunger'; neither is to be identified with the operation that makes an organism hungry and both are inferred from the effect of the operation upon behavior. In Tolman's system the notion of 'strength' is broken down into 'hypotheses,' 'demands,' and so on, according to the kinds of operations responsible for the strength. But the notion of strength itself or an equivalent is not clearly developed, probably because of the type of situation to which the system is typically applied. The maze is not a suitable instrument for the investigation of the dynamic properties of behavior. Even when we consider a single 'choice-point,' there remain two possible responses—turning right and turning left. No measure of the strength of either is provided by maze behavior, since a 'choice' reveals only the relatively greater strength of one. Instead of measuring behavior directly, Tolman is reduced to determining a 'behavior ratio,' which is of little use in following the various processes which are the principal subjects of investigation.

[2] The essential aspects of the present system which enter into the comparison were described in my paper on the concept of the reflex in 1931 (2).

That differences should arise over the question of fundamental variables at an early stage in the history of a science is neither remarkable nor alarming. The important thing is that the need for a system in the present sense (rather than in the sense of a classificatory vocabulary) is beginning to be realized. It would be an anomalous event in the history of science if *any* current system should prove to be ultimately the most convenient (and hence, so far as science is concerned, correct). The collection of relevant data has only just begun. But, paradoxically, the necessarily tentative character of any single current system cannot wholly excuse the prevailing multiplicity of systems. There are available criteria according to which a system may be judged. They are supplied principally by the usefulness and economy of the system with respect to the data at hand.

One outstanding aspect of the present book, which can hardly be overlooked, is the shift in emphasis from respondent to operant behavior. The definition of behavior as a whole given in Chapter One may not be altogether acceptable to the reader. By appealing to what the organism is doing to the environment a great deal of what is often called behavior is minimized or even excluded. Most of the responses of glands and smooth-muscle fail to act upon the environment in such a way as to yield the *conspicuousness* which is offered as a defining characteristic. Any definition of a scientific field is to a considerable extent arbitrary, but it is worth pointing out that, were we to make operant behavior a subject matter in itself, we should avoid many of these problems. Operant behavior clearly satisfies a definition based upon what the organism is doing to the environment, and the question arises whether it is not properly the main concern of a student of behavior and whether respondent behavior, which is chiefly involved in the internal economy of the organism, may not reasonably be left to the physiologist. Operant behavior with its unique relation to the environment presents a separate important field of investigation. The facts of respondent behavior which have been regarded as fundamental data in a science of behavior (Sherrington, Pavlov, and others) are, as we have seen, not to be extrapolated usefully to behavior as a whole nor do they constitute any very large body of information that is of value in the study of operant behavior.

Although a distinction may be drawn between the operant and the respondent field, there is also a certain continuity, which I have tried to indicate by beginning with respondent laws and by comparing conditioning of Type S (which is largely, if not wholly, respondent) with Type R (which is apparently wholly operant). A more important sort of continuity is manifested by the use of the term 'reflex' in both fields. This is to some extent a matter of controversy. In operant behavior the original figurative meaning of reflex is lost, since there is no stimulus to be 'reflected' in the form of a response. It is also true that from its being applied first to respondent examples the term has acquired incidental connotations (especially in its neurological use) which are opposed to the general use made of it here. But I have tried to show elsewhere (2) from a consideration of the history of the term that many of its connotations have sprung, not from the discovery of additional information, but from prejudices and preconceptions concerning the behavior of organisms. The simple positive fact of a correlation of stimulus and response has unnecessarily given rise to an elaborated negative definition of an action 'unlearned, unconscious, and involuntary.' Pavlov has extended the term into the field of 'learning' by showing that one can obtain the same kind of relation of stimulus and response in acquired behavior. The property of 'consciousness' is either irrelevant or ineffective in differentiating between two kinds of behavior. The remaining distinction between voluntary and involuntary (cf. 65) is probably closely paralleled by the operant-respondent distinction, but its traditional use in defining a reflex is more closely related to the question of predictability or freedom, which is of no significance here. A definition which respects the actual data may be derived from the simple observation of the correlation of stimulus and response. Somewhat more generally, the term applies to a way of predicting behavior or to a predictable unit. In this broad sense the concept of the reflex is useful and applicable wherever predictability may be achieved. Its range has steadily increased as more and more behavior has submitted to experimental control, and its ultimate extension to behavior as a whole is a natural consequence of an increasing demonstration of lawfulness.

One important practice has been observed in the traditional study of reflexes which is of paramount importance in the kind of system here set up and which supports the extended use of the term. The practice is that of referring to specific movements of parts of the organism. In spite of the generic nature of the term, the topographical reference has always been relatively narrow and precise. One reason why this is important is that the phenomena are then in a better position to be reduced to neurological terms. Such an argument may strike the reader as strange in view of the preceding chapter, but I agree with Carmichael (37) that 'those concepts which do not make physiological formulation impossible and which are amenable to growing physiological knowledge are preferable, other things being equal, to those not so amenable.' The principal significance of a sharp reference to behavior, however, is not that a neurological investigation is facilitated but that the descriptive value of the term is kept at a maximum.

This characteristic may be better understood by comparing a reflex or a law of reflex strength with a law or principle which describes the 'adaptive' or 'adjustive' nature of behavior or some other equally general property. Suppose, for example, that a principle is demonstrated from which it may be deduced that an organism facing a barrier in the path toward a goal will remain active until some response is made by virtue of which the barrier is surmounted. Granted the validity of the principle, we are still unable to say what the precise behavior will be. Similarly, a principle that enables one to predict that in a given situation behavior will have 'survival value' or will require 'least effort' may be valid enough so far as it goes, but it lacks the specificity of reference which the concept of the reflex presupposes. So far as I am aware, the reflex is the only important historical concept that has closely respected the actual movements of the organism, and the term may justifiably be preserved in a field in which that kind of reference is of first importance.

An obstacle in the way of a science of behavior is the failure to understand that behavior may be treated as a subject matter in its own right. The materialist, reacting from a mentalistic system, is likely to miss behavior as a subject matter because he wishes to have his concepts refer to something substantial. He is likely to regard conceptual terms referring to behavior as verbal and fictitious and in his desire for an earthy explanation to overlook their position in a descriptive science. Holt (45) adopts a modern position of this sort. His objection to such a term as 'instinct' seems to be reducible to the statement that you cannot find the instinct by cutting the organism open. A similar argument is commonly advanced against the concepts of 'intellect,' 'will,' 'cognition,' and so on, which have served in popular or philosophical descriptions of behavior for centuries. But the objection to such terms is not that they are conceptual but that the analysis which underlies their use is weak. The concepts of 'drive,' 'emotion,' 'conditioning,' 'reflex strength,' 'reserve,' and so on, have the same status as 'will' and 'cognition' but they differ in the rigor of the analysis with which they are derived and in the immediacy of their reference to actual observations. In spite of the conceptual nature of many of our terms we are still dealing with an existent subject matter, which is the behavior of the organism as a whole. Here, as elsewhere in the experimental sciences, a concept is only a concept. Whether or not it is fictitious or objectionable cannot be determined merely from its conceptual nature.

The traditional description and organization of behavior represented by the concepts of 'will,' 'cognition,' 'intellect,' and so on, cannot be accepted so long as it pretends to be dealing with a mental world, but the behavior to which these terms apply is naturally part of the subject matter of a science of behavior. What is wanted in such a science is an alternative set of terms derived from an analysis of behavior and capable of doing the same work. No attempt has been made here to translate mentalistic or philosophical concepts into the terms of the present system. The only value of a translation would be pedagogical. Traditional concepts are based upon data at another level of analysis and cannot be expected to prove useful. They have no place in a system derived step by step from the behavior itself.

The reader will have noticed that almost no extension to human behavior is made or suggested. This does not mean that he is expected to be interested in the behavior of the rat for its own sake. The importance of a science of behavior derives largely from the possibility of an eventual extension to human affairs. But it is a serious, though common, mistake to allow questions of ultimate application to influence the development of a systematic science at an early stage. I think it is true that the direction of the present inquiry has been determined solely by the exigencies of the system. It would, of course, still have been possible to suggest applications to human behavior in a limited way at each step. This would probably have made for easier reading, but it would have unreasonably lengthened the book. Besides, the careful reader should be as able to make applications as the writer. The book represents nothing more than an experimental analysis of a representative sample of behavior. Let him extrapolate who will.

Whether or not extrapolation is justified cannot at the present time be decided. It is possible that there are properties of human behavior which will require a different kind of treatment. But this can be ascertained only by closing in upon the problem in an orderly way and by following the customary procedures of an experimental science. We can neither assert nor deny discontinuity between the human and subhuman fields so long as we know so little about either. If, nevertheless, the author of a book of this sort is expected to hazard a guess publicly, I may say that the only differences I expect to see revealed between the behavior of rat and man (aside from enormous differences of complexity) lie in the field of verbal behavior.

The objection has been offered to the kind of system and method here described that it is not statistical. A number of meanings may be given to this term, and in considering the resulting forms of the objection, one or two other characteristics of the work as a whole may be pointed out.

In the simple sense of involving large numbers of measurements very little of the preceding work is statistical. The psychologist who is accustomed to dealing with fifty or a hundred or a thousand organisms may be disturbed by groups limited to four or eight. But large numbers of cases are required, if they are required at all, in order to obtain smooth and reproducible curves. The recourse to statistics is not a privilege, it is a necessity arising from the nature of many data. Where a reasonable degree of smoothness and reproducibility can be obtained with a few cases or with single cases, there is little reason, aside from habit or affectation, to consider large numbers. There are always limitations of time and energy to be considered, and one must inevitably compromise between the depth and breadth of an investigation. Before advancing to new problems I have tried to secure a reasonable degree of reproducibility or reliability, but the investigation has not been pressed beyond that point.

The records presented here must speak for themselves so far as orderliness is concerned.

The system of behavior proposed in Chapter One is statistical in another sense. In dealing with the behavior of what Boring has called the 'empty organism' the causal chain of events between stimulus and response is passed over in favor of the correlation of end terms. The substitution of correlation for cause is sometimes called statistical, but the distinction is trivial in view of modern theories of causality in which the correlational aspect is emphasized.

A third meaning of the term statistical is more important. There are at the present time two quite different modes of approaching the behavior of organisms which are hard to distinguish theoretically but which are clearly different in practice. The statistical approach is characterized by relatively unrefined methods of measurement and a general neglect of the problem of direct description. The non-statistical approach confines itself to specific instances of behavior and to the development of methods of direct measurement and analysis. The statistical approach compensates for its lack of rigor at the stage of measurement by having recourse to statistical analysis, which the non-statistical approach in general avoids. The resulting formulations of behavior are as diverse as the methods through which they are reached. The concepts established in the first case become a part of scientific knowledge only by virtue of statistical procedures, and their reference to the behavior of an individual is indirect. In the second case there is a simpler relation between a concept and its referent and a more immediate bearing upon the individual. It may be that the differences between the two approaches are transitory and that eventually a combination of the two will give us our best methods, but at the present time they are characterized by different and almost incompatible conceptions of a science of behavior.

It is obvious that the kind of science here proposed naturally belongs on the non-statistical side of this argument. In placing itself in that position it gains the advantage of a kind of prediction concerning the individual that is necessarily lacking in a statistical science. The physician who is trying to determine whether his patient will die before morning can make little use of actuarial tables, nor can the student of behavior predict what a single organism will do if his laws apply only to groups. Individual prediction is of tremendous importance, so long as the organism is to be treated scientifically as a lawful system. Until we are spared the necessity of choosing between the two approaches, we must cast our lot with a non-statistical investigation of the individual and achieve whatever degree of reliability or reproducibility we may through the development of techniques of measurement and control.

REFERENCES

A. Earlier Reports of Some of the Present Material.

(Note: These papers represent the gradual accumulation of facts and the progressive development of a formulation. So far as I am aware, no important contradiction or inconsistency has appeared in the experimental material. Several important changes in formulation have, however, necessarily been made, most of them contingent upon the notion of a non-elicited operant behavior. I have not recorded changes of this sort in the text. They should be obvious to anyone familiar with the papers and would only confuse the reader who is now approaching the subject for the first time.)

1. Skinner, B. F. On the conditions of elicitation of certain eating reflexes. *Proc. Nat. Acad. Sci.*, 1930, *16*, 433–438.
2. ———. The concept of the reflex in the description of behavior. *J. Gen. Psychol.*, 1931, 5, 427–458.
3. ———. Drive and reflex strength. *Ibid.*, 1932, *6*, 22–37.
4. ———. Drive and reflex strength: II. *Ibid.*, 1932, *6*, 38–48.
5. ———. On the rate of formation of a conditioned reflex. *Ibid.*, 1932, *7*, 274–286.
6. ———. On the rate of extinction of a conditioned reflex. *Ibid.*, 1933, *8*, 114–129.
7. ———. The abolishment of a discrimination. *Proc. Nat. Acad. Sci.*, 1933, *19*, 825–828.
8. ———. The measurement of spontaneous activity. *J. Gen. Psychol.*, 1933, *9*, 3–24.
9. ———. The rate of establishment of a discrimination. *Ibid.*, 1933, *9*, 302–350.
10. ———. 'Resistance to extinction' in the process of conditioning. *Ibid.*, 1933, *9*, 420–429.
11. ———. The extinction of chained reflexes. *Proc. Nat. Acad. Sci.*, 1934, *20*, 234–237.
12. ———. A discrimination without previous conditioning. *Ibid.*, 1934, *20*, 532–536.
13. ———. The generic nature of the concepts of stimulus and response. *J. Gen. Psychol.*, 1935, *12*, 40–65.
14. ———. Two types of conditioned reflex and a pseudo type. *Ibid.*, 1935, *12*, 66–77.
15. ———. A discrimination based upon a change in the properties of a stimulus. *Ibid.*, 1935, *12*, 313–336.

16. ————. A failure to obtain disinhibition. *Ibid.*, 1936, *14*, 127–135.
17. ————. The reinforcing effect of a differentiating stimulus. *Ibid.*, 1936, *14*, 263–278.
18. ————. The effect on the amount of conditioning of an interval of time before reinforcement. *Ibid.*, 1936, *14*, 279–295.
19. ————. Conditioning and extinction and their relation to the state of the drive. *Ibid.*, 1936, *14*, 296–317.
20. ————. Thirst as an arbitrary drive. *Ibid.*, 1936, *15*, 205–210.
21. ————. Two types of conditioned reflex: a reply to Konorski and Miller. *Ibid.*, 1937, *16*, 272–279.
22. Heron, W. T., and Skinner, B. F. Changes in hunger during starvation. *Psychol. Record*, 1937, *1*, 51–60.
23. Skinner, B. F., and Heron, W. T. Effects of caffeine and benzedrine upon conditioning and extinction. *Ibid.*, 1937, *1*, 340–346.

B. Other References.

24. Adrian, E. D. *The Basis of Sensation.* New York: W. W. Norton and Company, 1928.
25. Anrep, G. V. The irradiation of conditioned reflexes. *Proc. Roy. Soc.*, 1923, *94*, 404.
26. Bass, M. J., and Hull, C. L. Irradiation of tactile conditioned reflexes in man. *J. Comp. Psychol.*, 1934, *17*, 47–65.
27. Bayliss, W. M. *Principles of General Physiology.* Fourth edition. New York: Longmans, Green & Co., 1927.
28. Bethe, A. *Arch. f. microscop. Anat.*, 1897, *50*, 460.
29. Bousfield, W. A. Certain quantitative aspects of the food-behavior of cats. *J. Gen. Psychol.*, 1933, *8*, 446–454.
30. Bousfield, W. A., and Elliott, M. H. The effect of fasting on the eating-behavior of rats. *J. Genet. Psychol.*, 1934, *45*, 227–237.
31. Bousfield, W. A. Certain quantitative aspects of chickens' behavior toward food. *Am. J. Psychol.*, 1934, *46*, 456–458.
32. ————. Quantitative indices of the effects of fasting on eating behavior. *J. Genet. Psychol.*, 1935, *46*, 476–479.
33. Bousfield, W. A., and Sherif, M. Hunger as a factor in learning. *Am. J. Psychol.*, 1932, *44*, 552–554.
34. Brogden, W. J., and Culler, E. *Science*, 1936, *83*, 269.
35. Brücke, E. T. Refraktäre Phase und Rhythmizität, in *Handbuch der normalen und pathologischen Physiologie.* Band 9. Berlin: Springer, 1929.
36. Cannon, W. B. *Bodily Changes in Pain, Hunger, Fear, and Rage.* New York: D. Appleton and Company, 1929.*
37. Carmichael, L. A revaluation of the concepts of maturation and

* Second edition now published by D. Appleton–Century Company.

learning as applied to the early development of behavior. *Psychol. Rev.*, 1936, *43*, 450–470.
38. Carnap, R. *The Unity of Science.* London, 1934.
39. Creed, R. S., Denny-Brown, D., Eccles, J. C., Liddell, E. G. T., and Sherrington, C. S. *Reflex Activity of the Spinal Cord.* Oxford: Clarendon Press, 1932.
40. Crozier, W. J., and Hoagland, H. The study of living organisms, in *Handbook of General Experimental Psychology.* Worcester: Clark University Press, 1934.
41. Exner, S. *Pflüger's Archiv,* 1892, *28,* 487.
42. Forbes, A. *Quart. J. Exper. Physiol.*, 1912, *5*, 151.
43. Head, H. *Aphasia and Kindred Disorders of Speech.* London: Cambridge University Press, 1926.
44. Hilgard, E. R. The relationship between the conditioned response and conventional learning experiments. *Psychol. Bull.*, 1937, *34*, 61–102.
45. Holt, E. B. *Animal Drive and the Learning Process.* New York: Henry Holt and Company, 1931.
46. Hudgins, C. V. Conditioning and voluntary control of the pupillary light reflex. *J. Gen. Psychol.*, 1935, *12*, 208–214.
47. Hull, C. L. The goal gradient hypothesis and maze learning. *Psychol. Rev.*, 1932, *39*, 25–43.
48. ————. Learning: the factor of the conditioned reflex, in *Handbook of General Experimental Psychology.* Worcester: Clark University Press, 1934.
49. ————. Mind, mechanism, and adaptive behavior. *Psychol. Rev.*, 1937, *44*, 1–32.
50. Hunter, W. S. The behavior of the white rat on inclined planes. *J. Genet. Psychol.*, 1927, *34*, 299–332.
51. ————. The mechanisms involved in the behavior of white rats on the inclined plane. *J. Gen. Psychol.*, 1931, *5*, 295–310.
52. ————. The disinhibition of experimental extinction in the white rat. *Science,* 1935, *81*, 77–78.
53. Kantor, J. R. In defense of stimulus-response psychology. *Psychol. Rev.*, 1933, *40*, 324–336.
54. Keller, F. S. [Personal communication.]
55. Köhler, W. *The Mentality of Apes.* New York: Harcourt, Brace and Company, 1925.
56. Konorski, J., and Miller, S. *Podstawy Fizjologicznej Teorji Ruchow Nabytych (Fundamental Principles of the Physiological Theory of Acquired Movements).* Warsaw, 1933. (French Summary.)
57. ————. On two types of conditioned reflex. *J. Gen. Psychol.*, 1937, *16*, 264–272.
58. Lewin, K. *A Dynamic Theory of Personality.* New York: McGraw-Hill Book Company, Inc., 1935.
59. Liddell, H. S., and Bayne, T. L. The development of experi-

mental neurasthenia in sheep during the formation of difficult conditioned reflexes. *Am. J. Physiol.*, 1927, *31*, 494.

60. Mach, E. *The Analysis of Sensations*. English translation. Chicago: The Open Court Publishing Company, 1914.
61. Magnus, R. *Körperstellung*. Berlin: Springer, 1924.
62. ———. Croonian lecture. *Proc. Roy. Soc.*, 1925, *98*, 339.
63. Mowrer, O. H. [Paper presented at 1937 meeting of Am. Psychol. Assoc.].
64. Pavlov, I. P. *Conditioned Reflexes: An Investigation of the Physiological Activity of the Cerebral Cortex*. London: Oxford University Press, 1927.
65. Peak, H. An evaluation of the concepts of reflex and voluntary action. *Psychol. Rev.*, 1933, *40*, 71–89.
66. Pfaffman, C. Differential responses of the new-born cat to gustatory stimuli. *J. Genet. Psychol.*, 1936, *49*, 61–67.
67. Richter, C. P. Animal behavior and internal drives. *Quart. Rev. Biol.*, 1927, *2*, 307–343.
68. Sherrington, C. S. *The Integrative Action of the Nervous System*. New Haven: Yale University Press, 1906.
69. ———. Remarks on some aspects of reflex inhibition. *Proc. Roy. Soc.*, 1925, *97 B*, 519–545.
70. Thorndike, E. L. *Animal Intelligence*. New York: The Macmillan Company, 1911.
71. Tolman, E. C. *Purposive Behavior in Animals and Men*. New York: The Century Co., 1932.*
72. ———. Demands and conflicts. *Psychol. Rev.*, 1937, *44*, 158–169.
73. ———. The determiners of behavior at a choice point. *Psychol. Rev.*, 1938, *45*, 1–41.
74. Warden, C. J. *Animal Motivation, Experimental Studies on the Albino Rat*. New York: Columbia University Press, 1931.
75. Watson, J. B. *Psychology from the Standpoint of a Behaviorist*. Second edition. Philadelphia: J. B. Lippincott Company, 1924.
76. ———. *Behaviorism*. Revised edition. New York: W. W. Norton and Company, 1930.
77. Wendt, G. R. An interpretation of inhibition as competition between reaction systems. *Psychol. Rev.*, 1936, *43*, 258.
78. Wentink, E. The effects of certain drugs and hormones upon conditioning. *J. Exper. Psychol.*, 1938, *22*, 150–163.
79. Wright, S. *Applied Physiology*. London: Oxford University Press, 1926.
80. Young, P. T. Relative food preferences of the white rat. *J. Comp. Psychol.*, 1932, *14*, 297–319; 1933, *15*, 149–165.
81. ———. *The Motivation of Behavior*. New York: John Wiley and Sons, Inc., 1936.

* Now published by D. Appleton–Century Company.

82. Youtz, R. E. P. [Paper at the 1937 meeting of the Am. Psychol. Assoc.]
83. Zieve, L. The effect of benzedrine on activity. *Psychol. Record,* 1937, *1,* 393–396.

INDEX